Management, Systems, and Society: An Introduction

Management, Systems, and Society: An Introduction

Richard A. Johnson, R. Joseph Monsen,
Henry P. Knowles, Borje O. Saxberg
University of Washington

GOODYEAR PUBLISHING COMPANY, INC.
PACIFIC PALISADES, CALIFORNIA

Library of Congress Cataloging in Publication Data

Main entry under title:

Management, systems, and society: an introduction.

Includes bibliographies and index.
1. Industrial management. 2. Systems analysis.
3. Industrial organization. 4. Industry—Social
aspects. I. Johnson, Richard Arvid
HD31.M29385 658.4 75-13447
ISBN 0-87620-540-6

Current printing (last digit):

10 9 8 7 6 5 4 3 2 1

Y-5406-7

Printed in the United States of America

Contents

Preface

The manager of today performs a much different role from that of his predecessors of one hundred or even as recently as twenty-five years ago. Until about 1950, few constraints affected managerial decisions or actions other than those relating to maintaining a free-market economy. Businessmen tended to run their companies mainly by intuition. Today, however, conditions have changed along with the increasing complexities of our society. As a result of the emergence of giant corporations and a progressive dependence on world resources and markets, managers can no longer make decisions solely on the basis of competition or on the firm's profits and growth. The manager of today must consider the total human, social, political, and in many cases, international environments in which the organization operates.

Management, Systems, and Society is designed to describe the role of the business manager in this broad, new setting, but it pertains also to managing resources in a nonprofit organization. Directed toward the beginning student of management, this text presents an overview of the total system within which an organization must function. Once the student has acquired a perspective of the total picture, it is possible to expand this knowledge with specialized study in selected areas of the field of management. For example, analytical and mathematical approaches to decision making are an essential part of managing an organization and must, in time, be included in every manager's education. It seems more logical, however, that such specialized material should be presented only after a student has acquired a sound understanding of the basic philosophies and principles of organization and management. For this reason, we have deliberately excluded from this text mathematical approaches to decision making in order to avoid adding unnecessary complexity for the beginning student.

Throughout the book, the systems approach, including appropriate applications from the behavioral sciences, is used as a means of exploring both the organization's internal operations and also its relation to the environments in which it exists. We firmly believe that the systems approach to management will help the reader understand the interrelationships among the many parts of the organization and also help to relate the separate parts of the system to the organization's objective or mission. Although our primary focus is on business, the systems approach can be applied to all kinds of organizations, including governmental and educational institutions.

We continually encourage the reader to maintain a logical and positive outlook in assessing the present and future of business. Although the authors do not advocate drastic changes in our present political, economic, and social institutions, we do believe that there are ways to make these institutions more effective and responsive to human needs. For example, we believe we must develop social-action organizations to supplement present economic and political organizations. We believe also that, as a society, we should encourage managers of political-economic organizations to adopt a new social consciousness.

An ever-increasing number of women are entering the field of management. In recognition of this trend, we have often tried to avoid the use of "he" or "his" in writing this book. Unfortunately, we found this impossible without making many of the explanations clumsy or awkward. Therefore, we have reverted to the standard practice of using "he" or "his" as a neuter personal pronoun not implying reference to either sex.

We are aware that many students do not read the preface of a book. Therefore, we have opted to summarize the plan of this book at the end of Chapter 1. We have also elected to include a glossary of concise definitions for the convenience of those students who may be unfamiliar with some of the terms used before they are developed in subsequent chapters. The terms included in the glossary are indicated the first time they appear in the text by boldface type.

We wish to thank our many colleagues at the University of Washington who influenced or made contributions to this book. They include Cecil H. Bell, Jr., Wendell L. French, Fremont E. Kast, Richard W. Moxon, William T. Newell, and James. E. Rosenzweig. We would also like to thank our colleagues from other institutions who either contributed to the book or reviewed the manuscript and offered constructive suggestions. These include Gerald L. Bassford, Arizona State University; G. James Francis, Colorado State University; David D. Van Fleet, Texas A&M University; Roger C. Vergin, Simon Fraser University; and Thomas J. Von der Embre, Wright State University.

We also would like to thank Jane Smith, Nona Pederson, and Sandra Goodman for their help in typing the manuscript.

Richard A. Johnson
R. Joseph Monsen
Henry P. Knowles
Borje O. Saxberg

Part One
The Foundations of Management

The task of managing pervades all forms and all levels of **organizations**. In this section, we sketch managerial thought from its beginnings to the present. Although its roots reach back even further than the days of Adam Smith in the eighteenth century, most of the theory of management in current use was developed in this century. Acquiring a sense of this historical development will help establish a better understanding of organizational theory and behavior as they are viewed today.

One philosophy for classifying and categorizing the many fragments of thought that constitute the general theory of organizations is the **systems approach**. This approach is valuable as a conceptual philosophy, as a process for managing organizations, and as a model for analyzing **decision making**. The systems approach will be used to integrate the ideas developed in this book.

Today's manager cannot operate successfully within the traditional boundaries. Because the modern manager's decisions are affected by political, social, and economic systems, in this text we will identify and emphasize environmental influence.

Chapter 1
The Role of the Manager

Management is the art and science of making appropriate choices. To one degree or another, we are all involved in managing and are constantly making decisions concerning how to spend or use our resources. For example, we may need to choose between buying a television set and spending the money on a vacation, or between studying for an exam and going skiing. In such cases, we are managing scarce personal resources according to established preferences and needs. When the officers of a corporation decide between building a new plant and increasing dividends to stockholders, they are managing the company's resources according to a set of priorities. When the president of the United States presents the national budget to Congress for the coming year, he is serving as the manager of the federal economy. In other words, most of us act as managers when we find ourselves involved in choosing among alternatives or, as an economist would put it, in making decisions concerning "the allocation of scarce resources."

Like most things in our modern, changing world, the function of management is becoming more complex. In the past, simply having a knack for recognizing business opportunities was enough to achieve financial success. Management used to be thought of as an art, as the application of knowledge and skill gained from experience and from a kind of "gut feeling" for the market. Today, however, art is no longer enough. Managers for today and tomorrow need professional training for making decisions in our technically sophisticated and complicated economy—a systematic, scientific way of thinking about the intricate network of forces influencing modern business operations.

This book is about a systematic way to study the problems managers face. In this chapter, we will present a panorama of the field of management, discuss why studying management is important, and attempt to describe the excitement and challenge that lie ahead for the manager of the future. In the process, we will introduce many important terms and ideas that will be developed more fully in subsequent chapters. To acquaint the reader with unfamiliar terms, a glossary containing a concise definition is included at the end of the text. Those terms that are defined in the glossary are indicated by boldface type.

THE AMERICAN SUCCESS STORY

Because the emerging role of the modern manager has paralleled the historical development of the United States, it might be helpful to begin by reviewing some of the forces that contributed to the growth of this country. The question is often asked, What made America great? How did it become the industrial giant of the world, whose standard of living for its people has never yet been duplicated by any other country? Part of the answer to this question is related to the four basic factors necessary to produce goods and services: land, labor, capital, and good management. The abundance and availability of land and other natural resources directly affect a country's development. In America, the first settlers found rich farmland, ample water supplies, dense forests, and plentiful iron ore deposits. Second, the early American settlers were often skilled workers who had left their homelands to come to this "land of opportunity." A broad mix of educational backgrounds and talents formed the basis for a highly productive labor force. Third, surplus capital from England was invested in this country's development. Later, accumulated capital from the eastern part of America was used to develop the West.

The most important influence on the growth of America, however, was the ability of people to manage large organizations and vast accumulations of resources. Settlers coming to America possessed a pioneering spirit and were willing and able to accept change. Many of the first settlers were natural managers. Thus, America was abundantly endowed with the four prerequisites of production and seemed destined to become a great economic power.

Because the government of the United States supported private enterprise, the political environment was stable in relation to business, and because the industrial revolution coincided with the population's increasing need for more goods and services, destiny was served. Until the beginning of the twentieth century, economic growth was relatively slow. In the last seventy-five years, however, the acceleration has been spectacular. Viewing this country today with its sprawling cities, its teeming mazes of streets and highways, its grassy parks and paved plazas, it is easy to forget that only a little more than a century ago there were less than fifty million people living in the United States with more than two-thirds living in rural communities.

A Capsule History

The original American colonists settled near bodies of water, which could be used for commerce and transportation; near resources of timber, minerals, and good farmland; and in areas where they could defend themselves from conflicts with the indigenous Indian population. Nevertheless, the early settlers were faced with extreme climatic conditions and food shortages. In the 1800s, the American frontier shifted westward. Market towns, small industrial cities, and new, regional centers of government developed. Between 1825 and 1850, the first period of urban growth occurred. Communities such as Chicago and San Francisco grew from little villages to cities.

The Civil War left its scar on the nation for many years. Post–Civil War American cities were noted for their congestion, poverty, slums, and inadequate housing. Public education was introduced with an optimistic goal of assisting in the orderly economic and social growth of the cities. But poverty and crime flourished. In the city slums, water supplies and sewage disposal systems were desperately inadequate. Attempts to improve conditions were often obstructed by corrupt city administrators—the "big-city boss and his political machine." Not until the turn of the century did America become aware to what extent misery and blight had become a part of urban life.

During the Civil War era, the railroads became a vital part of the country's development. Railroads carried the coal necessary to expand the industrial power of the nation, transported cattle and grain, and helped distribute the goods that industry produced. The invention of the telegraph facilitated communication among the industrial centers. And mechanical reapers and harvesters, developed to replace the farmhands who had left their jobs to fight in the Civil War, greatly increased agricultural **productivity**.

In the early part of the twentieth century, mass production in this country received an assist from technological advances in the chemical industry, from the development and manufacture of light metals and alloys, and from the construction of high-speed mechanical steel mills. Industrial technology and managerial know-how were combined to mass-produce Henry Ford's first Model T, affectionately called the "Tin Lizzie." An increased number of automobiles affected employment opportunities for large numbers of the work force and influenced the way of life for the entire population. By the 1950s, many city residents were moving to the suburbs and building new homes; the automobile had helped solve the problem of transportation and had made the work force mobile. As we shall see later, there are currently forces at work that may alter this "automobile society" into a society patterned around mass transportation systems.

Today the United States is an urban society with approximately seventy percent of its total population living on little more than one percent of the land. This urban compacting was made possible by an integrated network of production, distribution, and consumption of goods and services. The concentration of population, however, has brought a corresponding pressure on the cities' facilities and organizations—freeways, streets, water and sewage systems, housing, and fire and police protection.

The Next Decade

Many experts have suggested that the coming decade will emphasize economic performance, with more material goods, savings, and profits the prime **objectives**. They base this forecast in part on the large number of young people who are now reaching the age of forming their own families. Traditionally, the young married group has generated increased demand for new homes, furniture, appliances, and cars—in addition to the desire to find productive work to support such demands.

From the late 1960s until 1971, the seventeen-year-old group was the largest single age group in this country. The youth revolution, which had its greatest impact in 1970–71, is partly explained by this phenomenon of population dynamics. Seventeen-year-olds are traditionally rebellious, for many of them are trying to discover their personal identity or purpose during this period of their lives, and they may often wish to be associated with a cause. In contrast, as young adults move into their twenties, they have generally become more conventional and less vehement about new ideas. Typically, people in their late twenties place primary attention on their job, the mortgage, financing new appliances, and the cost of living.

Today's young people will have a significant influence on our society during the next decade. Whether this group will be as oriented as their parents were toward material goods, and whether this country will have the resources to satisfy such desires is difficult to say. But we do know that there will be more people in the consumer category.

During the next decade, it will be more difficult to find jobs for many of the young people entering the labor market. Many of today's youth have delayed their entrance into the job market by attending college. Now there are increasing numbers looking for employment. Whereas in the past, the college diploma was practically a license to select from among an abundance of jobs, there is already an oversupply of college graduates. In the future, there will be more young people seeking jobs as well as significant shifts in the kinds of jobs available. For example, one of the largest sources of job opportunities for young people in the past decades was in the field of education but with the decreasing birthrate and the reduced number of young people attending schools and colleges, the demand for new teachers will be relatively small in the future.

The question, then, is who will hire the unemployed? Can a massive influx of new capital absorb this pool of labor, or will the government be required to create make-work programs? The particular solution that our political leaders choose to solve the unemployment problem will obviously influence the future environment in which businessmen must operate.

THE MODERN MANAGER

The role of the manager today is much different from what is was one hundred years, fifty years, or even twenty-five years ago. At the turn of the century, for example, the business manager's objective was to keep his company running and to make a **profit**. Most firms were production oriented. Thus products were developed which lent themselves to production **efficiency**, and there was only limited concern about the consumer's behavior. The typical strategy around 1900 was to expand whenever capital was available, and the exploding demand always seemed to catch up with any temporary overcapacity. Few **constraints** affected management's decisions. Governmental agencies imposed little regula-

Figure 1.1 The Manager's Role Fifty Years Ago

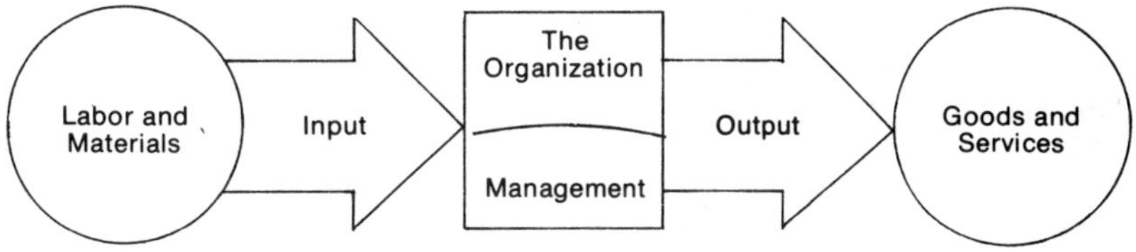

tion on business, and an ample and relatively unorganized work force made it possible for managers to hire and fire as demand for the firm's output changed. Businessmen ran their companies by a "gut feeling" of what they believed should be done.

The manager's role fifty years ago is diagrammed in Figure 1.1. The manager had access to ample inputs of labor and materials and could sell the firm's **output**, provided it was of good quality and at the right price. The manager's attention was directed primarily toward those factors within the organization that would improve quality and efficiency.

Today there are numerous constraints on the decisions made by business management. For example, unions have become a strong bargaining group and bring increasing pressure on management as their demands for higher wages, better working conditions, and more direct participation in the actual decision-making process continue to expand. In addition, government regulation over business has increased to such an extent that business must hire experts to understand the rules and guidelines set forth by the government, to process and report the taxes owed to the government, and to maintain a political voice, or **lobby**, to counter additional regulations. Next, environmental protection groups have organized in an attempt to control actual or potential environmental damage by business and industry and to encourage more governmental ownership of the nation's natural resources and key organizations (for example, the steel industry or the railroads). Also, consumer groups have been organized to monitor the quality and kind of output produced by industry. Finally the voices of women as well as of ethnic and racial minority groups, who have not received equal opportunities in the business establishment in the past, are now being heard. All of these forces are making the job of management more complex.

It has become apparent that the modern manager can no longer make decisions based solely on the effect of these decisions on the firm. The manager must now consider the environment in which the organization operates and be prepared to adopt a wider perspective. That is, the manager must have a good understanding of management principles, an appreciation of the current issues and broader objectives of the total economic, political, social, and ecological system in which we live, and he must possess the ability to analyze complex problems. Figure 1.2 illustrates the manager's role today and suggests how all the additional constraints and influences extend the need for greater knowledge, understanding, and expertise.

The modern manager must be sensitive and responsive to

Figure 1.2 The Manager's Role Today

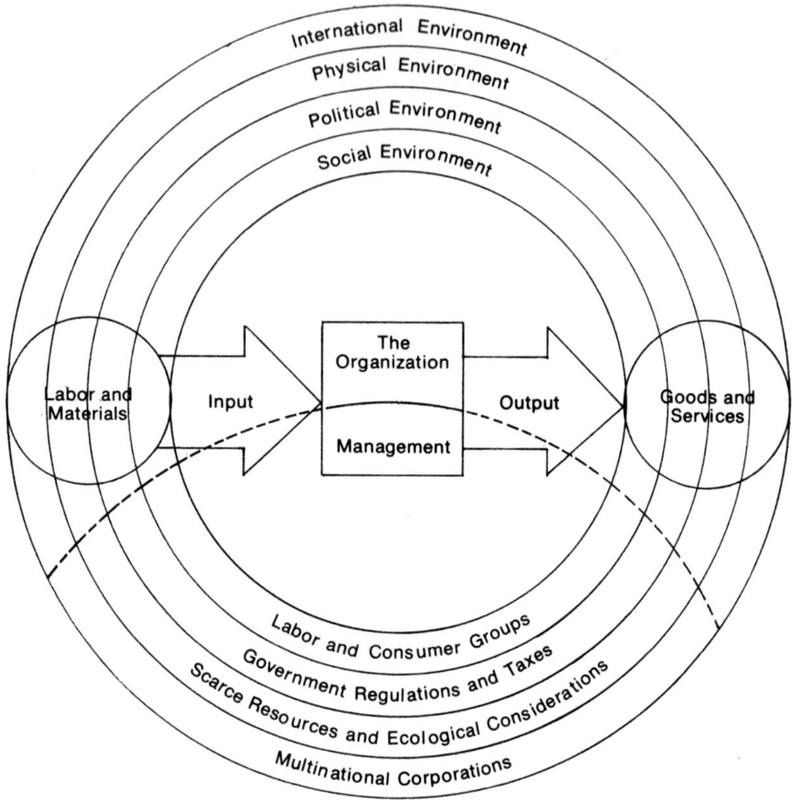

the environment—that is, he should recognize and be able to evaluate the needs of the total context in which his business functions, and he should act in accord with his understanding. Many people have wondered, for example, why the automobile industry continued to make large cars, when the demand for full-sized cars declined from forty to thirty-five percent between 1965 and 1972 and finally collapsed in the fall of 1973. Would not a responsive industry have anticipated the public's changing transportation needs and the imminent energy crisis? Would not more concern from industry about automobile quality and safety have made it unnecessary for people like Ralph Nader to become champions of the consumer?

There are many counter examples, however, in which individual companies have been concerned and consequently have made decisions that recognized an environmental need before it was expressed by the community. In one instance, between 1957 and 1960, the Bethlehem Steel Corporation spent millions of dollars renovating its Seattle plant. It replaced its old furnaces with new electric-arc furnaces and its old mills with new plate mills. But more important, a giant vacuum cleaner was installed to remove dust particles from the emission of the electric furnace, permitting only hot air to escape into Seattle's atmosphere. Unlike refining production equipment, these pollution-control devices did not increase earnings, improve the company's competitive position, expand production, or cut expenses—the new equipment was costly to maintain and operate. This simply illustrates a case where the managers of a business firm recognized problems and provided solutions to improve the environment long before **pollution** became a public issue.

Modern management must possess the ability to interact in an ever-more-complex environment and to make decisions that will allocate scarce resources effectively. A major part of the manager's job will be to predict what that environment needs and what changes will occur in the future.

The Choice among Alternatives

Current social and environmental conditions require some difficult decisions. For example, some suggest that it will be necessary to restrict the number of people who can live on this earth. Growing population is making more and more demands on the resources of this planet and causing increasing damage to the environment. For example, the amount of electrical energy required has doubled in the United States every ten years, millions of new automobiles are built each year and billions of gallons of gasoline are burned to operate these cars. A serious shortage of energy and other natural resources has already developed.

Statistics indicating a general expansion of our productive resources have been considered evidence of **progress.** The fact that more of everything was produced and at a faster rate than population has increased seemed to be proof that there was more for most people, if not for everyone. It is clear that the average citizen today is living more comfortably and more conveniently than he did either fifty or twenty-five years ago. There are more and better homes, new schools, better forms of transportation, and more time to enjoy recreational activities. It is easy to provide facts showing that the people in this country are enjoying a higher level of material wealth than has been experienced by any other society in either the present or in past history.

With improvements in the **standard of living** have come many problems related to the **ecology** and the way we live. In our cities, we have more crime, traffic problems, shortage of jobs, and racial, ethnic, and sexual discrimination. The natural environment has been used to meet the needs of an ever-expanding and demanding society, while waste from the production processes fouls the air, the water, and the land.

World Dynamics

Jay W. Forrester of M.I.T. studied these urban problems as a total social and biological system and considered the choices that we might make in the future.[1] His complex, highly interrelated system of analysis recognizes that the whole is more than a composite of individual problems. In his analysis, Forrester considered five

[1]Jay W. Forrester, *World Dynamics* 2nd ed. (Cambridge, Mass: Wright-Allen Press, 1973).

Figure 1.3 The Factors in World Dynamics

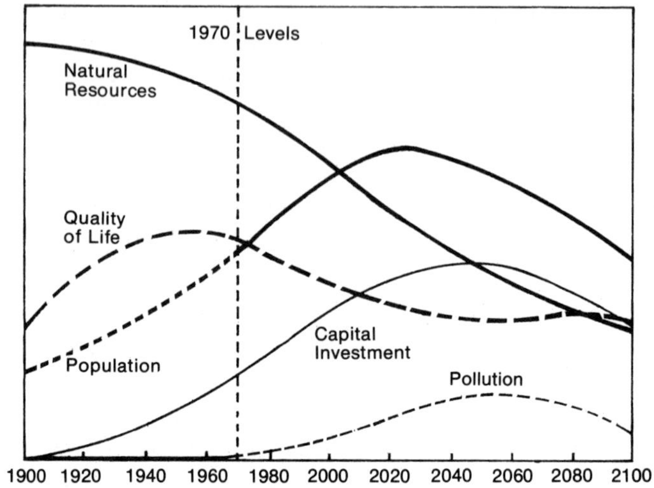

Source: Reprinted with permission from *World Dynamics*, 2nd ed., by Jay W. Forrester, p. 70. Copyright © 1973 Wright-Allen Press, Inc., Cambridge, Mass. 02142 USA.

basic factors: natural resources, quality of life, population, capital investment, and pollution. For example, the number of people living in a geographic region affects the consumption of natural resources, the quality of life, the capital investment necessary to supply the population, and the level of pollution generated. By projecting hypothetical levels over time for one of these factors, the effect on the other four factors can be plotted. For example as population increases, there will also be a tendency for capital investment and pollution to increase, and for natural resources and the quality of life to decrease. But this may be an oversimplification because sometimes the rate of increase or decrease is affected, and a lag, or delayed reaction, may extend beyond a change in the direction of one of the factors. Figure 1.3 shows one projection of these levels over time, graphing the different relationships and time lags among the five factors. It illustrates, for example, how resources will continue to decline even though there may be no change in the population level. If population continues to increase, the rate at which we use our natural resources will be even higher.

Such interrelationships point to the number of potential consequences of any social or economic decision and thus the necessity of analyzing each **tradeoff** carefully. As another example, if we reduce the birthrate and increase capital investment, the material standard of living will rise. But a more affluent population may then use natural resources much faster and cause a higher level of pollution (see Figure 1.4). Obviously, resource managers must take some positive, and maybe unpopular, actions if our destiny is to be controlled—if we are to manage our natural resources to ensure a reasonable quality of life for our future population.

Forging a new philosophy of life as well as new communities of living means transcending the politics of resistance and protest and taking constructive action. This pragmatic and structural

Figure 1.4 Variations in the Factors in World Dynamics

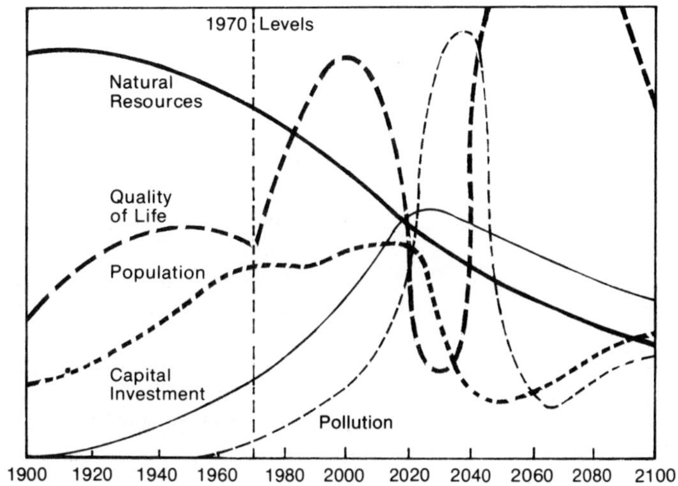

Source: Reprinted with permission from Forrester, *World Dynamics,* p. 111.

thinking suggests such things as the need to organize community-oriented centers around the country, and the need to model a new philosophy of life and living. Each community would play an active role in deciding **strategies** for the best way to provide the goods and services required by that social unit.

This country was founded on the premise that everyone has an opportunity for success. Horatio Alger, a popular American author of the late nineteenth century, wrote a series of books about boys, which focused on the famous "American Dream" of rising from poverty to riches and respectability. In one of his most famous novels, *Raggedy Dick*, the hero takes an initial step toward fame and fortune, which spurs him to achieve even greater success. This "rags to riches" theme symbolized the "American Dream" for young people for many years, and is only lately being questioned. America may be gaining maturity, self-awareness, and a new conscience.

PLAN OF THE BOOK

The plan of this book is to build on management concepts that have been developed and tested over time. The strategy is to use the systems approach as a framework for understanding management principles and the relation of organizations to the environment in which we live. The systems approach, combined with our knowledge about the role human behavior plays in management, is used as means to improve the output of organizations in relation to their objectives, to assist organizations in becoming more proficient in scientific analysis and decision making, and to help them achieve a more workable understanding and more sensitive use of the human resources that comprise their membership. We shall focus also on the environmental system in which we function and inspect the relationships between the environment and the organizations that must be managed.

Most books on management review general principles, and a few include a taste of the systems approach. This book goes further. It includes discussions of the general principles of management, some important applications from the behavioral sciences, and describes the interactions of the organization within the context of the larger social, economic, and political systems. All of these facts of modern management are then brought together and explained within the framework of the systems approach.

Part One of this book introduces the field of management and discusses the systems approach. Part Two covers the human aspects of administration and the time-tested principles of managing organizations. Part Three considers the changing society and how societal values and **pressure groups** contribute to these changes. Part Three also considers several of the pressing issues that managers face today. And Part Four attempts to forecast some of the developments that may affect business organizations in the future.

SUMMARY

Some people believe that this country became great solely because everyone had a chance to achieve success. In addition, the United States was endowed with a wealth of natural resources, talented people, who believed in self-improvement through education, capital to invest in goods, and innovative managers. Moreover, the government has been stable and supportive of private enterprise. The country grew from an agricultural and rural society to a nation with large cities and a massive industrial complex.

The modern manager must consider the total environment in which the organization operates. Today management has a more difficult role to play. The manager must have a good understanding of management concepts, an appreciation of the objectives of the total system in which we live, and an ability to analyze complex issues. Some of the choices he must make will be very difficult. Nevertheless, the effective manager must possess a positive and constructive approach and great **sensitivity** in order to meet the human needs of the workers, consumers, and society in general.

DISCUSSION QUESTIONS

1.1. Why do you think that the automobile industry in this country continued to build large cars? Assuming that you agreed with the need to build smaller cars, analyze the strategy of placing a tax on cars according to gas consumption as a way to discourage the use of large cars.

1.2. If we could limit the growth of our cities to their present size, what effect would this have on our way of life?

1.3. Does the increasing number and kinds of constraints placed on management increase or decrease the need for long-range planning? Discuss.

1.4. List those factors that you believe should be included in determining the standard of living for the United States. How could each of these factors be measured?

APPENDIX
The Case of the Independent Service Station

Tom Lloyd is the owner of an independent service station. He receives his gasoline from an independent distributor who, in turn, receives gasoline from a major oil company. During the spring of 1973, the oil company had cut the distributor's allocation to 84 percent of its previous year's sales, and a corresponding reduction in deliveries was passed on to Tom Lloyd. During the following months, deliveries continued to the Lloyd station at 84 percent of the previous year's sales and then, in November, the distributor notified Tom Lloyd that he would be getting no more deliveries in the future.

Tom Lloyd is a black businessman who had received a Small Business Administration loan, and when he notified the bank of his dilemma, an investigation was made. It developed that the oil company had been supplying the distributor with 105 percent of its previous year's sales, except for the one month mentioned earlier. The distributor also owned and operated several retail stations, and it was apparent that the gas was being cut off to Mr. Lloyd because the distributor could make more money by selling it through his own retail outlets. When the major oil company was contacted by the Small Business Administration, its managers proclaimed their concern and willingness to talk to the distributor but suggested that once they sold the gas to the distributor they had no jurisdiction over how it should be allocated. The management of the oil company admitted that they had subscribed to the program to see that all parties received fair and equal distribution but were emphatic about their inability to do anything about this situation.

1. What would a "consumer-oriented" oil company do in a situation of this kind?
2. As a consultant to Mr. Lloyd, what action would you recommend he take?
3. What monopolistic implications (Sherman and Clayton Acts) will the major oil company face if it controls the dealer's operations after delivering the gas?

SELECTED READINGS

Ehrlich, Paul R. *The Population Bomb*. New York: Ballantine Books, 1968.

Falks, Richard. *The Endangered Planet*. New York: Random House, 1971.

Mintzberg, Henry. *The Nature of Managerial Work*. New York: Harper & Row, 1973.

Shaw, Irwin. *Rich Man, Poor Man*. New York: Delacorte Press, 1969.

Toffler, Alvin. *Future Shock*. New York: Bantam Books, 1971.

I will build a car for the Masses...

Chapter 2
Traditional Management

Management, as we have said, is the process of making appropriate choices. A moment's thought, however, tells us that there is more to management than sorting out **alternatives**. Management must also involve others because no manager, whether the head or as a part of an organization, makes all the decisions. The manager makes a few decisions and delegates to others the **authority** to make some, but most often he helps to make decisions after consulting with others. Further, the manager cannot act alone to carry out decisions. To a very great extent he must depend on the help of others to do this, and thus he becomes involved in relationships and patterns of behavior that facilitate the work of others and that will ultimately gain the cooperation and support of his employees.

Behind the manager's behavior are attitudes about people and situations which are derived from underlying **values**. These attitudes and values are combined in a more-or-less consistent philosophy, or system of thought, that determines the manager's style and, to an important extent, the way he performs various managerial functions, such as organizing, directing, controlling, deciding, and so on.

In this chapter and the next, we will outline some of the important concepts that have contributed to the art and science of management, and that form the foundation for developing new management theory to meet the needs of our changing environment.

ORGANIZATIONS

Although the function of managing has its roots in antiquity, only during this century has management begun to make the transition from a technology based on experience and intuitive reasoning to a science based on a system of verifiable laws. A significant portion of managerial technology has been handed down from the experience of the Roman Catholic church, the military, and ancient political states. The church and the military, in particular, have bequeathed to us a legacy of administrative techniques that still remain a powerful influence on managers and the organizations they control. The idea of a **hierarchy** of authority, the division of organizations by function, and the differentiation of

responsibility of the line (the doers) and the staff (the advisors) are all inventions of the church and of military leaders who were faced with the need to manage large aggregations of human, technical, and material resources. We shall draw on this heritage as we describe what an organization is and trace the emergence of the scientific conduct of its affairs.

Management and organization are two terms that often appear together. In one sense they are inseparable; one cannot exist without the other. Since we believe that what a manager does can best be understood with the help of the language of organizations and organizational theory, we will discuss organizations first and later the development of managerial philosophy, practice, and style.

Growth of Large, Complex Organizations

Organizations range in size from small, informal work groups such as a family to large formal organizations such as a university, a hospital, a manufacturing concern, a union, or a political group like the Republican party. It is useful to imagine a continuum ranging from individual activity at one extreme to formalized groupings at the other, and with a wide variety of social organizations in between. In all instances, these social organizations consist of: (1) **goal-oriented** groups (people with a purpose), (2) **psychosocial systems** (people working together), (3) **technical systems** (people using knowledge and **techniques**), and (4) an **integration** of structured activities (people coordinating their efforts).[1]

The family is man's earliest form of organization. For centuries, the family has served as a cooperative system for accomplishing certain individual and collective objectives. Out of this first form of social organization developed the tribe, the village, larger political states, and nations, among other types of social groupings. Each of these separate **sociocultural subsystems** has its own goals, behavioral norms, and hierarchical relationships, along with some differentiation or specialization of skills or labor.

During medieval times, the feudal system was the primary social system to which individuals belonged. A dramatic change occurred with the advent of the industrial revolution. The essence of this revolution was the need to concentrate resources and to produce goods on a larger scale. This necessity fostered the development of the large economic organization. Parallel to the growth of economic organizations was the creation of massive government bureaucracies concerned with formulating and enforcing the laws of commerce and industry.

The size and complexity of organizations is closely related to the wealth of a society, the level of literacy, the degree of population concentration in urban centers, and the level of consumption of goods and energy. As countries pass through different phases of industrialization, they tend to evolve more complex organizational systems. India, for example, is developing from a family-oriented

[1]Richard A. Johnson, Fremont E. Kast, and James E. Rosenzweig, *Theory and Management of Systems*, 3rd ed. (New York: McGraw-Hill, 1973), pp. 42–44.

society to one that contains more industrialization and a system of centralized government. In China, the breakdowns of the traditional family unit and the local social structure are dramatic examples of the evolution of strongly centralized organizational systems instituted by the Communists. This trend toward larger and more complex organizations would seem to be common to all human society.

TRADITIONAL ORGANIZATIONAL CONCEPTS

It is not easy to say what an organization is in a few well-chosen words. Such units are not concrete entities but possess a number of different properties, both substantive and abstract. U.S. Steel, for example, includes within its groupings tremendous physical assets such as production machinery and equipment and real assets such as a huge accumulation of working capital, a trained labor force, and so forth. At the same time, this organization includes social relationships, attitudes, and aspirations that, although they cannot be seen, are a vital part of the company's organization. The management of U.S. Steel functions as the motivating and coordinating force that orchestrates and conducts this vast aggregation of parts and attempts to ensure that it achieves its corporate **goals** and objectives.

Structure

Structure is the arrangement of the various functions or activities in an organization. It provides the basic framework that relates the various parts of the organization and within which they operate. From the broad objectives of the organization, more specific goals are established, and these in turn are distributed among the appropriate parts of the organization for accomplishment. The structure is usually symbolized in an "**organization chart**." As illustrated by the typical organization chart in Figure 2.1, the structure creates a framework for communicating information to and from each subunit, for assigning responsibility to the subunit, and for specifying authority for making decisions. The structure serves both as a general skeleton to hold the organization together and as a network for the flow of information.

Hierarchy

Closely related to structuring in the organization is the concept of hierarchical relationships. Almost any organization—the federal government, the university, or a business—is divided into subsets of the whole, which are subdivided into smaller units, which are in turn subdivided into still smaller segments. This concept is known as the **scalar principle** and refers to the vertical division of authority and responsibility and the corresponding assignment of duties to the various levels of organizational units.

Almost everything in nature has a hierarchical structure. For example, the universe is made up of a complex array of systems and **subsystems**. One of the systems of the universe is the galaxy containing our suns and planets. The earth is a subsystem of that galaxy, and it includes countless subsystems down to the smallest independent particle of an element, the atom. The hierarchical structure, then, is a phenomenon of nature. Similarly, in human organizations there is a fundamental structural principle that suggests divisions into levels or segments of the whole. A city's public educational system may be designed in a hierarchical chain relative to grade levels and geographical subdivisions. A large corporation is often organized in a tier of levels composed of a centralized headquarters, regional divisions, branch headquarters, and individual sales units.

Division of Work

Decisions regarding how to divide the work are integral to setting up an organization. As a small organization expands, the owner-president is likely to review which particular functions should be delegated to somebody else. This process is called **functional organization** and results in creating labels for the various activities, such as manufacturing, marketing, financing, purchasing, and personnel (see Figure 2.1). The labels attached to the divisions immediately below the president's level and the way in which these activities are separated may reveal the format for the division of work. For example, as shown in Figure 2.2, in a product organization, each product may be given independent organizational existence or, in other instances, the division of work may be based on the customer, the manufacturing process, or the geographical territory.

Two examples can illustrate the importance of the basis of division of work. Cluett, Peabody is famous for having ownership rights to the Sanforizing antishrinking process of textiles. In addition, it manufactures men's clothing under its brand name of Arrow. The organization was originally set up on a functional basis: manufacturing, finance, sales, and so forth. A salesman who called on customers was expected to represent the whole line of Arrow products, ranging from undershirts, ties, and dress shirts to fashionable sports shirts. The company found that salesmen tended to concentrate on selling those products from which they made their major commissions. For other goods, they merely acted as order takers. Therefore, Cluett, Peabody decided to reorganize based on the product. Each product was set up with its own organization including its own sales force.

In contrast, the Hotpoint division of General Electric was organized on a product basis. Thus a salesman represented only one product of the Hotpoint appliance line. The appliance retail dealer might be visited by a Hotpoint salesman for electric ranges, but perhaps his inventory was low in freezers and refrigerators. Unfortunately, another salesman handled that line. Feedback from the dealers concerning the excessive time they were forced to spend with Hotpoint salesmen finally dictated the need for a reorganization. In this case, the reorganization was from a product-based organization to a functional organization, giving each salesman responsibility for the complete line of appliances corresponding to those sold by the appliance retailer.

Figure 2.1 Functional Organization Chart

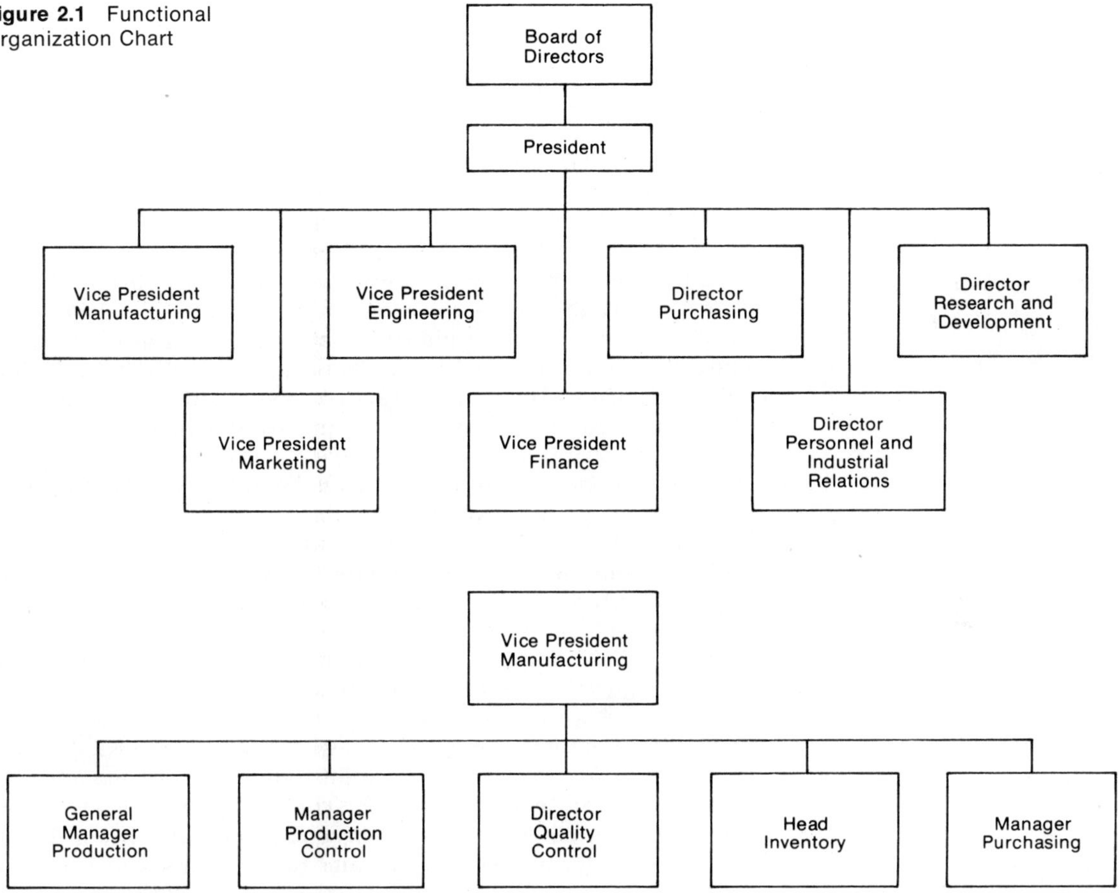

Figure 2.2 Territory and Product Organization Plans

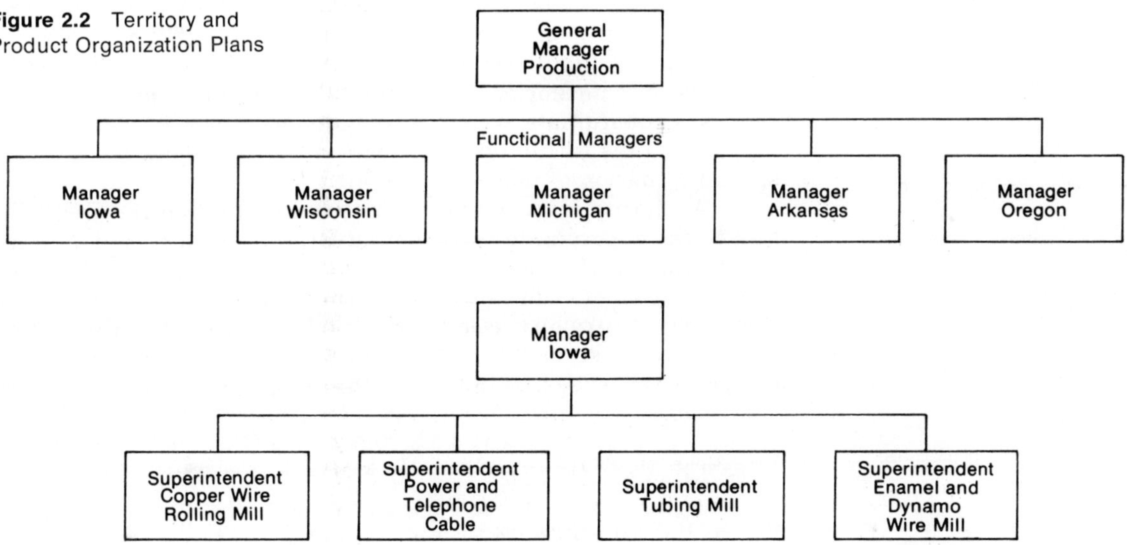

Authority and Responsibility

In the classical view, the concept of authority as a central source of **power** is of prime importance. The superior "has the right to command someone else and the subordinate has the duty to obey the command. This is implied in the notion of official legitimacy, legal in nature rather than social and informal."[2] Authority, therefore, is the right to control subordinates on the basis of position in the organization hierarchy and the ability to disperse the rewards and sanctions that go with the formal position. Authority is a power concept and forms the framework for much traditional organization theory. It is a dispersal of command prerogatives associated with each position of the organization's hierarchical levels. Authority supplies direction to the activities of the organization and is integrated with its objectives and goals.

This description seems to imply that members of the organization will follow the directives or commands promptly and efficiently, and that the organization will achieve its goals. Obviously, such is not always the case. Chester Barnard (discussed in more detail in Chapter 3) was one of the first to suggest that the force of traditional authority may be overstated: "Authority is the character of a **communication** (order) in a **formal organization** by virtue of which it is accepted by a contributor or member of the organization as governing the action he contributes" [boldface added].[3]

In other words, the possession of authority is meaningful only if it is accepted by those who receive the commands. Management has no effective authority unless and until the members of the organization confer it upon their supervisors. Barnard suggests that a subordinate will accept a communication as authoritative only when (1) he can and does understand the communication, (2) he believes that it is consistent with the purpose of the organization, (3) he believes it is compatible with his personal interest as a whole, and (4) he is mentally and physically able to comply with it. Thus, if the chief cashier of a bank at closing time on Friday afternoon ordered an employee to leave the bank vault unlocked, this might result in rejection, that is, nonacceptance of the superior's authority by the employee. In the normal course of events, a subordinate accepts management's exercise of authority because it falls within his set of expectations, or because it represents a situation of no concern to him.

Authority is delegated or distributed through the organization to the subordinate managers at the various levels in a scalar chain. That is, the people in the highest authority positions delegate a portion of their authority to those managers directly below them, who in turn do the same with their subordinates. Delegation does not suggest that top management *gives away its authority*, because higher-level managers retain general authority to carry out all necessary functions and grant authority to subordinates as

[2]John M. Pfiffner and Frank T. Sherwood, *Administrative Organization* (Englewood Cliffs, N.J.: Prentice-Hall, 1960), p. 75.

[3]Chester I. Barnard, *The Functions of the Executive* (Cambridge, Mass.: Harvard University Press, 1938), p. 163.

a form of trust only within prescribed limits. The distribution of authority may be substantial in some organizations while minimal in others, as we shall see.

When a subordinate manager accepts authority he also accepts responsibility to carry out the function to which he has been assigned. A subordinate manager should not be held accountable to perform a particular function unless he has the authority to see that it is accomplished. In other words, both the concepts of responsibility and authority flow from the top down, but the final responsibility to accomplish the duties cannot be shifted to a subordinate by higher management. Thus responsibility cannot be shifted to a subordinate by delegating authority. To put it another way, management may delegate authority but can never delegate responsibility, in the sense that the manager is never relieved of the responsibility associated with accomplishing the assignment. All managers, therefore, from the upper to the lower echelons, must continue to exercise responsibility as managers by supervising, controlling, and evaluating the performance of those subordinates who have been assigned authority for specific tasks.

Responsibility and authority should be equal. Whenever a subordinate has been granted the authority to carry out an assignment, he should also be responsible to see that it is accomplished. On the other hand, he cannot be responsible to a degree greater than his authority to see that the task is performed. Sometimes an organization may function well with an imbalance between authority and responsibility. For example, some subordinates may be willing to accept assignments beyond the scope of their authority, but most managers agree that a balance between authority and responsibility is more desirable.

Authority Relationships in the Organization

Organizations can be classified according to the type of authority relationships that govern. Frederick W. Taylor (discussed again later in this chapter) introduced the idea that a foreman's task was too demanding. Traditionally, the shop organization was similar to the lower-level military organization—that is, a **line organization** in which each individual has one person supervising his performance. Taylor suggested that the foreman's task should be divided among eight different foremen, with the purpose of separating **planning** activities from those having to do with direct production, or *performance*. In this design, his four foremen on the planning side included a cost clerk, a route clerk, an instruction clerk, and a disciplinarian. On the performance side he had a gang boss, a speed boss, a repair boss, and an inspector (see Figure 2.3).

Taylor's functional foremanship plan did not get much support. For a workman to have eight line bosses led to much confusion and buck-passing. Taylor's general idea, however, has stood the test of time. But instead of designating the various separate activities as line activities, today they are designated as *staff*, or *advisory, activities* (excluding the position of the gang boss). Thus the cost clerk has become the cost-accounting function, the route

Figure 2.3 Taylor's Functional Foremanship Plan

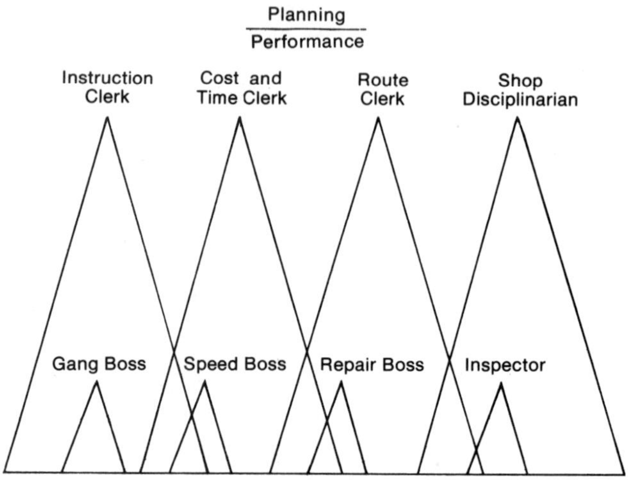

clerk is production control, the time clerk is the payroll department, and the disciplinarian is sometimes located in the office of personnel and industrial relations. Also, the speed boss has become the industrial engineer and the time-and-motion study, the repair boss is known today as maintenance, and the inspector remains as inspection and quality control.

That the line executive needs the assistance of specialists has become a generally accepted principle. The specialists are attached to the line position, serve at the discretion of the line, and exercise only advisory authority, at least in theory. In practice, **staff authority** frequently goes beyond advisory authority, as specialized competence is implicitly accepted by the organization at face value and thus staff is likely to wield considerable *functional authority*. Consequently, employees increasingly find themselves working for more than one boss. For example, an individual may serve as an accountant reporting directly to the division manager who initially hired him. However, this accountant also may report to the accounting organization at headquarters. For pay raises, recommendations for promotion, or transfers, the accountant may be dependent on favorable recommendations by both his line executive and the accounting organization (staff executive) in the home office.

A **committee organization** refers to an arrangement in which committees make important decisions. Though it might be difficult to find an organization that has decision-making committees at all levels, many companies use a top-management structure that operates through decision-making committees. A story is told about the DuPont executive committee, in which the president like the other executives had only one vote, and decisions had to be unanimous. A reporter asked the president if he did not find it embarrassing to think of the possibility that he might be outvoted. The president repied, "No, I take care to state my position first!"

Nevertheless, committee decision making, properly utilized, is often beneficial because it provides an opportunity to exchange ideas and judgments and can result in a higher level of acceptance and cooperation from those who participated in formulating the decision.

Finally, there have been efforts to introduce a **free-form organization** where a fixed organizational structure tends to be unwieldy. The assumption in such an arrangement is that employees are versatile and available for whatever tasks need to be done at whatever moment the need appears. As an example, the team organization, such as that of the Systems Group of TRW (Thompson, Ramo, Wooldridge), is based on a professional research and development environment in which individuals are expected to find and join appropriate groups with a matching of interests and personalities.

The free-form organization suggests flexibility and timeliness of organizational arrangements. Today, a number of research and engineering organizations involve their members in projects and programs that have multiple authority relationships, which change in response to the particular needs of projects in progress. One variation of this, the **matrix organization**, may use a traditional functional arrangement in which engineers and scientists have an organizational home base in their departments. However, as the company prepares bids on projects and has their contracts accepted, the engineers and scientists may then be assigned to work on one or several projects. The project manager responsible for the particular project will have a budget that enables him to purchase the services of the engineers and scientists he needs. The functional managers, in turn, are dependent on this income flow to maintain their departmental operation (see Figure 2.4).

INFORMAL ORGANIZATIONS

It would be impossible to design a formal organization that incorporated all the procedures and rules and all the conditions necessary to provide employees with answers to any situation that might arise. In the final analysis, the organization must rely on the spontaneous behavior and ingenuity of the men and women who have joined it. These individuals find themselves spending a large proportion of each working day in the organizational environment and following its formal requirements. But their special requirements for getting the work done, and their needs for protection, security, maintenance, and survival are not completely provided for by the formal organization. Consequently, there emerges a natural, self-acting **informal organization** to serve the additional needs of the members.

It is possible to identify numerous aspects of informal organizations. There are structures of power, friendship, functional expertise, and communication. These cut across the formal lines of authority reflected in the organization chart and can be pictured as overlays on the formal organization. Thus a structure of functional expertise may bind together the bookkeepers, accountants,

Figure 2.4 Matrix Organization

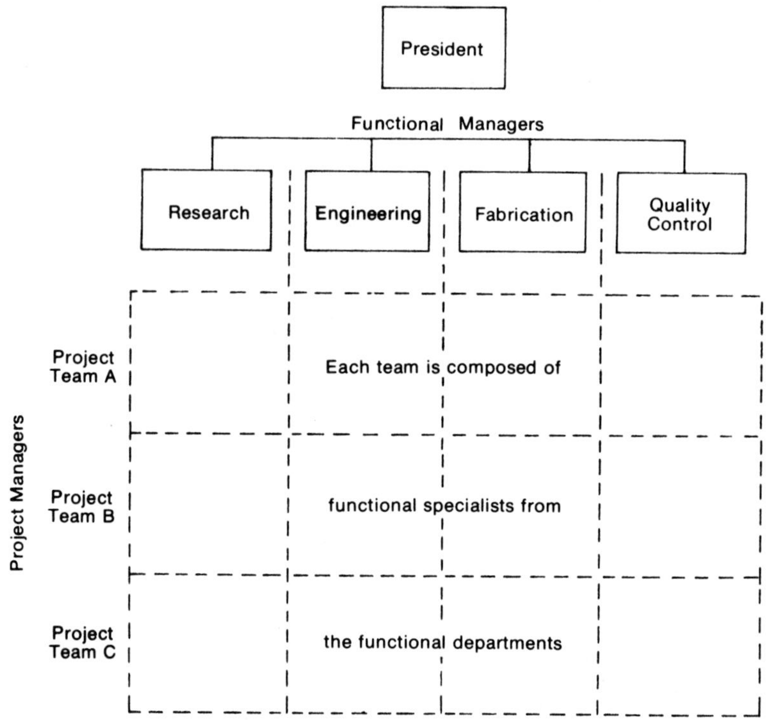

controllers, the vice-president of finance, and the treasurer—based on a common background and a shared set of cultural values. A structure of communication, better known as the **grapevine**, provides information without which we would have very limited knowledge of what really goes on in an organization. Research suggests that, contrary to common opinion, the grapevine is accurate most of the time. Clearly, management should be connected to the grapevine. If the grapevine circulates inaccurate information, management could help to correct it.

Each group structure in the informal organization has its own cultural values and established norms of performance. It can demand conformity to these behavioral norms by applying discipline in the form of group pressure, or social control. Thus the informal group supports the general morale of the organization. Although it is not always the case, good morale in the informal organization will often support achievement of the objectives and goals of the formal organization. On the other hand, the informal organization may also take on the role of critic and at worst, virtually destroy the formal organization by refusal to cooperate.

The relation between activities in the formal organization and in the informal organization is demonstrated in appointments to managerial positions. The formal organization may designate and appoint an individual to become a chief or head of an organization or its subunit and accord him the authority of the position. However, he can become a leader only to the degree that he gains acceptance from members of the informal organization. This acceptance corresponds to his personal capacity and ability. If the chief or head is unable to command personal ability and capacity commensurate with his position, he becomes merely a figurehead.

MANAGEMENT

As indicated earlier, the concepts of management and organization are difficult to separate. However, the foregoing discussion of organizational concepts should help you to understand the concept of management. Management can be considered goal setting, and also allocation and proper use of such resources as people, money, and machines, or people and money—according to whether output consists of products or services. At the same time, the organization is management's means of achieving these ends. We turn now to a discussion of some early attempts to define the function of management.

Scientific Management

Around the turn of the century, Frederick W. Taylor was the motivating force for introducing into practice prevailing ideas on manufacturing management, in addition to adding his own contributions to their development. An eye problem forced Taylor to give up a career in law, and instead he decided to work for the Midvale Steel Company, which was owned by friends of his family. Taylor soon discovered that the factory environment was characterized by a tug-of-war between management and workers. Foremen attempted to increase workers' output, while workers contended that they were already doing their best. Taylor saw men performing the same job by a variety of different methods. He was particularly bothered by the amount of **soldiering** (loafing) he saw. It appeared to him that some workmen were by nature lazy and not overly keen on doing their best. In addition, he observed systematic soldiering, in which production was limited by agreement among the workers and in defiance of management's desire to increase output.

Taylor, who had completed a degree in engineering, developed an interest in getting things done easier and faster. His experiences convinced him of the need to introduce science into management, to eliminate rule-of-thumb methods and management by what he called "drift." He therefore adopted efficiency as his cause. His work *Scientific Management* (still in print) mentions almost every aspect of management known today. He was a tireless worker and researcher and was also instrumental in inventing high-speed tool bits. In addition, based on experiments in the machine shop, he helped develop a slide rule for determining the speed, feed, and depth of metal cuts. By the time of his death he had more than a hundred patents to his name. This scientific attitude, called **scientific management**, was reflected in his whole approach to management.

Taylor's Experiments. Taylor conducted a number of experiments to determine the best method for doing work. He believed that method study was necessary to establish the right way to do a job. Management should then enforce this method so that all workmen doing the same job would, in effect, be adhering to mangement's way. The method selected was the result of a careful analysis directed toward streamlining the job and eliminating unnecessary movements. Thus there was a considerable saving of time and a corresponding increase in productivity. In addition, the newly streamlined job was probably easier to perform than it had been.

As an illustration, one of Taylor's experiments concerned loading pig iron. He established what he believed to be the right method to use and then asked a workman named Schmidt to do exactly as he was told. In return, Schmidt was to receive increased pay. Taylor recognized, just as the economist Adam Smith had noted more than a century earlier, that the nature of this assignment required "a man of the type of the ox, a rare specimen of humanity, difficult to find and therefore very highly prized . . . a man so stupid that he was unfitted to do most kinds of laboring work."[4] The detailed analysis of the job of handling pig iron made improvements possible and established managerial control over the method. Schmidt was this kind of a man and he was able to increase his daily loading to 47 tons of pig iron from the usual 12.5 tons. His wages increased according to Taylor's promise, from $1.15 to $1.85 per day. And the work was no harder than before, for Schmidt now had a schedule of preestablished rest periods.

Once Taylor had introduced a standardized method, it was possible to apply time studies to arrive at a standard output. Thus, if a job took five minutes to perform, it was now possible to establish a **standard** of twelve pieces per hour. However, Taylor recognized the need for soliciting the cooperation of the workers in the new method. He therefore proposed a differential piece rate. Under this system, the worker would be paid a low piece rate if he failed to reach the quota, whereas quota and above would yield a high piece rate. Taylor believed that only the best workmen would remain with him, through a process of self-selection. In addition, his system would give an individual early feedback about his ability to match the requirements of the job. Taylor was convinced that every individual could find a job where he would be a "first-class man."

Taylor's Writings. Taylor's writings on management reflect his two major preoccupations: accomplishment of the job task and a general concern for the employee. He believed that there was a commonness of purpose for management and workers. Management was interested in high profits and workers were interested in

[4]Frederick W. Taylor, *The Principles of Scientific Management*, in Frederick W. Taylor, *Scientific Management* (New York: Harper & Bros., 1947), p. 62.

high wages; both would be able to reach their goals through cost reductions resulting from increased productivity. Taylor spelled out precisely that the critical ingredient of the art of management is obtaining the workers' cooperation.

Taylor characterized scientific management in the following way:

> *Management must assume responsibility for knowledge to apply to the work, or the development of a science of management to replace rule-of-thumb knowledge.*
>
> *Management must assume responsibility for the scientific selection of workmen and their training and development.*
>
> *Management must come together with the workmen in order for science and the scientifically selected and trained workmen to be brought together.*
>
> *Management must assume almost equal responsibility with their workmen in the division of the work and the responsibility for its completion.*[5]

The older, traditional forms of management had led to strife and deceit, not to mention low productivity. Under Taylor's system, the manager and the worker would be cooperating in harmony, with the workman receiving rewards corresponding to his efforts.

Unfortunately, many managers were anxious to reap the advantages of the system by using the stopwatch, standardized methods, and the standard time study—all of which Taylor had introduced—but they failed to take into account the human aspects of the system. As a result, unions became increasingly antagonistic toward scientific management. They saw it as management's way of reducing employment opportunities. The stopwatch became management's snooper. As a result of this controversy, federal legislation prohibited the use of stopwatches in government operated ordnance plants. This legislation remained in force until 1949.

In practice, Taylor violated many principles of his system. His autocratic ways caused difficulties with his clients. He insisted that if they did not accept every bit of his advice, he would not be responsible for the results. His writings were clearly ahead of the times, but management at the turn of the century was not yet ready to accept his entire philosophy.

Taylor's Associates. Taylor had a number of associates who became versed in scientific management. Henry Lawrence Gantt introduced a task and bonus system for wage payments that was somewhat less strict than that designed by Taylor. It reflected Gantt's continual awareness of the manager's responsibilities in training workers and in creating an industrial **democracy**. His most significant contribution to the field of management was the Gantt

[5]Frederick W. Taylor, *Taylor's Special Testimony before the House Committee*, in *Scientific Management*, pp. 41–45.

Chart, the predecessor of all charts used by management to maintain a performance review in relation to planned performance.

Another associate, Frank Gilbreth, became interested in the details of time-and-motion study. He was a construction engineer and had at one time learned how to lay bricks. He had discovered that however hard he tried, he could not match his teacher's output. With his second teacher he did a little better, but it was only with his third teacher that he realized that the brick mason used a different method for laying bricks depending on whether he was going at a slow, an average, or a fast pace. Later Gilbreth devised a set of standard elements by which he believed any job could be analyzed and described. He called these elements *Therbligs* (from *Gilbreth* spelled backward), which were the result of his work with time-and-motion studies, in addition to direct observation and an analysis of the light patterns made by hand movements in the workspace of an individual.

It was fortunate for Gilbreth's career that he married a psychologist. Lillian Gilbreth complemented her husband's activities. She studied the effect of fatigue on labor and suggested a separate allowance for fatigue in establishing a quota for a job. Their combined work emphasized the importance of approaching motion study qualitatively before timing a job and establishing a quota.

Today, when we speak of scientific management, we tend to refer to the historical period from the turn of the century through the 1920s, and to the work of Taylor and his associates. Frequently, their work has been characterized as a **mechanistic view** to manufacturing management, which equates workers with the machines on the factory floor. As we have indicated, their concern transcended this narrow focus.

The character of Taylor's era and the prospect that the alternative to work was starvation created a setting in which financial incentives and management's drive for productivity tended to hold the upper hand. Scientific management was originally applied to repetitive tasks in the manufacturing plant. Although the early studies dealt with factory work at the shop level, the system was later applied to other levels of the organization.

Recently, for example, the ratio delay-time study has been used by defense contractors. To increase productivity, the observer walks through the engineering department on a random time schedule, noting the number of engineers who are at their desks or in the area, presumably working, and the number of engineers who are absent, presumably not working. This ratio reflects the presumed efficiency of the department. The figures are accumulated and returned to the respective engineering department supervisors. The engineers as well as the supervisors frequently protest, however, that they often leave their desks to consult with machinists or draftsmen.

The Managerial Process and Principles Approach

We have seen that much of the early study of management was done by practitioners. These practitioners included many who explored the dimensions of management in the Taylor tradition while attempting to arrive at a set of principles that they believed characterized successful management. These principles frequently involved managerial functions or processes and seemed to exist apart from other managerial functions or philosphies.

A Frenchman named Henri Fayol was one of the first contributors to the view that managerial functions had a separate existence from other managerial theory. Fayol was originally a geologist who, through his technical proficiency, worked himself into a position of management and finally to president and chairman of the board of a large industrial mining complex in France. As he reviewed his career and compared it with that of others, he realized his need to learn more about management. He recognized that technical men were often given managerial positions without sufficient preparation for these responsibilities. However, to provide a proper education in management, there had to be a body of knowledge available, and he saw little of this.

Upon his retirement, he wrote the first of a projected three-volume work, called *Principles of Industrial and General Administration*, published in 1916. In this he suggested the need to recognize the varying mixture of managerial knowledge that was required at different levels of management. A manager should have a combination of technical, accounting, financial, commercial, security, and managerial knowledge. The lower in rank a manager was, the more he had to rely on technical knowledge. As he advanced to higher levels of responsibility, managerial and administrative knowledge per se became more important.

Fayol discussed the critical functions in which he believed every manager was involved: planning, organizing, commanding, coordinating, and controlling. By *planning* he meant the manager's responsibility for developing a vision of the future—setting a course for the organization over a future time horizon. By *organizing* Fayol meant establishing the organizational structure and its resources to enable the manager to have the work accomplished according to the applicable authority relationships, procedures, and rules. By *commanding* Fayol referred to the manager's exercising supervisory and leadership skills. *Coordinating* meant the efforts a manager must put forth to balance the various interests in the organization and establish communication among the units and members. **Controlling** encompassed the multiple aspects of comparing progress to plan and checking and verifying the experience of the organization with regard to personnel, materials, and decisions.

Fayol elaborated further by establishing a set of fourteen principles of management: division of work, authority, discipline, unity of command, clarity of direction, subordination of the individual interests to the general interest, remuneration, centralization, chain of command, order, equity, stability of tenure of personnel, initiative, and **esprit de corps**. He explored these principles in detail, emphasizing that they must be applied with flexibility and moderated by judgment. It was significant that Fayol gave strong attention to the human element in the organization.

Fayol agreed with Taylor in considering management a science that should be studied and taught. However, he took issue with Taylor on two points. First, Fayol believed that the technical aspects of business, and particularly the mathematical engineering-production preparation, were vastly overemphasized in view of broader needs in management. He took exception also to Taylor's emphasis on division of labor, especially the division of the duties of the foreman, believing that this violated his principle of unity of command.

Other Contributers

Though Fayol's work first appeared in 1916, an English translation did not become widely available until 1949. In the meantime, an English management consultant, Lyndall Urwick, helped formulate the strategic functions of the managerial process as they are generally listed today—planning, organizing, staffing or assembling resources, directing (implying motivating, communicating, and exercising leadership), and controlling. Most executives, if asked to talk about their role as managers and the work they do, feel comfortable using the managerial process approach.

In the United States, the 1931 publication of *Onward Industry!* by Mooney and Reiley gave additional support to the direction Fayol's work had taken. Mooney stated that, "organization is an art, and as such, it must have its technique based on principles."[6] Understanding the correlation of these principles became a focal point for their work and typified much managerial thought in the 1920s. The managerial process and principles approach also contrasts Taylor's production orientation with the emerging concern for management and organizations as such.

Mooney's principles included **coordination** as the orderly arrangement of group effort, the scalar principle as related to the formal hierarchy of the organization; and **functionalism**, or departmentation, as concerned with the definition of jobs by process and purpose. Functionalism gives rise to specialization of labor and forms the basis for creating a staff function to support the work of the line organization. Coordination was the superordinate principle to which the others were subject, since it included consideration of authority as the source of coordination, the moral phase of mutual service, the definition of the objectives and goals of the enterprise under its doctrine, and finally, discipline as a special managerial responsibility.

[6]James D. Mooney, *The Principles of Organization*, rev. ed. (New York: Harper & Bros., 1947), p. 4.

Several women were actively engaged in the management consulting field during this phase of scientific and classical management. We have already noted the contribution made by Lillian Gilbreth and her emphasis on the psychology of work. In addition, Mary Parker Follett was involved in exploring the quality of **leadership** in the setting of "the mutual dependence of men in a complex social organism."[7] Follett's career spanned community activities, social work, vocational guidance, and also business and industrial organizations. Her work with the management of New England Telephone and Telegraph gave her an early appreciation of the challenges facing managers in their daily activities. She was a forerunner in emphasizing that management must be supportive of the employees in the organization and must proceed from a level of gaining the consent of the governed, to true democratic participation.

Follett's most famous contribution was to regard **conflict** as an opportunity for achieving progress. She suggested that we have the three choices of domination, compromise, and integration in solving conflicts. **Domination** obviously indicates that one side wins, sometimes very easily, but perhaps not in the long run. **Compromise** is frequently resorted to but never fully appreciated by those involved, because it means that each side must give up something. Ideally, management should seek integrated solutions in which all parties can find satisfaction. None of the parties then believes that it has had to make an undue sacrifice. The key to integrated solutions is to review the facts of the conflict in new ways in order to expand the number of possible solutions. Through open discussion, it may become clear that the facts in the situation can lend themselves to a creative solution on another plane that had not been considered.

We should note that Fayol, Urwick, Mooney, and Follett carried out their discussions on management without restricting themselves to business organizations. Rather they emphasized that the approaches to management and the principles they developed were applicable to military organizations, hospitals, educational institutions, and any other organization setting.

Bureaucratic Management

Around the beginning of the twentieth century, Max Weber, a German scholar whose work spanned the fields of comparative law, sociology, political science, comparative religion, and economic history, made a landmark contribution to managerial and organizational theory. His concepts are introduced at this point because, although many of his ideas belong with the classical school, they seem to bridge the gap between the traditional concepts of organization and management and more modern ideas (the subject of the next chapter). In some respects, Weberian **bureaucracy** is based on mechanistic assumptions about human nature in the pre-1930s

[7]Harold F. Smiddy and Lionel Naum, "Evolution of a 'Science of Managing' in America," *Management Science*, vol. 1, no. 1 (October 1954), p. 16.

tradition, whereas in other respects, his concepts match the modern scholar's search for general theories and understanding and the idea of the ultimate predictability of organizational behavior.

In Weber's view, the bureaucratic form was the most efficient type of organization possible in modern society. Basically, he wanted to construct an ideal organization that reflected rational human behavior. This attitude obviously differs from the label "bureaucratic" often used negatively to describe large-scale, cumbersome governmental or business offices. Speaking of the technical advantages of bureaucracy, Weber said:

> *The decisive reason for the advance of bureaucratic organization has always been its purely technical superiority over any form of organization. The fully developed bureaucratic mechanism compares with other organizations exactly as the machine with the nonmechanical modes of production.*
>
> *Precision, speed, unambiguity, knowledge of the files, continuity, discretion, unity, strict subordination, reduction of friction and of material and personal cost—these are raised to the optimum point in the strictly bureaucratic administration, and especially in the monocratic form.*[8]

Weber emphasized that the theory of bureaucracy was relevant to all forms of organization. He stated that its use could increase efficiency in such organizations as business enterprises, governmental units, military operations, and labor unions. To develop the "ideal" bureaucracy, certain conditions or policies should be followed: (1) a division of labor based on functional specialization, (2) a well-defined hierarchy of authority, (3) a system of rules covering the rights and duties of positional incumbents, (4) a system of procedures for dealing with work situations, (5) impersonality of interpersonal relations, and (6) promotion and selection for employment based on technical competence.[9]

Weber's contention was that man is unpredictable, often emotional, not necessarily rational, and that he possesses characteristics that tend to interfere with efficient organizational performance. In accordance with this philosophy, he set forth the depersonalized form of organization as an ideal that would minimize the effect of dysfunctional human behavior.

Rather than diminishing in importance, as have some aspects of Taylor's scientific management, the bureaucratic organization seems to be holding its own. The bureaucracy emphasizes predictable institutional performance and positive control, perhaps the two most critical managment variables in large, complex organizations. As we shall point out in our discussions of contingency theory in the next chapter, a high degree of structure and

[8]H. H. Gerth and C. Wright Mills, *From Max Weber: Essays in Sociology* (New York: Oxford University Press, 1946), p. 214.

[9]Richard H. Hall, "The Concept of Bureaucracy: An Empirical Assessment," *American Journal of Sociology* (July 1963), p. 33.

centralized authority both seem appropriate when markets and technology are relatively stable. Although such conditions were more characteristic of conditions in the United States during the period between the Civil War and World War II, nevertheless, the complexity of many large organizations today and the relatively limited amount of innovation required in some industries continue to provide a basis for using Weber's ideas of management and the ideal organization.

SUMMARY

This chapter reviewed some of the traditional ideas that have contributed to the managerial function. Noting that many of these ideas originated with early church, military, and political leaders, the basic elements of organization and management—structure, hierarchy, the division of work through departmentation, and the concepts of authority and responsibility were described.

The early twentieth-century contributions of such practitioners and theorists as Taylor, Fayol, Mooney and Reiley, and Follett to managerial processes and principles were outlined and the early scientific work of Max Weber in organization theory was noted.

This period marked the beginning of a transition in managerial thought from a principles approach and management by intuition to a more scientific analytical handling of the problems of management and organization. In the next chapter, the exploration of this transition will be continued by reviewing the growing contributions of the behavioral sciences, which emphasize the human dimension of the managerial problem.

DISCUSSION QUESTIONS

2.1. If you compare Taylor's approach to management and organization with that of Fayol and Mooney, what similarities and differences do you discover?

2.2. What were the features of the bureaucratic organization as defined by Max Weber?

2.3. What features of Weber's model are still used today by large organizations?

2.4. Compare the Taylor and Weber approaches to improving organizational efficiency.

2.5. In what ways is the informal organization important in a business firm, a government office, or in any type of organization?

2.6. Is authority a part of job assignment or is it earned? Discuss.

2.7. How does responsibility relate to authority?

APPENDIX
Non-Linear Systems, Inc.—A Reorganization

Some years ago, an electronics company in California experienced a major reorganization. The President, Mr. Andrew F. Kay, had become acquainted with the writings of many authorities on contemporary management and organization. He became convinced that for this small company, there was a potential benefit both in human and business terms to change the form of organization.

Non-Linear Systems was in the business of making electronic instrumentation products and test systems. The products were protected by patents. The company was owned by Mr. Kay; he controlled all of the stock. Annual sales in the 1960s amounted to about $6 or $7 million. The market was quite competitive.

One day Mr. Kay announced a change to his employees. He discontinued the use of time cards, disbanded the assembly line, introduced a salary base for everyone, and indicated to the employees that they were free to organize their work as they saw fit. With one specialist on electronics acting as team leader, sixteen production teams were established. Under its team leader, each group of six employees was able to decide how to organize its effort.

The company's organization incorporated the traditional functions of sales and finance, as well as supporting staff functions such as engineering and accounting. The accounting department was eliminated and the accountants were reassigned to areas such as personnel, shipping, and purchasing. The departments now kept their own books and reported to the treasurer of Non-Linear Systems. An attempt was also made to get rid of the drafting room. Teams were created for engineering and research, consisting of an engineer, a draftsman, a designer, and a technician. However, the need for a central drafting department became too strong as the pooling of drafting talent turned out to be a better use of this resource. For the salesmen, the requirement of filing expense reports with the office was eliminated, and each salesman was given an allowance to cover his selling expenses. This plan also was scrapped as selling expenses became higher than under the old system.

The executives immediately under Mr. Kay were given new titles indicating functional responsibilities for such matters as "productivity," "profitability," "performance and development," "market standing," "physical and financial resources," and "public responsibility." The press reported, however, that these members of the executive council of Non-Linear Systems were uneasy about their new titles. They desired more specific definition of their responsibilities. Confusion was also created because the next level of executives now had to report to each member of the executive council. Discussions carried out under a specialist in sensitivity training were unsuccessful. The executives reverted to their old titles after a year.

1. What aspects of our discussion on organization can you find illustrated in the Non-Linear Systems experience?
2. Why was Non-Linear Systems unsuccessful in changing the staff and managerial organization?

SELECTED READINGS

Drucker, Peter F. *The Practice of Management*. New York: Harper & Row, 1954.

Fayol, Henri. *General and Industrial Administration*. London: Sir Isaac Pitman & Sons, 1949.

George, Claude S., Jr. *History of Management Thought*. Englewood Cliffs, N. J.: Prentice-Hall, 1968.

Hutchinson, John G. *Organizations: Theory and Classical Concepts*. New York: Holt, Rinehart and Winston, 1967.

Mooney, James D., and Reiley, Alan C. *Onward Industry*! New York: Harper & Bros, 1931.

Taylor, Frederick. *Scientific Management*. New York: Harper & Bros., 1947.

Wren, Daniel A. *The Evolution of Management Thought*. New York: Ronald Press, 1972.

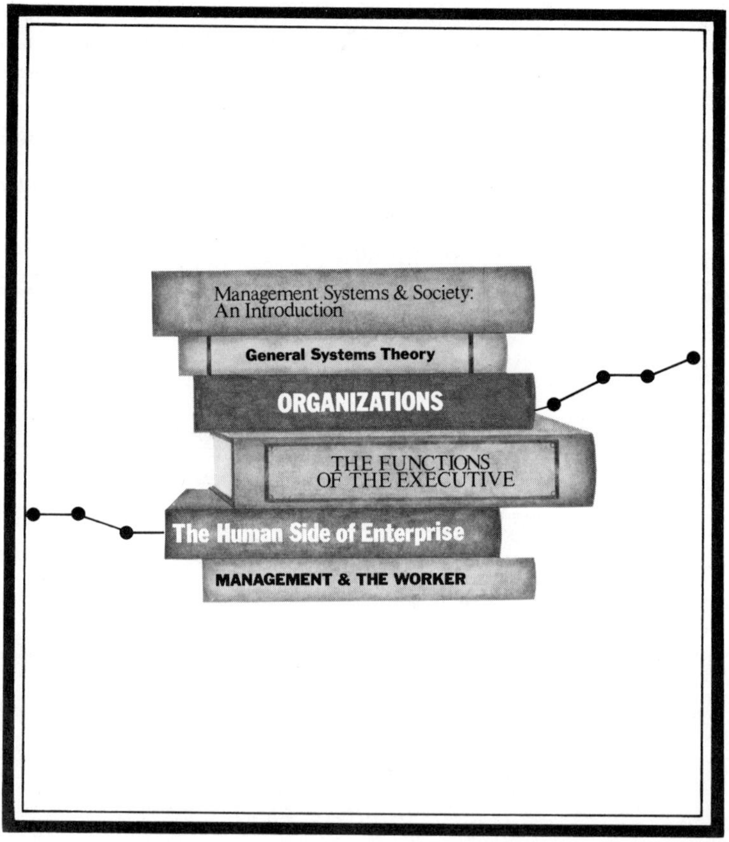
Management Systems & Society:
An Introduction
General Systems Theory
ORGANIZATIONS
THE FUNCTIONS
OF THE EXECUTIVE
The Human Side of Enterprise
MANAGEMENT & THE WORKER

Chapter 3
Transition in Management

During the 1930s, several important events changed the course of managerial thought. In 1933, Kurt Lewin, a brilliant **behavioral scientist** from Germany, began a systematic study of the effects of internal and external forces on human behavior. Also in 1933, Elton Mayo published the first report on a series of investigations, the Hawthorne experiments,[1] which raised serious questions about the mechanistic assumption associated with the earlier work of Taylor and Weber (discussed in Chapter 2). These experiments indicated that social and psychological factors in the work environment appeared to be of equal or greater importance to the workers' productivity than money, work methods, or the hours and conditions of work. The Hawthorne experiments triggered what has become a growing and continuing effort among behavioral scientists and practicing managers to unravel the riddles of worker **motivation.**

In this chapter, we will explore the transition in managerial thought as we continue our survey of the practical and scientific backgrounds of organization and the functions of management.

THE HAWTHORNE STUDIES

In the late 1920s, the management of the Hawthorne plant of the Western Electric Company, Cicero, Illinois, became interested in studying the fluctuations in productivity of their assembly line workers. They thought these fluctuations were related to working conditions, and that if such a relationship could be established, productivity could be controlled. The management called in a group of researchers from the Massachusetts Institute of Technology to make the study. The original experiments involved illumination—that is; the research team attempted to establish a simple correlation between the experimental group's output and the relative intensity of lighting. The researchers maintained a control group for whom working conditions were not changed. Productivity increased in the experimental group regardless of

[1]Elton Mayo, *The Human Problems of an Industrial Civilization* (New York: Macmillan, 1933; reprinted by Division of Research, Harvard Business School, 1946). A more complete and comprehensive account of the Hawthorne research is found in F.J. Roethlisberger and William J. Dickson, *Management and the Worker* (Cambridge, Mass.: The President and Fellows of Harvard College, 1939; reprinted by Harvard University Press, 1956).

whether illumination was increased or decreased, and to the surprise of the researchers, output increased in the control group as well. Having no explanation for this, the researchers abandoned the experiment. It had become apparent that other things besides working conditions were causing the fluctuations in productivity.

A new experiment was organized at the Hawthorne plant under the supervision of Elton Mayo, who was affiliated with the Harvard Business School and whose associates included Fritz Roethlisberger and George Homans. This team used a significant new research methodology—they approached the analysis of the factory as if they were sociologists or anthropologists studying a preliterate tribal village.

Elton Mayo's research team recognized the need for exploring more dimensions than merely the physical working conditions. For their investigation they established an experimental setting involving a relay-assembly test group of women and an observer. In this group they proceeded to change wage payment plans, frequency and duration of rest periods, hours of work per day, and lunch arrangements. The conditions of the experiment related to physical working conditions and were changed in consultation with the workers who were assigned on this experimental assembly line. In addition, the researchers incorporated in their notes information about all aspects of the events affecting the women, both inside and outside the working environment. They recorded a general increase of output, and the conclusions that were drawn about the increases were related not only to the work place but also to other conditions surrounding the work. In brief, the women had responded with greater effort to such incentives as the friendly atmosphere they sensed in the test room, the belief that what they were doing was important, the less rigid supervision, and the opportunity to participate in decisions affecting their work.

Another famous experiment involved a group of employees engaged in installing electrical wiring in a manufacturing operation. The researchers observed that the workers split themselves spontaneously into one group of individuals who considered themselves an **elite** group, and another group composed of the remaining workers who believed that they had lower status. The social organization that emerged in the wiring room determined the entire group's output, their relationships with superiors, and the relationships between individual members and the cliques within the group. The group's general objective was to protect the members as well as to enforce conformity to the group's norms. The group presented a united front resisting suggestions of change from supervisors. This experiment demonstrated the power of social pressure to alter supervisory demands and to influence the behavior of individuals in work groups.

The Human Relations Movement—Organizations as Social Systems

The Hawthorne studies identified the need to focus on the organization as a social system that has both a formal and an informal structure. People are individuals who have psychological as well as social needs. Group pressure (social control) determines the norms to which members of the group must adhere. By this means, the group can protect its members against excessive control by management and against pressure to increase output through change. These studies found that the interests of management and employees may indeed be different and that employees must be provided with psychosocial, in addition to economic, means to gain satisfaction on the job.

Elton Mayo pursued a particular cause. He perceived that industrialization had brought with it *anomie*, or normlessness, to the industrialized, urban working world. Individuals had been uprooted and turned into impersonal, mobile, competitive units, isolated in the city culture and given little chance to feel secure in the thought that they belonged to a human community.

Mayo believed that the established society had been displaced by the adaptive society of the industrialized urban community, and so he advocated that administrators should assume responsibility for creating working conditions that would give workers a sense of belonging. The Hawthorne studies appeared to show that people worked for several reasons: (1) financial security to meet present needs, sickness, and old age, (2) good working conditions, which included attention to physical facilities and friends, quality of supervision, and other aspects of the working environment, (3) opportunity to advance, which did not imply a requirement to advance but rather a desire to match capabilities with the job requirements and the possibility for growth and development of these capabilities, (4) recognition in the form of feedback expressing the importance of their contribution to the success of the total enterprise, and (6) participation, which represented a belief in a democratic society and the right of people to have a voice in matters that affect them.

Mayo interpreted the Hawthorne findings as proof of his theory that returning to the ideology of an idealized medieval rural community would yield beneficial results both for workers and for management. Mayo believed that a harmonized utopia in the industrial workers' world was possible.

Other researchers have accused the proponents of human relations of glad-handing and using the image of "contented cows" as a standard for working conditions. The critics have pointed out that the Hawthorne and similar studies gave little attention to the union movement. As a result, the researchers involved with the Western Electric Company and subsequently behavioral and social scientists generally have been accused, perhaps unjustifiedly, of "selling out" to management.

In any case, the **human relations movement** was responsible for introducing the social and behavioral sciences into business and industrial management. This trend has continued, and today we see sociologists, social psychologists, cultural anthropologists, and political scientists pursuing their research in all types of organizations. The original bias of the Hawthorne researchers for the priority of human values has remained the benchmark for the work of many of today's social and behavioral scientists.

EARLY MANAGEMENT THEORY

Chester Barnard was one of the most important contributors to the theory of management during this pioneering period. He was an executive with New Jersey Bell Telephone Company and had a great interest in the social sciences. He had become acquainted with members of the Harvard faculty and followed the work of Elton Mayo and his associates. Combining his knowledge of the social sciences with his own practical experience as a manager, Barnard developed a classical theoretical statement on management in his book, *The Functions of the Executive*, first published in 1938.[2] This book generated hypotheses that could be tested and verified through research. Barnard's work later yielded new and revised managerial theory, thus ultimately helping to create the possibility of prediction and control in management.

The organization can join the physical capabilities, reasoning, ideas, and other talents of single individuals to create a planned *coordinated* effort. What one person could not accomplish alone could easily be achieved through an organization relying on the *cooperation* of its members. Barnard viewed the organization as a social system that had been created by man to overcome human biological limitations. One of these social systems is the impersonal organization, in which the members' efforts are coordinated toward achieving certain objectives and goals. But successful coordination depends on the goodwill of the individuals and groups involved. Their cooperation was needed to enable the organization to contend with the environment and to help assure satisfaction for individual members. This result was labelled the *efficiency* of the organization, in Barnard's terminology.

As mentioned in Chapter 2, Barnard also introduced the notion that authority flows both up and down in the organization. Authority signifies that the manager has the right to issue orders, to decide, and to act. However, as we have said, this authority is nullified if an employee does not accept it. Before the manager can gain acceptance, the employee must believe that the order conforms to the objectives and goals of the organization and, in addition, must be physically able to perform as directed. The employee must also understand the meaning of the order and must believe that it is in his personal interest to carry it out.

In Barnard's terms, the theory of management hinges on the balance of exchange, or the equilibrium achieved, when employees weigh their contributions to the organization against the inducements provided them. The way an executive functions is particularly critical in attaining and maintaining this stability. Barnard believed that the manager's duty was to provide moral leadership, to infuse the organization with an ethical code and thus overcome the pull of selfish behavior for the ultimate goal of achieving the interests of all. Because of its logical theoretical scheme, Barnard's *The Functions of the Executive* has proved very influential in sub-

[2]Chester I. Barnard, *The Functions of the Executive* (Cambridge, Mass.: Harvard University Press, 1966).

sequent academic study and research in management and organization and ranks with Taylor's *Scientific Management* as one of the most important early contributions to the theory of management.

Centralization and Decentralization

During the 1920s, the managements of DuPont and General Motors were exploring the type of management that would complement their large and enormously complex business units. As DuPont at this time had acquired more than twenty percent ownership of General Motors in return for helping that company out of financial difficulties, managers of both corporations were exchanging ideas about organization. In *My Years with General Motors*, Alfred Sloan has related his more than forty years' experience as an executive with that company and has also reflected on the importance of his early study of decentralization for the company throughout its growth. His ideas generated a unique theme for managers of industrial organizations.

As long as the founder-owners of organizations maintained control of their companies, authority tended to be **centralized.** The company was in effect their personal property, and they dealt with it as an extension of their personal assets (and sometimes their liabilities). The objectives and goals of the **entrepreneur**-owner were also the objectives and goals of the company. Little authority and responsibility were delegated to others.

However, the immensity of companies in certain industries made centralization and centralized control increasingly difficult to maintain. The term **decentralization** was used originally to suggest the dispersion of physical facilities over a large area, as companies spread out geographically. The growth of organizations in size also made it difficult to rely on a functional division of work. Work was traditionally divided into manufacturing, selling, financing, purchasing, and other functions. This became cumbersome for a large organization that had units located throughout the nation, many of which were larger than most of the other companies in the area. Thus Alfred Sloan and his associates at General Motors and DuPont proposed that division of work be based on product. The product division managers would report to the chief executive. This **divisionalization** or departmentation by product also was referred to as decentralization.

It soon became clear that it was impossible to manage the separate product divisions from a central headquarters and also control the day-to-day operations. It was necessary to delegate authority and responsibility to the division managers, who were charged with running their divisions as if they were separate enterprises. This policy of a company was implemented only to the degree that it adopted a *managerial philosophy of decentralization*, or maximum delegation of authority and responsibility to the point of action, supported by a fundamental belief in the high quality of subordinates and co-workers.

The president of a very successful bank once said: "I believe in decentralization. The branch managers run their offices as independent businessmen in the towns where they are located."

Later, in discussing the maximum dollar amount that a branch manager could authorize, he remarked: "He is authorized to sign for $250. Anything above this requires central approval." This clearly suggests the importance of having a managerial philosophy of decentralization to assure that actual behavior is consistent with stated policies. Alfred Sloan labelled the General Motors approach as "decentralized management with centralized control," maintained by unique financial and accounting reporting systems. DuPont contributed the concept of independent profit centers in which managers are responsible for a specific return on investment for the unit over which they have charge.

The concepts of centralization and decentralization remain important in managerial thought today. Most large national and multinational corporations pattern their management structures after the decentralized models introduced by General Motors and DuPont. The most commonly cited advantage of decentralization is the opportunity to make maximum use of individual skill and drive. In addition, it disperses knowledge about the organization and its work to managers in the subunits.

Ralph Cordiner, president of General Electric in the early 1950s, stated that unless "we can put the responsibility and authority for decision making close to the scene of the problem, where complete understanding and prompt action are possible, the Company will not be able to compete with the hundreds of nimble competitors who are, as they say, able to turn on a dime"[3] On the other hand, the increasing use of centralized information systems may lead to efforts to recentralize management, as communication no longer presents the barrier it used to in setting limits on growth and size.

Finally, let us illustrate the concepts centralization and decentralization with a classic example from the annals of corporate history. There are two famous mail-order houses that have national and international distribution of their retail services and mail-order centers, and thus have widely dispersed physical facilities. If you entered the store of one company and found an electric range with a corner damaged, you might propose a reduction in price to the salesman. He would very likely respond: "I have to see the manager about that." The manager would inspect the damage and agree that a reduction may be merited but add: "However, I will have to write to headquarters to get authorization." Disappointed, you might decide to try the other firm's store. After a diligent search, you find a damaged range and make the same proposal: "How about a $25 cut in price?" The salesman for this store could bargain with you and acccept or reject your offer. The differences in the effect on customers and customer relations are not hard to see in these contrasting policies on the locus of authority.

[3]Ralph J. Cordiner, *New Frontiers for Professional Managers* (New York: McGraw-Hill, 1956), pp. 45–46.

BEHAVIORAL SCIENTISTS—THE REVISIONISTS

Many investigators in the behavioral sciences during the 1930s developed experiments and theoretical schemes that were later incorporated into the study of organizational behavior and management. One of the leading researchers in this field was Kurt Lewin. Continuing in the United States with work he had begun earlier in Germany, Lewin developed his **"force-field concept"** of human behavior. His force-field theory conceived of behavior as the resultant of **vectors** (forces) of different strengths and directions impinging on the individual and determining his behavior. These vectors can be described as all forces, both internal (psychological and biological) and external (social and physical) operating within the individual's "life space," or total sphere of experience. Conflicts arise when the individual encounters opposite forces of equal magnitude.

Lewin later extended his force-field concept to analyzing social change.[4] It still remains a useful way for managers to analyze the variables that maintain the present state of an organization or organizational subsystem and to evaluate and isolate those forces that, if altered, would help bring about a desired change in the system or subsystem.

Figure 3.1 illustrates the application of this concept to a hypothetical problem—increasing the level of production in a boat yard. The present state of the system, that is, the level of production, is held in what Lewin would call quasi-stationary equilibrium by two sets of opposing forces—those increasing and those decreasing. The algebraic sum of these forces is zero. The production level can be changed by changing the balance of forces—either by decreasing any or all of the forces (*a*) through (*d*), or by increasing any or all of the forces (*e*) through (*h*). The vertical scale at the left of the figure helps management *estimate* the relative importance, or strength, of each of the forces involved (represented graphically by the lengths of the arrows) and provides an order of priority for dealing with them. For example, if forces (*a*) and (*d*) could be increased, the production level would improve. Similarly, if supervision, force (*g*), could somehow become more effective, production could also be expected to increase. Note, however, that changing the strength of some forces in the force field could result in increasing the existing *tension* level in the system without changing the level of production.

Lewin's notion that a field of force governs behavior called attention to the significance of studying the individual in relation to the total organization of which a person is a part, and also of studying how he or she is related to other individuals and objects. Expressed algebraically, behavior (B) is a function of individual personality (P) and the environment (E), or $B = f(P, E)$. This classic expression serves to remind us of the sometimes neglected fact that behavior can be influenced or changed through a change in an

[4]Kurt Lewin, "Group Decision and Social Change," in E.E. Maccoby; T.M. Newcomb; and E.L. Hartley, *Readings in Social Psychology*, 3rd ed. (New York: Holt, Rinehart and Winston, 1958), pp. 197–211.

Figure 3.1 Force-Field Analysis in a Boat Yard

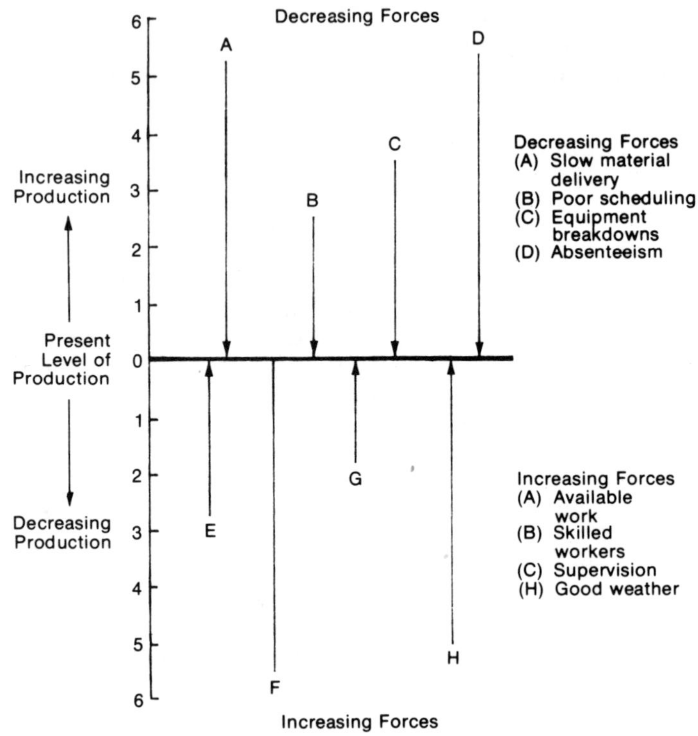

individual's relevant environment as well as through a change in the individual himself. The organization can be viewed as a system of position-roles, each surrounded by other people and objects. The definition of any one position-role is a reflection of the individual's own definition of the role combined with the expectations of all the other members of the organization, functioning in their position-roles, and of the influence of the material and ethical environment.

Lewin went on to explore the dimensions of leadership behavior and the processes of accomplishing change in the organization, as we shall see in later chapters. His work also provided a basis for further explaining the results of the Hawthorne studies, that is, the importance of the social control exercised by the group over its membership. This interest in group behavior has continued and is commonly known as **group dynamics.**

Other Behavioralists

A large number of behavioral scientists working in the 1950s concluded that management should assume a double responsibility. Not only should the organization successfully realize its objectives and goals by adhering to the logic of cost efficiency, but it must also recognize that employees need to achieve their personal objectives and goals in the process. The *logic of efficiency* had to be combined with the *logic of sentiment*, or emotions.

Lewin's work and the Hawthorne studies described the problems associated with poor management during World War II, and for the postwar years demonstrated the value of studying management and organizational theory in universities and of recognizing the potential for considering management an independent profession. In addition, the employment of behavioral, physical, and natural scientists in the war effort (both in Great Britain and in the United States) underscored the need for further research and development in the operation of organizations and in the control of the constituent economic and human resources

Argyris and McGregor

The behavioral scientists' contention that organizations cannot achieve their objectives at the expense of the employees has persisted. This revisionist approach to management and organization is well stated in the work of Chris Argyris, who explored these ideas extensively in his book, *Personality and Organization.* Argyris proposes that individuals proceed through a developmental process from infancy to adulthood—from a state of passivity to a state of increasing activities, from a state of dependence to a state of relative independence, from being capable of behaving in a few limited ways to behaving in a variety of ways, from having a few shallow interests to having deep interests. In addition, individuals progress from possessing a short-term perspective to a long-term perspective, from a subordinate position-role in the family setting to an expectation of equality and even superiority, and from a lack of awareness of self to awareness and control of one's self. In contrast to this, Argyris believes, when the individual becomes an employee in an organization, he or she is expected to conform, to do what is ordered, and to behave very much like a child! The organization represents an ordered world where predictability is assured through employees who submit to the demands of a specialized task, to the requirements of lines of authority, and to the characteristics of the organizational structure, including taking orders from a boss who exercises additional control by having only a limited number of subordinated reporting to him, and for whom he is responsible.

As an illustration, Argyris investigated an industrial plant and found a situation that appeared to reflect a detached and uncommited relationship between the employees and their work environment. In addition, he noted an absence of friendly interaction among the employees and very little interest in the objectives of the organization and its management. At the same time, management regarded its plant as a productive unit manned by employees who caused very little trouble. Using Argyris's assumptions, however, we can understand why he would consider this company to be in trouble. It represented a situation of stability based on management's limited desire for growth and development and the employees' low levels of expectation and aspiration.

Argyris urges that all employees should have a maximum opportunity to participate and express themselves at work. He used the term **fusion** initially to refer to a situation in which both the organization and its employees achieve the highest state of satisfaction of needs. Through the term *fusion* Argyris conveyed an ideal mix of company and individual objectives, in which the goals of the two have become literally indistinguishable from each other. Later Argyris used the term *integration*, which appears to suggest that the objectives of the organization and those of the employees remain separable, though intermeshed. The personal and social conflicts that occur when this intermeshing or integration is not achieved are explored further in Chapter 11.

Continuing the revisionist approach to management, Douglas McGregor stated: "If employees are lazy, indifferent, unwilling to take responsibility, intransigent, uncreative, [and] uncooperative,"[5] the difficulty probably stems from the use of traditional assumptions and methods of management and control. McGregor summed up his position by saying that the "essential task of management is to arrange organizational conditions and methods of operation so that people can achieve their goals *best* by directing *their own* efforts towards organizational objectives."[6]

Clearly, the revisionists were strongly concerned with the idea of employees' satisfying their needs in their work environment. This idea reveals an interesting underlying assumption—that individuals who work for a living in business and industry or other organizations look mostly to their work environment for satisfaction. Management is asked, in effect, to believe that employees seek to satisfy their highest motivations within the organization that employs them, and management must therefore create conditions at work that will enable employees to achieve this satisfaction.

Whyte, Likert, and Bennis

William F. Whyte, Jr., and Rensis Likert have elaborated further on these ideas. Whyte relies on the notion that, in addition to activities and sentiments, interactions are important in creating satisfaction. Interaction may refer to written or oral communication, the silent language of gestures and facial expressions, or activities determined by position in the work group.

In an investigation of restaurant work, Whyte analyzed the relationships between the customer who gives an order to the waitress, the waitress who gives the order to the pantry or store room boy, who in turn passed on the order to the chef. This interaction flow, according to Whyte, is contrary to the status ranking or hierarchy of the individuals involved. The chef is in fact ordered about by the least important person though he has the highest rank. As a result, conflict and animosity develop. Whyte's findings indicate that using an impersonal order holder or spindle, which

[5]Douglas McGregor, *The Human Side of Enterprise* (New York: McGraw-Hill, 1960), p. 48.

[6]McGregor, "The Human Side of Enterprise," *Management Review*, vol. 46, no. 11 (November 1957), pp. 22–28, 88–94.

leaves the chef free to organize his orders as he sees fit, preserves a work flow that corresponds to the status hierarchy, thus eliminating the basic conflict.

Rensis Likert, long-time head of the Institute of Social Research, University of Michigan, emphasized measurement of social data. He and his associates studied employees in an insurance company's office force, railroad section gangs, earth-moving equipment plants, and a nationwide moving van service to explore the relationship between productivity, leadership behavior, employee satisfaction, and other dimensions of the work environment. He used questionnaires and survey instruments that made it possible to report findings in quantifiable terms.

Organizational psychologists, such as Likert, suggest that the authoritarian leadership style of traditional management should be changed into a democratic, employee-supportive leadership style. Likert has attempted to demonstrate that a supportive and participative leadership style results in improved organizational performance and also serves the needs of employees for professional development and growth.

The notion of a reality-oriented approach to management—in the sense that it takes into account the position of the organization as well as the position of the employees—has been set forth particularly by Warren Bennis, who with McGregor has forcefully argued for a shift from a style that the latter labelled Theory X philosophy in management to Theory Y, also a McGregor invention. Theory X assumes that an individual is innately lazy and unreliable; consequently, organization and control must be based on authority imposed from the outside. Theory Y assumes that an individual can be basically self-directed and creative; management is assumed to have an accepted leadership position.

The recent work of Warren Bennis, based on his experiences in an administrative assignment at the New York State University at Buffalo, indicates that these thoughts are still under development. In evaluating his experiences during the late 1960s, Bennis concludes that he has revised his thinking regarding managerial decision making by consensus.[7] He argues that this is not always feasible in the day-to-day realities of the organization, in which there is the inevitable sense of excitement and commitment as people play for power and control. This revised attitude seems particularly relevant today, as management expands its responsibilities beyond the boundaries of the organization to include the social and natural environment in its decision-making analyses.

THE VALUE APPROACH TO MANAGEMENT

Management makes decisions today under the careful scrutiny of numerous **interest groups**. In the past, management could narrow its focus to the affairs of the organization and make decisions from the point of view of what best served its objectives and

[7]Warren G. Bennis, "A Funny Thing Happened on the Way to the Future," *American Psychologist*, vol. 25 (July 1970), pp. 595–608.

goals. But in the present, we see emerging a client- or consumer-oriented society.

Generally when we talk about science we think of its dependence on precise and logical knowledge that can be expressed in quantitative or mathematical terms. In contrast, ethical knowledge is philosophical and depends on personal values. It must be acquired firsthand by an individual and applied directly to choices or decisions in everyday life. Research and investigations increase scientific knowledge, which is therefore cumulative as each successive scientist draws on the results of those who worked in the field before. However, in regard to ethical knowledge, each person must learn anew a framework of value attitudes that will be unique to him alone. Ethical knowledge, therefore, is noncumulative, though it is likely to reflect the shared values of society in general. The higher up in the organizational hierarchy a decision is made, the greater the role the ethical component plays. Managers cannot claim that their decisions are only economic decisions derived through strictly scientific methods; they are also ethical decisions and as such must reflect the manager's personal values.

The **value approach** to management assumes that the organization exists in a social, political, and economic context. Management takes place within this environment and must recognize that the decision to participate or not to participate is a reflection of the organization's moral responsibilities, which will affect society's sanction for the organization to exist within it. Society's norms include an expectation that organizations will be sensitive to the opinions of consumers and clients—to a greater extent now than ever. If managers ignore this expectation, the public view will ultimately prevail nevertheless, through laws and regulations passed to protect the position of consumers, customers, and clients.

Management is recognizing the inadequacy of defining business or progress according to the logic of cost efficiency, profits, sales, or share of the market. It must prepare itself for the question: What contribution is the organization making to society, and what values is it supporting? We will elaborate in later chapters on management's widening social responsibilities.

Decision-Making Model

How can managers make good decisions when they must consider the organization's interests, its employees, and the environment simultaneously? One group advocates focusing on the decision-making process as the key to improving management. These writers have rejected most of the traditional concepts, and emphasize human problem-solving processes and decision mechanisms as primary forces in determining organizational behavior.

According to Herbert Simon, who further developed the ideas originally introduced by Barnard, organization members should not be viewed as robots but instead as individuals with wants, motives, aspiration levels, drives, and perhaps a limited rationality and capacity for problem solving. Thus management theory ought to revolve around decision making. Though economists had assumed that man acts rationally as a decision maker, Simon suggested that no individual has access to all the necessary information. All that can be assumed is that individuals act as rationally as they can, given limited access to knowledge, a limited time horizon, and thus, limited choices. Within this bounded rationality, Simon argued that individuals will engage in **satisficing**, or roughly adequate behavior, rather than striving for wholly satisfying behavior.

Simon thought that each decision could be divided into a *factual* component and an *ethical* component. The factual component could be judged for correctness or incorrectness according to tangible data. The ethical component could be judged only in terms of good or bad; only by hindsight would information be available for judging how correct or incorrect the ethical components of a decision were. As a manager rises in the organization to higher levels of responsibility and to more encompassing decisions, this ethical component becomes correspondingly larger.

Simon used the term *decision making* as though it were synonymous with *managing*.[8] In this context, decision making has three principal stages: (1) **intelligence**—searching the environment for conditions calling for decisions, (2) **design**—inventing, developing, and analyzing possible courses of action, and (3) **choice**—selecting a particular course of action from the available alternatives.

Herbert Simon and James March expanded the concept of decision making and incorporated it into a more general theory of organization.[9] Their thesis is that organizational structure and functions are derived from the characteristics of human problem-solving processes and rational human choice. Because it is impossible for the human intellect to capture and analyze the complexities of problems that involve individuals and organizations, rational behavior suggests the need for simplified **models** that represent the main features of the problem without involving all of the complexities.

Richard Cyert and James March developed these ideas further in an elaborate theory of business behavior whose focus centered on organizational decision making. They proposed that: "In order to understand contemporary economic decision making, we need to supplement the study of market factors with an examination of the internal operation of the firm—to study the effects of organizational structure and conventional practice on the development of goals, the formation of expectation, and the execution of choice.[10] Within the general framework, the organization is viewed as an adaptive system with many interest groups integrated in-

[8]Herbert A. Simon, *The New Science of Management Decision* (New York: Harper & Row, 1960), pp. 1–4.

[9]James G. March and Herbert A. Simon, *Organizations* (New York: John Wiley & Sons, 1958), p. 169.

[10]Richard M. Cyert and James G. March, *A Behavioral Theory of the Firm* (Englewood Cliffs, N.J.: Prentice-Hall, 1963), p. 1.

to a loose coalition. The organization then develops mechanisms for avoiding uncertainty, searching for problems, learning through experience, and seeking satisfying, rather than optimal decisions. Because organizations exist in different situations and in different stages of development, what may be appropriate for one institution may not work in another instance. These ideas have led to the development of the **contingency** model of management and organization.

The Contingency Approach

Early in the century, Taylor made a distinction between the industrial work environment, which produces great quantities of the same product, and the shop, which produces special products to customer order. In each case, he assumed that since the technological conditions were different, the particular manpower used was different, and therefore management had to be different. In other words, the type of management and organization employed by a company must be contingent on its particular situation. Not much was made of this observation until the 1960s, when Joan Woodward of England revealed that her research findings on management and organization made sense only when the firms were classified by the technology that was used in the production process.

Thus Woodward distinguished between the single-unit, production-to-order shop, the mass-production factory, and the continuous process plant, such as a chemical plant or an oil refinery. In each case, the type of management and organization—defined by the number of levels in the organization, the number of subordinates reporting to the president and the proportion of payroll devoted to direct labor versus indirect labor—were related to a particular technology.

Two other researchers in Great Britain, Burns and Stalker, have shown that a change in the environmental features of the market will bring internal changes that correspond to the existing organizational and managerial patterns.[11] A business firm facing change in the environment is served better by an **organic organization**. Such an organization is characterized by a managerial philosophy of decentralization, or delegation of authority and responsibility based on technical expertise and knowledge rather than on position. In an organic organization, levels of hierarchy would be few, and the visual representation would look flat rather than like a pyramid. Because of the effects of continual change on their organization, managers would exercise leadership by virtue of their knowledge and by persuasion rather than by their position of authority. The organic managerial system is expressed in a network of communication rather than hierarchical control, and lateral rather than vertical interactions.

In contrast, the *mechanistic* view would appear to characterize an organization with a relatively stable market and few innovations in technology and technical processes. Such a system

[11]Tom Burns and G.M. Stalker, *The Management of Innovation* (London: Tavistock Publications, 1961).

has a rigidly prescribed organizational structure. In these situations, the task, including the methods, duties, and powers, are attached to each functional role and are defined precisely. The interactions within the managerial system tend to be vertical, between an employee and his boss, and consequently form a strong hierarchy of command and authority relationships.

The **contingency approach** to management has been supported further by the American researchers Paul Lawrence and Jay Lorsch.[12] The degree to which the organization has to deal with change affects whether its members hold a long or a short term perspective, favor formality or informality of organizational structure, and become either more or less involved with interpersonal relations.

Evidence collected by Lawrence and Lorsch in their study of organizational change, illustrates how the varying production technologies and the customers' changing specifications for products would suggest a short term perspective and a low degree of formality in the organization, accompanied by a high degree of interaction and communication. This natural interpersonal involvement along with a personal commitment to their changing roles helps to assure a company's successful performance in a highly competitive industry. The internal organization may also reflect differing management approaches between departments. The sales department would be more likely to be aware of change and the need for change, whereas the production department would set its sights on long production runs to serve the goal of efficiency.

Lawrence and Lorsch point out that these differences within the organization imply a corresponding need for integration, in order for the organization as a whole to be successful. For this purpose there are departments or personnel who act in the role of integrators. They may be called "expediters" on the shop floor, or "project directors" in the management offices.

The contingency approach to management is not a radically different approach. Rather it serves as a reminder that each firm, each office, each organization must analyze and diagnose its own situation, and in light of such an evaluation choose a pattern for management that appears to have merit, given such factors as type of product, technology, degree of stability, quality of employees and managers, and future prospects.

TOWARD A SYSTEMS APPROACH TO MANAGEMENT

Traditional organizational theory focuses on the parts and segments of the organization and attempts to separate activities into tasks or operational units. Unfortunately, it gives us limited information about the interrelationships among the parts or about how to integrate the separate activities into total organizational objectives. The human relations movement did not correct this de-

[12] Paul R. Lawrence and Jay W. Lorsch, *Organization and Environment* (Boston: Division of Research, Graduate School of Business Administration, Harvard University, 1967).

ficiency. The primary aim of these theorists was to inject considerations of human motivations, aspirations, desires, and limitations into the traditional, mechanistic models. The value approach points to the need to consider the external environment as well as internal organizational objectives, but it does not show us how to integrate these two environments. The decision-making approach describes the difficulty and importance of making proper choices but does not furnish a structure for accomplishing this task. The contingency approach suggests the need for a versatile framework that can be used in a variety of situations in a changing society but it, too, stops short of suggesting what such a framework should be.

Each of these approaches makes a valuable contribution to management theory, but they all share the common need for a conceptual scheme to integrate what is useful in each approach into a meaningful whole. The systems approach meets all of these requirements. First, this approach is a philosophy, a way to visualize and to understand; second, it is a way to manage an organization and finally, it is a means to analyze information and to make meaningful decisions. In this view, an organization is a unit consisting of an integrated whole with inputs of energy, materials, and information. A process transforms these inputs according to a plan into an output of useful goods and services. An organization is also a social system designed to obtain a particular goal, and each organization operates as part of a larger system, the society.

In our next chapter, we will discuss what a system is and how the process of management and organization fits into this concept.

SUMMARY

The pioneering work of behavioral scientists challenged traditional, mechanistic ideas about managing organizations through the famed Hawthorne experiments of the late 1920s and early 1930s. Around the same time, the work of Kurt Lewin and a parade of experimenters, researchers, and writers emphasized the human side of the organization and demonstrated that there is not necessarily one best way, but a variety of possible ways, to structure and govern human productive effort.

World War II gave a strong boost to experience in management and organization, both in military and nonmilitary activities. For the first time, the behavioral sciences were recruited to perform research that could be useful in the war effort. The emphasis on employees as supporters of the organization's objectives and goals continued after the war. However, increasing importance was attached to management's side of an implicit psychological contract requiring management to commit itself to satisfying its employees' personal objectives and goals. This revisionist approach received further elaboration from ideas that would make decision processes synonymous with management.

The basic question concerning management's responsibility is: What should its contribution be to society and the quality of life? Understanding the interrelationships that confront management and managers helps suggest some answers. We need to remember that conditions are not always the same and the choice of a particular type of management and organization must be related to the organization's situation, that is, be contingent on how and to what extent the situation reflects stability or change.

The multiplicity of viewpoints and theories that are emerging in the wake of growing scientific interest in the art and science of management calls for an integrative scheme to act as a synergizer of ideas for students and practitioners. Such a philosophy is the systems approach, in which the organization is treated as a system of mutually dependent parts and variables.

DISCUSSION QUESTIONS

3.1. What were the most important contributions of the Hawthorne studies to the theory and practice of management?
3.2. What is your understanding of Kurt Lewin's contribution to management theory?
3.3. What did Chester Barnard contribute to the field of management and organization?
3.4. In what way can you relate Barnard's work to centralization and decentralization of authority and responsibility?
3.5. Should a manager be more responsible for the corporation or for the people working for that organization?
3.6. Do you agree with Chris Argyris, that the business firm is likely to ask its employees to revert to behavior more characterisic of children than of adults? Give an example.
3.7. In what way do you relate the revisionists' approach to management to the Theory X and Theory Y approach?
3.8. What are the components in Herbert Simon's philosophy of management and organization, which uses decision making as a base?
3.9. What ideas support the development of a contingency approach to management?
3.10. Why do you think that values have become important elements in deciding on issues and problems facing the organization today?

APPENDIX
Non-Linear Systems, Inc.— Another Look

Mr. Andrew F. Kay, the president and controlling stockholder, had provided the workers and first-line supervisors in his small electronics company free reign in setting up their work-team organization (see Appendix, Chapter 2). The **response** had taken some time but finally came into full operation. The old assembly-line environment for putting together the company's electronics products became a thing of the past, as Theory Y took hold.

The results were very encouraging. Inspection as a special function became incorporated into the work teams, whose general level of skill was substantially increased. Rejection rates dropped, and customer complaints all but disappeared. Though it took nearly three months for the new organization to reach the former production rate, it continued to improve. New model changeovers could now be handled in a few weeks' time, compared to the months required before. Flexibility had increased as a result of the team effort. Workers displayed an amazing array of formerly hidden talents.

A year or so later, however, a recession hit the electronics industry and Non-Linear Systems discovered that it had little business, costs were way out of line, and the entire organization was on the verge of bankruptcy.

1. In what way do you find the Non-Linear Systems illustrating a revisionist approach to management and organization discussed in this chapter?
2. Would you recommend that the approach taken by Mr. Kay should or could be used by another organization?
3. Why do you think Non-Linear Systems was so vulnerable to the economic recession?

SELECTED READINGS

Argyris, Chris. *Personality and Organization.* New York: Harper & Bros., 1957.

Bennis, Warren. *Beyond Bureaucracy.* New York: McGraw-Hill, 1973.

Blau, Peter M., and Scott, W. Richard. *Formal Organizations,* San Francisco: Chandler, 1962.

Chandler, Afred D., Jr. *Strategy and Structure,* Cambridge, Mass.: M.I.T. Press, 1962.

Dale, Ernest. *Planning and Developing the Company Organization Structure.* New York: American Management Association, 1952.

Homans, George C. *The Human Group.* New York: Harcourt, Brace & World, 1950.

Johnson, Richard A., Kast, Fremont E., and Rosenzweig, James E. *The Theory and Management of Systems,* 3rd ed. New York: McGraw-Hill, 1973.

Kast, Fremont E., and Rosenzweig, James E. *Organization and Management,* 2nd ed. New York: McGraw-Hill, 1974.

Knowles, Henry P., and Saxberg, Borje O. *Personality and Leadership Behavior.* Reading, Mass.: Addison-Wesley, 1971.

Lawrence, Paul, and Lorsch, Jay. *Organization and Environment.* Boston: Harvard University, Harvard Graduate School of Business Administration, 1967.

Litterer, Joseph A. *The Analysis of Organizations.* New York: John Wiley & Sons, 1973.

McGregor, Douglas. *The Human Side of Enterprise.* New York: McGraw-Hill, 1960.

Mayo, Elton. *Human Problems of an Industrial Civilization.* New York: Harper & Row, 1947.

Perrow, Charles. *Organization Analysis: A Sociological View.* Belmont, Calif.: Wadsworth, 1970.

Roethlisberger, F. J., and Dickson, William J. *Management and the Worker.* Cambridge, Mass.: Harvard University Press, 1956.

Seiler, John A. *Systems Analysis in Organizations.* Homewood, Ill.: Irwin-Dorsey, 1967.

Simon, Herbert. *Administrative Behavior*, 2nd ed. New York: Macmillan, 1957.

______*The New Science of Management Decision*. New York: Harper & Row, 1960.

Whyte, William F., Jr. *Human Relations in the Restaurant Industry*. New York: McGraw-Hill, 1948.

Woodward, Joan. *Industrial Organization: Theory and Practice*. Oxford: Oxford University Press, 1965.

PRATT WHITNEY
JT 8D-7-9-15
BOEING

Chapter 4
The Systems Approach

Laurence Lessing, an editor of Fortune, has asserted that systems management is one of the greatest inventions coming out of World War II.[1] He has commented that innovative industry has used this approach to achieve its goals in the shortest possible time, and that the concept makes managing more straightforward and easier to comprehend. Lessing has suggested that systems management strips away the traditional structure that has been superimposed on the function of managing and directs its attention to what is important—the task of analyzing complex situations and making effective decisions. In this textbook, we take a systems approach to management.[2]

The term **systems** has become a common word in our vocabulary. Most people use the word as an everyday expression to describe what they do and how they live. For example, we have educational systems, transit systems, and health care systems. Note, however, that the phrases "systems philosophy," "systems management," and "systems analysis" are not synonymous, although it is convenient to categorize all the concepts relating to systems under the general classification "systems approach." Within this framework, we may include the various subconcepts of **systems theory** (the set of related concepts or the body of knowledge that is fundamental to all systems), **systems philosophy** (a way of thinking), **systems management** (the design and operation of organizations as systems), and **systems analysis** (the application of modeling and problem solving). The relationships among these terms are illustrated in Figure 4.1. Notice that the items flow from theory to practice—from the conceptual to the analytical, and from theory to day-to-day operations. Throughout this model of the systems approach, the idea of systematic (that is, logical, thorough, and methodical) thinking is emphasized.

When systems theory is applied as a philosophy, the viewpoint is conceptual and the method is cogitative (thinking or reflecting). In systems management, the viewpoint is pragmatic and the method is synthetic (building an organization by assembling or

[1]Laurence Lessing, "Where the Industries of the Seventies Will Come From," *Fortune* (January 1967), p. 98.

[2]Many of the concepts in this chapter are adapted from: Richard A. Johnson, Fremont E. Kast, and James E. Rosenzweig, *The Theory and Managment of Systems*, 1st ed. (New York: McGraw-Hill, 1964).

Figure 4.1 The Systems Approach

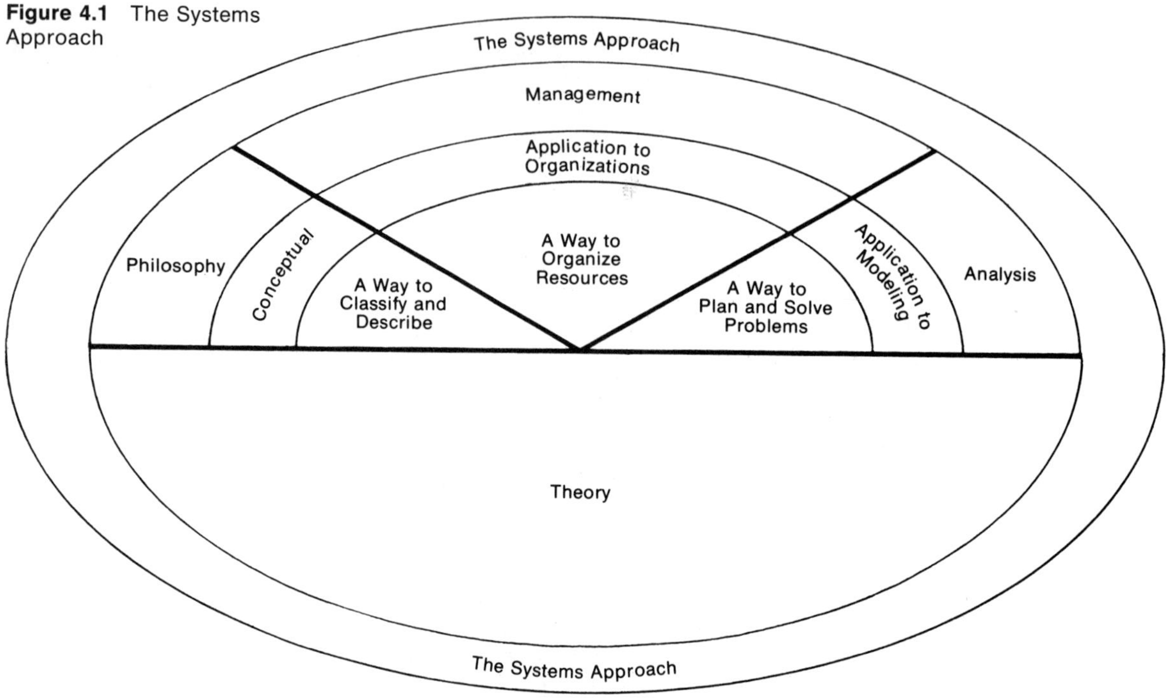

combining parts to form a whole system). The organizational subsystem is coordinative and the task is to integrate operations and achieve objectives through design. In systems analysis, the viewpoint is optimizing or problem solving and the method is modeling.

SYSTEMS THEORY

Before we describe each of these segments of the systems approach in more detail, we first need to define systems according to the general theory underlying the approach.

Systems Defined

The term *system(s)* is defined and applied in different ways. Some writers associate systems with science—which they describe as a systematic body of knowledge; an array of essential principles or facts arranged in a rational dependence or connection; a complex of ideas, principles, and laws forming a coherent whole. Scientists endeavor to develop, organize, and classify materials into interconnected disciplines, or bodies of knowledge. Two classical works, Darwin's *The Origin of the Species* and Keynes's *General Theory of Employment, Interest, and Money*, are attempts at integrating a large amount of related material. In his theory of evolution, Darwin integrated all life into a "system of nature," in which the myriad of subsystems were all interrelated. Keynes, an economist, integrated the complicated natural and man-made forces that influence an economy. Both men created a major impact on man's thinking, because each in his field was able to conceptualize interrelationships among complex phenomena and integrate them into a systematic whole.

When used in this context, the term *systems* connotes plan, method, order, and arrangement—the development of a systematic, theoretical framework for describing general relationships of the empirical world. It is possible to construct a systematic, theoretical model applicable to many disciplines, whether physical, biological, behavioral, or social. A distant goal of systems theory is to develop a framework (or system of systems) that will ultimately tie all disciplines together into an understandable, cohesive whole.

The *whole* is fundamental in systems theory. This point of view was set forth by the **vitalist theory** of deductive reasoning, according to which: (1) the whole is primary and the parts are secondary; (2) integration is the condition of the interrelatedness of the many parts within one; (3) the parts constitute an indissolvable whole such that no part can be affected without affecting all other parts; (4) parts play their role in light of the purpose for which the whole exists; (5) the nature of the part and its function is derived from its position in the whole and its behavior is regulated by the whole-to-part relationship; (6) everything should start with the whole as a premise and the parts and their relationships should evolve.[3] Now think of your body as the whole, and then reread points (1) through (6). All of the parts, or subsystems, of your body exist because of you as a whole, and if any of these parts fails to function properly, the whole will be affected. A system renews itself constantly through a transposition process; the identity of the whole and its unity are preserved, but the parts may change. This process continues endlessly; sometimes it is planned and observed, other times it may occur without notice.

Let us define a system as "an organized or complex whole, an assemblage or combination of things or parts forming a complex or unitary whole." This usage covers an extremely broad spectrum of concepts: in the world around us, there are mountain systems, river systems, and the solar system. The human body, as we have already suggested, is a complex organism made up of a skeletal system, a circulatory system, and a nervous system, among others. As mentioned before, we come into daily contact with transportation systems, communications systems, and economic systems.

When relating systems to organizations, it may be even more meaningful to define a system as "an array of components designed to accomplish a particular objective according to plan." This definition contains three significant parts: first, there is an objective, or a purpose, that the system is to perform. Second, there is design (and sometimes construction) of components in a particular arrangement. Third, there are **inputs** of information, energy, and materials allocated according to an operating plan, from which output results. According to this definition, Figure 4.2 can be called an input-output model of a basic system. Planned resource inputs

[3]L. Thomas Hopkins, *Integration: Its Meaning and Application* (New York: Appleton Century Croft, 1937), pp. 36–49.

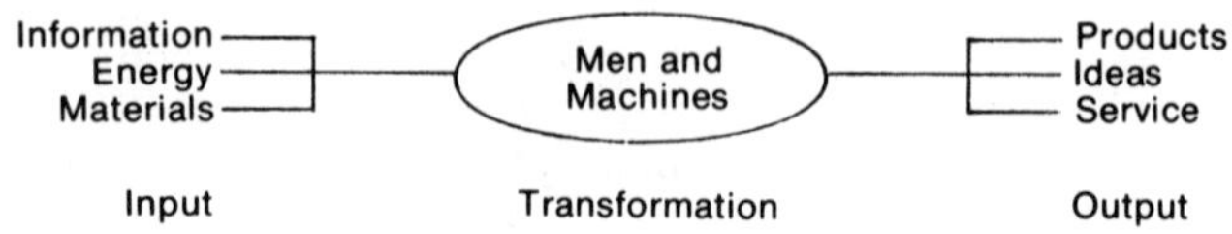

Figure 4.2 Input-Output Model of a Basic System

of information, energy, and materials are transformed by man and/or machines to produce output of products or services. The output, if the system is **effective**, accomplishes the system's objective.

To illustrate, a system whose objective is to produce automobiles must be designed, which requires information concerning the people and machines to be used and the arrangement of the transformation process, energy to build the plant and its machines, and materials to construct the plant, equipment, and machinery. After the system is designed and constructed, inputs of operating plans (information), human and electrical energy, and raw materials are transformed by people and machines into an output of automobiles.

Systems Classified

Systems may be classified and compared in various ways—for example, a natural system such as the solar system may be contrasted to a man-made system such as a transportation system. Those relationships between objects and sequences of events observed in their natural setting that have meaning to us constitute natural systems. Such systems behave according to the laws of nature and their input-output relationships are scientifically predictable. In contrast, man-made systems are those designed and operated by man. They utilize inputs from natural systems but retain little of the scientific predictability found in nature, largely because man is somewhat unpredictable. In this book, our discussion will be limited to man-made systems.

It is important to distinguish between a **flexible system** and a **rigid system.** In a flexible system, the structure or design of the system adjusts continually to maintain a balance, or equilibrium, between the system and its environment. Most life forms as well as economic, political, and social systems are examples of flexible systems. In contrast, a rigid system is not a self-maintaining unit, and its structure does not adjust to changes in the environment, at least not in the short run. A highway is an example of a rigid system—once constructed, little can be done in the short run to adjust to a change in the traffic pattern. Man tries to build some flexibility into every system he constructs. For example, a building might be designed so that it can be converted to some other use if the primary need for the structure changes.

Another classification of systems pertains to the degree of human involvement in relation to the machine. At one end of the spectrum, we can visualize a system utilizing few, if any, machines or tools. A group of natives weaving straw hats illustrates a manual system of production. At the other extreme, an automated refinery programmed to produce petroleum products without human involvement during the transformation process is an example of an automatic machine system. The man-machine relationship in most systems falls somewhere between these two extremes. Some industries have a high capital investment in machines (for example, the steel industry), whereas in others the human factor is more significant (for example, service industries).

The classifications "natural versus man-made," "flexible versus rigid," and "man versus machine" have all pertained to the design or structure of systems. The nature of a system's output forms still another basis for classification: systems that produce goods or products in contrast to systems that produce information or some other service for clients. The manufacturing department of the Chevrolet Division of General Motors is a subsystem of that corporation, which is designed to produce automobiles. Man and machines follow predetermined instructions to achieve the operating goals outlined by the company. The essence of a *product-oriented* organization or system rests in the processes by which resources are converted or transformed into measurable output.

The fundamental purpose of *service-oriented* systems is to serve clients. Service may be performed by distributing products or ideas for people. For example, a retail store does not produce a product as such, but makes goods available to customers and therefore produces a service. Similarly, a doctor or lawyer produces a service by supplying medical or legal assistance for people. Many organizations produce both a product and a service.

The classifications presented above illustrate but a few of the various hierarchies, or levels, of systems and should suggest the difficulty of making distinct and meaningful groupings. The classifications are presented chiefly to help the reader become acquainted with systems and relate the abstract concepts of the systems approach to the real world.

Systems, Subsystems, and Components

The "total system" is determined by definition. For example, a single university could be defined as a total system, within which subsystems and components would be described. In another instance, this same university might be organized, operated, or analyzed as a subsystem of a much larger educational unit. Outlining the boundaries of each system according to the system's objectives is crucial to analysis and understanding. When the total system is determined, and the inputs of each segment can be specified relative to the objectives of the total, it is possible to analyze the functions and relationships among the parts and those of the parts in relation to the whole.

The boundaries of a system are a function of the scope of the organization or system, the extent to which control can be exercised, and the nature of the problem requiring analysis. The secretary of the Department of Health, Education and Welfare would include as boundaries to the systems under his jurisdiction all of the institutions associated with the function of his department.

Figure 4.3 Hierarchy of System and Subsystems

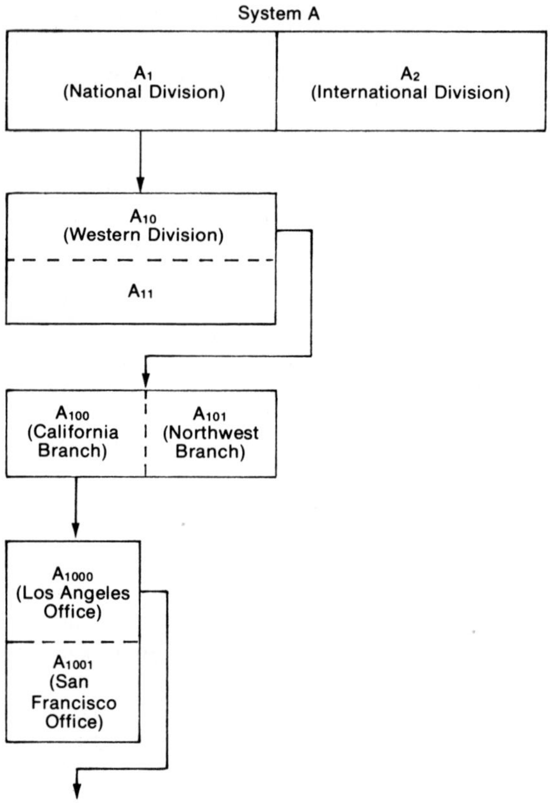

The president of a university would specify all of the parts of that institution under his control as the total system.

Within each system there are parts, or subsystems. In a university there are departments or colleges, each performing as a subsystem under the general direction of the president of the university. Figure 4.3 shows the hierarchical relationship of systems and subsystems. For example, when A is defined as the total system, there are subsystems at levels A_1, A_{10}, A_{100}, and A_{1000}.

To illustrate, System A may have two major subsystems, a national A_1 and an international operation A_2. The national system may also have two subsystems as represented. One of these subsystems, perhaps the Western Division, may have two subsystems, A_{10} and A_{11}. When we establish additional levels, or subsystems, of the "total system," we do so because it is important to define and analyze the inputs and outputs of these subunits and to determine how the transformation process occurs in each operation. Let us suppose that we are not concerned about what happens within the subunits at the A_{100} level, but only in the input and output of that grouping. In this instance, these subparts would be considered **components**.

In many kinds of analysis, the inputs and outputs of a system are more significant than the actual transformation that occurs to produce the output. In such instances, we can say that the transformation process occurs in a "black box"—and we know only what goes in and what comes out, nothing else. We can use this "black-box" approach to demonstrate the role of components. A basic unit, or "black box," which performs, or provides the facility for performing, some part of the defined transformation process will be defined as a component. For our analysis, there is little value in subdividing or describing this part of the system beyond input and output. To illustrate, a classroom building may be defined as a component in an educational system, when there is no particular reason to detail the analysis of the building according to its own subsystems, such as heating systems, lighting systems, and so forth. A_{100} in Figure 4.3 could symbolize two area offices that would be defined as components of subsystem A_{100}, even though we know that each of these subunits could be a system with subsystems in another situation.

Systems, subsystems, and components, then, are determined by definition. If conditions or the need for analysis changes, what has been analyzed as a subsystem may become the total system, or what has been considered a component may be analyzed as a subsystem.

SYSTEMS PHILOSOPHY

If we regard the world in which we live as a total system and picture the many natural and man-made subsystems with their countless levels of interaction, and if we begin to picture things relative to the input-output model, then we are adopting a systems philosophy. We are using this philosophy as a way of thinking and of seeing the world around us and are stressing the idea of logical, thorough, and methodical thinking. The systems philosophy becomes a way to relate complex ideas, principles, and laws so that they become meaningful.

For example, let us consider the cross-relationships of some of our economic, political, and social problems relating to pollution. Figure 4.4 shows how pollution (waste) is a product of both production and consumption. Even more basic, the amount of pollution depends on many factors relating to many systems: (1) the total population and its rate of growth, (2) the demand of the population for goods and services, (3) the kinds of goods and services that are produced as determined by priority, (4) the degree to which resource transformation can occur without waste, and (5) the degree to which consumption can occur without waste.

When population expands, the demand for goods and services grows, and waste increases. Eventually, waste becomes excessive and is regarded as a negative factor in evaluating general living conditions. By adopting the systems approach as a philosophy, it is possible to identify and understand these relationships.

SYSTEMS MANAGEMENT

When organizational units are arranged and operated as systems, we refer to this as *systems management* or *management*

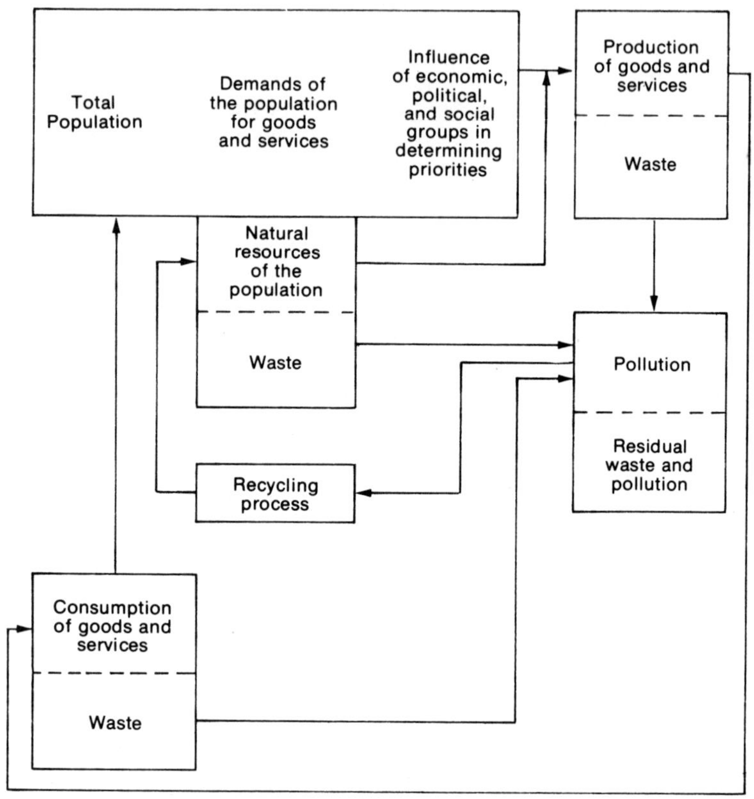

Figure 4.4 A Systems Approach to Pollution

by system. Each segment of the total, or each subsystem, is viewed as a distinct entity, and its relationship or contribution to the next level in the hierarchical structure is programmed and measured, but always in consideration of the total system.

There are four characteristics that apply to a philosophy of systems management: (1) *goal orientation,* which continually emphasizes objective or goal achievement (effectiveness), (2) *total-system orientation,* as the decision strategy stresses optimizing the total system rather than the subsystems, (3) *responsibility orientation,* for each manager is given a specific assignment with measurable inputs and outputs, and (4) *people orientation,* as workers are given challenging assignments that are associated with output (achievement is recognized and rewarded). Although work assignments may be more specific in systems management, the ability of systems to adjust to changing demand encourages employees' professional growth and development.

There are four decision stages in the management of systems; each is distinct but closely related to the others: (1) systems determination, (2) design and creation, (3) operation and control, and (4) review and **evaluation.**

Systems Determination. First a decision is made to create a system. This is called "systems determination"—the decision an entrepreneur makes in deciding to open a corner grocery store, the decision a large industrial organization makes in establishing a new product division, or the decision a government makes in creating an agency to help solve the problems of the cities.

Design and Creation. Next, a system must be designed; that is, subsystems and components must be arranged in some combination to transform inputs and produce output. For example, a system designed to prepare invoices for mailing would probably include both billing machines and machine operators. These parts of the system would be arranged in a planned order to achieve the optimum output. Remember that it requires information during the design stage to determine the exact and proper way to organize the system so as to accomplish the objectives, it requires materials to construct the plant and facilities, and it requires inputs of energy to create the system.

Operation and Control. After the system has been designed it is ready to operate. Again there will be inputs of information (for example, customer data, quantities, prices, discounts, and delivery dates), inputs of materials (for example, invoice forms and machine ribbons), and inputs of energy (human energy exerted by the workers and electrical energy supplied for the machines). The operating inputs are allocated according to predetermined decisions (plans) or by on-the-spot decisions made by operating managers. For example, a supervisor may determine which kind of invoice to use, when various customers should be billed, when the operators should work overtime, and when the machines should be repaired.

It is possible to eliminate parts of the decisions required during operating by designing systems with predetermined input and transformation activities. Such structuring permits the system to operate in a more calculated fashion and with more predictable results. The more the operation of the system is preplanned, the more automatic it becomes—but it also loses the flexibility to adapt to unusual inputs.

Basic to the theory of systems is the premise that given certain inputs, the processor will give certain outputs or operate within established limits. However, an organization is not a structured or predictable system. Its equilibrium cannot be determined by equation, and it will change (within limits) as the components of the system are rearranged or according to how the inputs are allocated.

In more advanced form, a system will have some means of control. The control subsystem will include a **sensor** for measuring output or related characteristics, a means of comparing the measurement with a standard, and an activating unit to adjust inputs to correct the indicated deficiencies. The objective is to control certain variables so that the system will tend to stabilize near its ideal equilibrium point. This objective is possible only if a desired standard can be determined and if the significant operating values can be measured. Control and control decisions are discussed more fully in Chapter 13.

Review and Evaluation. The final stage of decision making pertains to how well the system has operated. Has it been effective and/or efficient? A system is *effective* if it accomplishes its objectives, whereas efficiency is the ratio of the resource input to the output. A system can be effective while wasting resources (inefficient); conversely, a system may transform inputs efficiently without achieving the objectives as outlined (for example, more shoes per man-hour of input but all for the left foot, or a product few people wish to buy). The answer is to determine a balance between effectiveness and efficiency whenever the two are opposing each other. Adjusting this tradeoff correctly defines the best course of action, and we say the system tends to be optimized. To the extent this balance does not occur, there is a nonoptimal condition. For example, there is always a question pertaining to the quality of the product and the cost of production. We would attempt to achieve the greatest quality (effectiveness) and still keep costs (efficiency) acceptable to a competitive environment.

Review and evaluation occur at periodic times during the life cycle of a system and may lead to design changes in the present system or recommendations that may be incorporated into future systems. Information for the review and evaluation process is often gathered as control data during operations. The feedback of operating information furnishes evidence of how the system is doing and whether or not design changes are required. A review or audit analysis may bring changes in design, whereas action pertaining to control causes changes to occur in operating input.

Decision making and planning take place at all four stages, although the kind and extent at each stage will vary with the nature of the system. For example, the operation of an automated oil refinery (rigid system) must be carefully preplanned, with the operating decisions programmed into the system. In contrast, a department store (flexible system) establishes more general policy and plans and uses managerial decision making in the operating subsystem to adjust the organization to various environmental influences.

Illustrative Model

Obviously a single model cannot describe the detailed activities of every organization; each organization is unique. Figure 4.5 illustrates the major activities of an organization that applies the systems approach. How these activities would be grouped and detailed in a particular situation depends entirely on the pertinent objectives and conditions associated with each case. In our example, the master planning committee would consider inputs relating to the demand or need for the product or service, the present state of research and development technology, the resource capability of

Figure 4.5 A Systems Model of an Organization

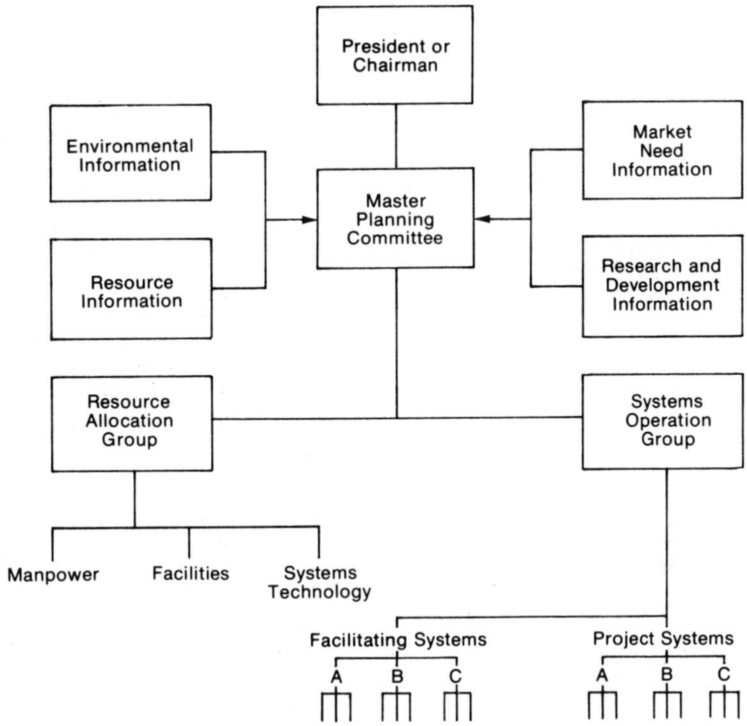

Source: Richard A. Johnson, Fremont E. Kast, and James E. Rosenzweig, *The Theory and Management of Systems,* 3rd ed. (New York: McGraw-Hill, 1973), p. 128.

the organization, and other influences generated by the environment. At this level, decisions would be made concerning the selection of new programs and the expansion or discontinuation of existing programs.

Once such decisions were made, the actual design and creation of the system would be delegated to the resource allocation group. Specific inputs of manpower (energy), facilities (materials), and systems technology (information) could be combined as required to plan and assemble new systems or to make major revisions in existing systems.

Two types of operating systems can be created—project systems or facilitating systems. *Project systems* are the major programs of an organization, and their output is directly related to organizational objectives. In contrast, *facilitating systems* are created to service project systems. Some may classify facilitating systems as subsystems of the major projects, but it is more meaningful to consider them as distinct systems, inasmuch as they usually serve more than one project. For example, a hospital's organization may include major programs for maternity, heart, and elderly patients and also have a facilitating system to provide medicines for all three programs (although the objective of a hospital is not to dispense drugs).

Generally, it is better to provide each program with all of the subsystems and components it needs to accomplish its goals. Such a strategy maintains closer responsibility, reduces problems of communication, transportation, and scheduling, and simplifies the process of administration. However, when the service may be unique or specialized, or when the cost of operating a facility cannot be justified for each project system, it is proper to create a facilitating system. But even in such cases, the goals of the service system should be specified and inputs and outputs measured.

SYSTEMS ANALYSIS

Analyzing systems is closely related to all stages of decision making in systems management. Questions related to the exact **mission** of the system, the kind or nature of facilities to use, and the arrangement of the facilities and the people who use them all involve detailed study. During a system's operation, it is important to analyze the various units of input and to decide on the combination that will provide the most effective and efficient output. During this process, an analysis is made of the timing of inputs relative to outputs (scheduling) and of material requirements at the input, process, or output levels (inventory management). An analysis of the operation relative to the system's effectiveness and efficiency also may be conducted at this time. At every state of the process, relevant information should be furnished to accomplish the appropriate analysis.

Decision Making

The process of systems analysis results in decision making, which is more than making the final choice among alternative courses of action. Decision making involves outlining the possible alternatives, determining the factors that are pertinent to each alternative, and establishing value coefficients for the variables. This process is discussed in some detail in the next chapter.

The hierarchies of decisions were illustrated in the discussion of systems management by pointing out that some decisions involve organizational strategy, others involve design and construction of systems, others relate to operation and control, and some involve the review and evaluation process, whereby changes in design may be indicated.

Model Building

Model building, a more-or-less abstract representation of a system through an analogous form, is one important way to analyze systems and improve the quality of decision making. A model is used to capture the essence, but not necessarily the detail, of the system. It permits experimentation among various decision strategies to test the results of assigning different values to the

Figure 4.6 Behavioral and Technical Subsystems of an Organization

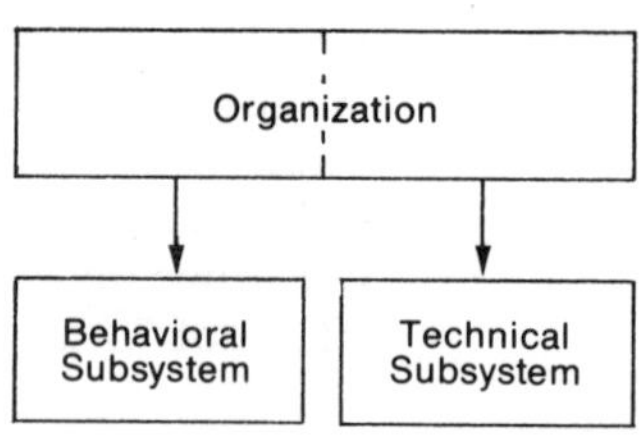

variables. The more appropriate the model to the actual situation being analyzed, the more valuable model building is as a tool in analysis (see Chapter 5).

THE BEHAVIORAL SUBSYSTEM

An important subsystem of any organization is the system of variables that characterizes the human element and its response to the total system. If we think in terms of levels, the **behavioral subsystem** appears as a subpart at the organizational level, the other major subpart being the technical subsystem. These major subparts of the organizational system are shown in Figure 4.6.

The function of the behavioral subsystem is shown in detail in Figure 4.7. There are three general types of variables involved: (1) those that represent inputs, or the *determinants* of organizational behavior, (2) those in the transformation stage, which are called *emergent influences*, and (3) those that comprise the output, or *results*.[4] **Feedback loops**, to provide measurement data, are the system's self-correcting mechanisms. These mechanisms keep the organization's inputs and emergent processes in line with expected outputs, usually through managerial action.

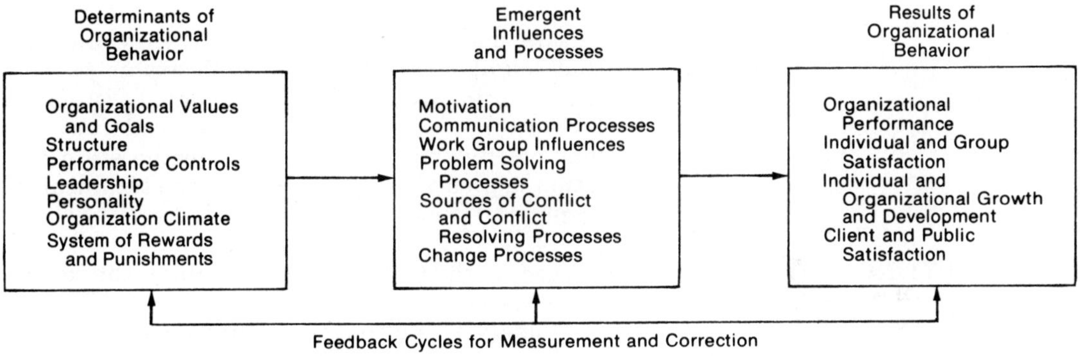

Figure 4.7 Behavioral Subsystem of an Organization

The determinants, or inputs, are *causal* with respect to emergent processes and the latter, in turn, are causal with respect to results, or output. Thus, when output is not meeting standards, the human causes must be looked for in the preceding processes and will ultimately be found in such management-controlled inputs as organizational structure, controls, and leadership, and in their effects on the behavior of organizational members. In the behavioral subsystem, the *input* stage is of primary importance, because these are the variables that management can alter directly. As Likert points out, many managers make the mistake of trying to deal directly with aspects of emergent behavior or output.[5] (Output, in particular, may be easily measured—for example, as pro-

[4]This analysis is an adaptation of Likert's systems description of the organization. His terms for the three classes of variables are *causal, intervening,* and *end-result.* See Rensis Likert, *The Human Organization* (New York, McGraw-Hill, 1967), pp. 26–29.

[5]Likert, *The Human Organization*, p. 130.

ductivity, grievances, absenteeism, and so on.) The *causes* of these results, however, can be traced back to managerial inputs into the system.

Systems analysis may be applied to constituent elements of the behavioral subsystem. For example, Knowles and Saxberg utilize systems theory to analyze the interpersonal communication process.[6] The sender and receiver of data are viewed as interdependent parts of a system engaged in achieving mutual understanding. Feedback between the sender and the receiver has the effect of either stabilizing or changing the relationship, according to its quality and purposes in a given case. In another example, Lawrence and Seiler devised a conceptual scheme for describing work-group behavior based on systems concepts and the work of George Homans.[7] In this scheme, certain "background factors" and "required behavior" (work methods), when combined are analogous to "determinants of organizational behavior" (Figure 4.7.) This leads to "emergent behavior" and to "consequences" of group activity—which may cause management to make changes in such inputs as technology (managerial assumptions and practices) in an effort to restore the work-group system to a satisfactory level of activity.

There are many other examples of behavioral subsystem analysis that might be cited.[8] In all such examples, recall that changes that occur in any part of a system or subsystem will affect other parts. For this reason, managers must be alert to the system ramifications of any change that occurs in the state or level of activity or that is prescribed in any part of the technical, social, or human organization, or in managerial practice.

PUTTING IT ALL TOGETHER

After reading the material in this chapter, the reader should have a general idea of what is meant by systems philosophy, management, and analysis. Beyond these fundamental concepts rests the philosophy of application—the idea of relating the materials in this book to the input-transformation-output model.

This philosophy can be used to help us understand the real world and to perform more effectively as members of resource-allocating organizations. First, it is helpful to view everything as a system having inputs and outputs. Second, it is necessary to establish specific boundaries or to define each system precisely. Third, each organization should be directed toward optimizing its total system rather than its parts. Finally, an orderly approach should be used to study alternative courses of action.

The systems approach will train you to organize bodies of knowledge into logical disciplines. It will also help you see the rela-

[6]Henry P. Knowles and Borje O. Saxberg, *Personality and Leadership Behavior* (Reading, Mass.: Addison-Wesley, 1971), pp. 98–101.

[7]Paul R. Lawrence and John A. Seiler, *Organization Behavior and Administration* (Homewood, Ill.: Irwin-Dorsey, 1965), pp. 154–64.

[8]See, for example, Chapter 12 of this text. See also Lewin's force-field concept, described in Chapter 3.

tionships among disciplines. From this integrated perspective, the relations existing among political, social, and economic systems may become more meaningful. The influence of each system on a manager's function in allocating resources is significant input in the decision-making process.

SUMMARY

The systems approach is used as a broad classification that includes other subconcepts; systems theory (a fundamental body of knowledge), systems philosophy (a way of thinking), systems management (the design and operation of organizations as systems), and systems analysis (the application of modeling and problem solving).

The term *systems* is defined as an array of components designed to accomplish a particular objective according to plan. Inputs of information, energy, and materials are transformed by man and/or machines to produce outputs of products, ideas, or services.

Systems may be classified according to their design (for example, man-made or natural, flexible or rigid, man or machine), or according to their output (for example, product or service). The scope and hierarchical arrangement of systems depends on a definition of a particular system's boundaries and on a determination of the need to subdivide in order to acccomplish operational or analytical objectives.

Organizations can be viewed as having two subsystems—technical and behavioral. The behavioral subsystem is fundamental for the very nature of the organization, its values and goals, and its involvement in operations and change are dependent upon this segment of the total system.

DISCUSSION QUESTIONS

4.1. List and describe three different examples of a system using the definition of a system developed in this chapter.

4.2. List the resource inputs, transformation components, and outputs of a bank.

4.3. How would you classify the following systems?
   a. A hospital
   b. A public market
   c. A guided missile
   d. A rubber plantation

4.4. To what extent might environmental inputs affect planning at the project level? Give an example.

4.5. What is systems analysis? How does it relate to decision making?

4.6. The systems approach was used successfully by the National Aeronautics and Space Administration (NASA) in the manned lunar landing program. What advantages did this approach have in this particular application?

4.7. What problems in our country might be alleviated if we had zero population growth? What changes might occur?

APPENDIX Emergency Ambulance Service

Emergency ambulance service can be viewed as a portion of a larger system, including communication, transportation, and medical treatment. The communication subsystem is the way in which emergency medical treatment is summoned for a patient, the screening and ranking of calls as they acccumulate, and the means of communicating among dispatchers, ambulances, and hospitals. Calls for medical aid may be made via telephone directly from the public to an ambulance dispatch center or to the police or fire departments' communications sections. Requests for ambulance service may also be made by police officers investigating an accident.

The transportation subsystem provides a way of conveying a patient to a medical treatment center or of transporting medical facilities to a patient. This can include regular ambulance service, air ambulance service (helicopters or fixed-wing aircraft), rescue service vehicles, aid cars, and specialized vehicles such as mobile intensive coronary-care units. Factors to consider in the transportation subsystem include the boundaries of ambulance districts, the location of ambulances and medical treatment centers, and the number of vehicles available for service.

The medical treatment subsystem encompasses the nature, speed, and adequacy of emergency medical treatment—that is, the qualifications of personnel, their prompt availability, the organization, procedures, and equipment in an emergency room, the equipment carried in an ambulance, and the possible utility of first aid stations. Improvements in the transportation subsystem would be to no advantage if a doctor were not available immediately after the patient was delivered to the hospital.

The primary objective of emergency ambulance service is to minimize the time between the occurrence of an accident or sudden illness and the moment at which medical treatment can be supplied. Medical evidence suggests that the duration of time between the event of injury and application of medical aid directly corresponds to the degree of recovery and length of hospitalization of the victim.

Examining the sequence of events that takes place between the occurrence of an injury or sudden illness and the delivery of the patient to a hospital (and the eventual return of the ambulance to its home base) can determine where time savings might be achieved. At first glance, it may appear that by moving the location of the ambulance dispatch center to a point where the most calls for service originate, the round-trip time might be cut in half. However, the decrease in time is not proportional to a decrease in the distance between the location of the ambulance, the patient, and the hospital.

Figure 4.8 presents the sequence of activities following an accident or sudden illness.

The activity "delay before call" is the time lapse between the moment of the accident and the call for an ambulance. This length of time varies widely, depending on the circumstances surrounding

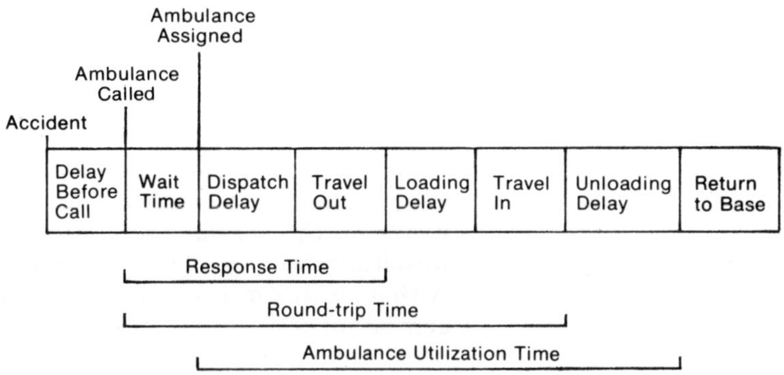

Figure 4.8 Sequence of Activities from the Time of an Accident until the Ambulance Returns to Its Home Station

the accident. For example, if the accident involves an automobile leaving the road in a mountainous area when there is no other traffic in the vicinity, the accident may not even be noticed for some time. Furthermore, if the accident is discovered by a passerby, this person may not know whether to leave the scene to call for aid or to stay to give aid to the victim and wait for another passerby to call for an ambulance.

In another case, a witness to an accident may call for an ambulance immediately, and the time delay would be negligible. However, it would be difficult, if not impossible to reduce the time delay between the accident and the call for an ambulance because it depends heavily on the circumstances of the accident.

Furthermore, the "wait time" that exists between the receipt of a call and the assignment of an ambulance probably could not be shortened, either. During this time, the dispatcher must decide whether the call requires an ambulance, the priority of the call relative to other calls that are waiting, and which ambulance is closest to the scene. Unless a queue of calls has occurred, however, the time between the receipt of a call and the assignment of an ambulance would be insignificant.

The "dispatch delay" is the time between the assignment of an ambulance and the moment it begins its trip to the scene of the emergency. If an ambulance is returning from a previous run and the radio dispatcher assigns it to the current run, the ambulance would begin its trip to the accident immediately. If the call comes when the ambulance is stationed at its home base and the crew is sleeping or lounging, it may take several minutes before the ambulance departs. Again, there seems to be little room for improvement in elapsed time.

The "travel-out" time is simply the length of time enroute to the scene of the emergency. This time depends to a large extent on the spatial relationship between the location of the ambulance and the location of the patient requiring aid. It also includes the transportation network linking the two. That is, it may take as long to travel one mile along residential streets as it takes to travel three miles along a freeway.

A "loading delay" involves the time required to administer first aid and load the patient into the ambulance. The length of time involved will vary with each call but generally cannot be reduced.

"Travel-in" time is the time it takes to transport the patient from the scene of the accident to the hospital (or emergency medical aid treatment center). This time, like the travel-out time, depends on the spatial relationship between the acccident and the hospital as well as on the transportation network between the two. Although the location of existing hospitals is fixed, a certain percentage of accident victims will not require hospitalization and can be treated at neighborhood health clinics or first aid stations that may be closer to the scene.

The "unloading delay" is the time that elapses between arrival at the hospital and until the patient is received and the ambulance attendants can be released. In some cases, this delay is inordinately long, which allows some possibility of time reduction. Improved admission-acceptance procedures will increase the effective availability of ambulances.

Finally, the "return to base" is the time it takes between leaving the hospital and arriving at the home station. This period of time is not necessarily relevant to the problem, because during this time the ambulance is available to be assigned to another call.

1. Diagram the emergency ambulance system to show each activity and the relationships among the subsystems.
2. Illustrate an alternative system that might be more effective in achieving the stated objectives.
3. To what extent is your proposal limited in application because the different subsystems are under the jurisdiction of different managers?
4. Would your proposal be more efficient than the one described in the text? In what way?

SELECTED READINGS

Churchman, C. West. *The Systems Approach.* New York: Delacorte Press, 1968.

Johnson, Richard A., Kast, Fremont E., and Rosenzweig, James E. *The Theory and Management of Systems*, 3rd ed. New York: McGraw-Hill, 1973.

McCleland, David I., and King, William R. *Systems, Analysis, and Project Management*, 2nd ed. New York: McGraw-Hill, 1975.

Part Two
Managing Organizations

A manager makes decisions to determine what systems to create, how they should be designed, and how they should be operated and controlled. In carrying out this task, the manager must seek a balance between the technological achievement of operations and the human needs of the organization's members. Information generation, communication, and motivation are vital concepts to consider in establishing a proper balance.

People are an organization's most important resource. It follows that the manager must have an understanding of the principles of human motivation and behavior and of the relationship between personal and organizational goals. Leadership is the integrating force that can accomplish the organization's mission and also effectively support and manage the company's human resources.

Conflict and change are inevitable parts of managing organizations. With sensitive and creative management, however, conflict and change can become important motivators for renewal and development.

Part Two describes the practice, functions, and principles of good management.

PLAN YOUR
WORK WORK
YOUR PLAN
Coca-Cola

Chapter 5
Decision Making and the Planning Process

Much has been written about decision making and the decision-making process. There are also many books that describe planning. What do these two concepts have in common, and how are they different? To begin with, most writers refer to planning as the managerial process of deciding in advance what to do and how to do it. This process includes first determining objectives and goals and next developing programs for achieving them. The outputs from this process are called *plans*, or predetermined courses of action.

A decision has been described as a resolution among alternative choices. Some writers suggest that a decision is not a plan because it need not involve either action or the future, however, when decision making is considered a process—from making a choice among alternatives to developing a program of action—then it is identical to the planning process. For our purposes, we will consider decision making and planning synonymous (although decisions designated for future action will be called plans).

THE DECISION-MAKING PROCESS

The decision-making process is a procedure for solving problems and implementing action. Although there is no single procedure used by managers, there are certain elements that are common to the process. They include:

1. Problem definition.
2. Development of alternative courses of action.
3. Evaluation of alternatives.
4. Selection of best course of action.
5. Implementation of the decision.
6. Follow-up.

Careful consideration of each of these elements is essential. Model building is often used at each stage to capture the essence of the entire process.

The Role of Modeling

A model is an abstract representation of a system. It is used to capture the essence, but not every detail, of the system. Models are often used to represent organizational systems since it is seldom possible to experiment with the actual system to determine preferred decision strategies. The extent to which a model is able to

Figure 5.1 Modeling and the Decision Process

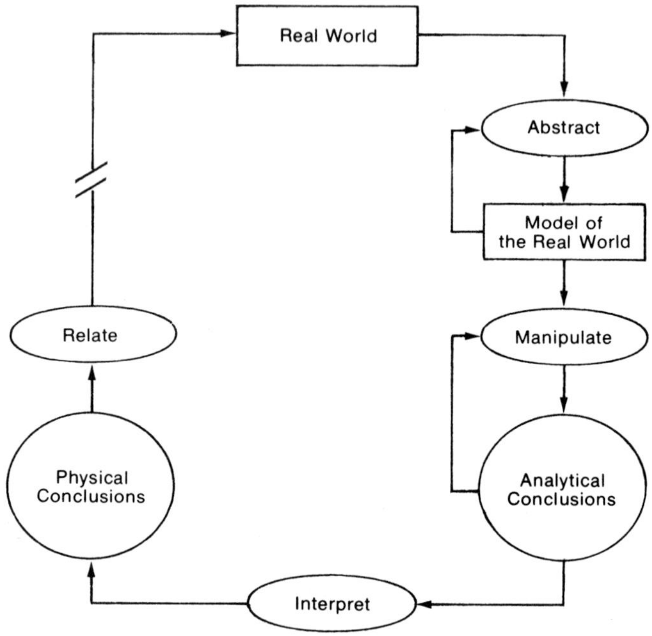

predict a system's behavior depends on how accurately it represents the underlying system structure. A model that successfully captures the essence of a system can be used for experiments to discover the effects of various decision strategies.

Figure 5.1 illustrates how problems from the real world can be abstracted in a form suitable for analysis, testing, and evaluation. Starting with the *real world*, the model builder *abstracts* those factors that represent the behavior of the system. The initial abstractions are of the major variables, making little attempt to detail or refine the model. Once the general model has been developed, it is much easier to expand the details to make the model a more realistic representation of the actual problem.

It is important that the model accurately represent the system. It should outline the various parts of the system and show their relationship to the other parts. Building realistic models is one of the keys to good decision making, despite the fact that it may take some practice to represent the system or the problem in the real world accurately. The decision maker must continue to improve and refine the model as more information becomes available through the analytical process.

The model can be *manipulated* by solving for unknowns based on known values, or hypothetical values can be given to the unknowns to test the reaction of different values on the total system. This process is called **simulation**. As a result of these simulations, certain *analytical conclusions* are drawn. Note that there are many instances when the validity of analytical solutions may be questionable. Such conclusions should be *interpreted* by managers, who can add the dimensions of experience, common sense, and good judgment to the analysis. All of this points to the unavoidable fact that the model cannot capture the full complexity of reality. The analyst who recognizes the limitations and interprets the conclusions carefully is making proper use of the modeling process. His experienced interpretation should produce realistic and workable physical conclusions.

Once the actual decision has been made and implemented, it is *reviewed* in a follow-up exercise to judge whether or not it was the best solution. Obviously, the interpretation of what is "best" depends on the decision criteria that have been established. If the most important criterion is to maximize profit, then the level of profit will be the basis on which to judge the merit of the particular decision. All of this presumes that the analyst can measure the outcome of other alternative courses of action for a point of comparison.

Management is turning increasingly to model building, because evidence shows that formal models can improve decision making. In some areas, such as important strategic decisions, model building is at a low stage of development. However, for many important tactical decisions, such as inventory control, model building is quite advanced.

ILLUSTRATING THE DECISION OR PLANNING PROCESS

Although there may be no single sequence of events that managers follow in the decision or planning process, following a systematic or orderly approach is a valuable habit to acquire. Figure 5.2 illustrates one framework or model to use for making decisions. It describes an orderly method of reviewing and appraising alternative ways of using scarce resources to accomplish a particular objective. In view of the cause-and-effect relationships among the elements of a problem, the constantly changing environment in which decision making and planning must be formulated, and the limited resources and time available to analyze a problem, such a logical approach can assist and refine the decision-making process.

The following steps can help the practitioner to make a sound decision: (1) determine the objectives of the system or the problem to be solved; (2) describe the system or define the problem in detail; and (3) try to be completely objective, not making a judgment or hazarding a solution until you have completely studied the problem. (Often, studies are made to justify a predetermined decision rather than beginning with an unbiased analysis of the system and its requirements.)

In many ways, the model shown in Figure 5.2 is similar to the scientific method. For example, the model includes stages where requirements and constraints are carefully outlined, alternative avenues of achieving the objectives are analyzed in detail, tradeoff studies are made of the criteria selected to determine the best approach, and a decision plan is developed to satisfy the specified objectives or stated problem.[1]

[1]P.G. Thome and R.G. Willard, "The Systems Approach—A Unified Concept of Planning," *Aerospace Management*, General Electric Company, Missile and Space Division (Fall–Winter 1966), pp. 25–44.

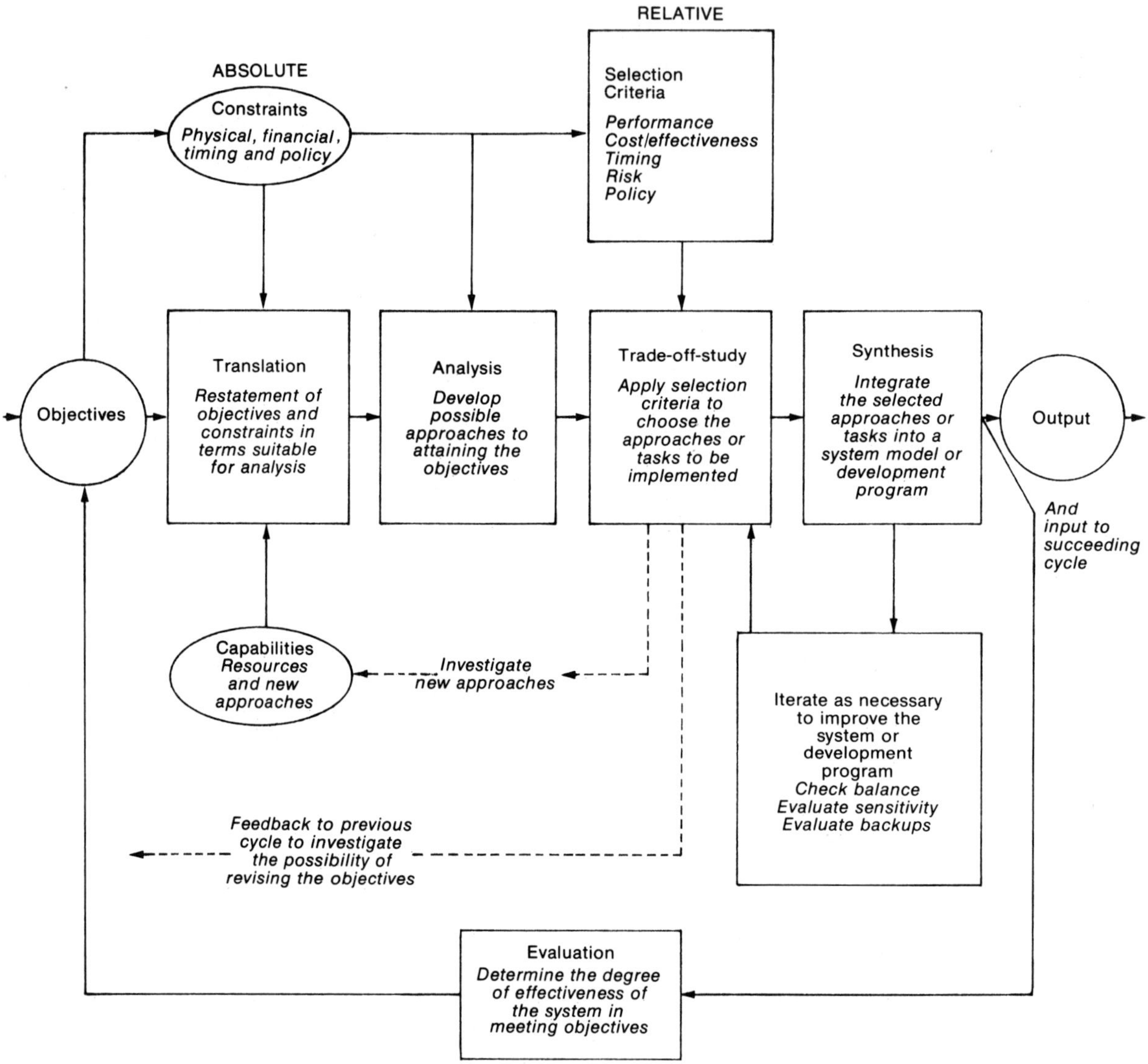

Figure 5.2 A Model of the Decision-Making Process

Source: P.G. Thome and R.G. Willard, "The Systems Approach—A Unified Concept of Planning," *Aerospace Management.* General Electric Company, Space/RESD Library, 1, No. 3 (Fall–Winter 1966), 31.

The logic of the model follows the deductive-inductive sequence. That is, the process starts with the objectives or general statement of the problem and then develops those details or data relevant to solving the problem or outlining a plan of action. This pattern of thinking produces a cycle that creates (as nearly as possible) a closed loop for continuous feedback with periodic inputs. The process illustrated in Figure 5.2 goes through several stages of development, starting with general objectives or problem definition, translation of objectives or problem detailing, constraints, and analysis. Following this, the process includes development of alternatives, criteria for selection, tradeoff analysis, and synthesis. Note the close relationship between objectives and selection criteria, as shown in the figure, and also the need to relate the plan, or decision, back to the objectives in an evaluation procedure. To understand how this decision-making process works, let us discuss each of the parts in more detail.

Objective or Problem Definition

Objectives serve as the starting point in the decision-making process, but these must be translated into specific terms. The more specific the statement of objectives, the easier it is to focus the analysis. It is not always possible at the beginning of a study to have a clear and precise definition of objectives. Therefore, it may be necessary to refine the general definition as more information becomes available. An analysis of the various factors in the study will often clarify the mission and permit the analyst to define the objectives more precisely than was possible at the start.

When decision making involves problem solving, there is a similar need to define the problem precisely. For example, let us consider some of the difficulties facing decision makers engaged in studying traffic problems. Their problem is to provide the means for moving people from their homes to other locations as necessary for work, school, shopping, and play. Further, materials must be transported to distribution centers within the city. Planners must consider some of the following questions: How serious is the problem of traffic? Are people being inconvenienced to the point where something must be done? Of the criteria of speed of movement, convenience, safety, and cost, which is the most important?

First, decision makers must make sure that a traffic problem exists and then identify the size and nature of the problem. There is always the possibility that the wrong problem may be solved. For example, to avoid congestion on the highways, it seems logical to build more highways to speed traffic. However, when more and better highways are constructed, more people drive their cars and go farther to work, shop, and play. Consequently, traffic builds more rapidly than highway capacity, and congestion increases again.

The objectives of a particular subsystem must be considered in relation to the total system so that optimizing the subsystem does not reduce the effectiveness of the total system. If the part is improved at the expense of the whole, we say that the total system has been "suboptimized." For example, it may be possible to build more freeways if less money is spent on entrance and exit ramps. When the flow of traffic on and off the freeway is not smooth, however, the entire system may become ineffective.

Constraints

Constraints are factors that limit the number of feasible solutions or alternative plans of action. They establish the boundaries within which a solution must be found. Constraints may be technical (a limitation on the state of the art), economical (cost or capital restrictions to a particular solution), political (legal restrictions resulting from laws or ordinances), educational (limitations on the ability of consumers to understand or to initiate certain actions), social (restrictions imposed by religious groups or by informal organizations), or limitations on time (requiring that a solution be found within a prescribed period and consequently eliminating from consideration certain long-range solutions).

Everyone would like to maximize all parts of a system and do so with little, if any, cost. Obviously, this is not possible, for every time we maximize something, something else must be sacrificed. When priorities (value functions) are established constraints are automatically created.

Using our example again, since the minimum cost of producing a better transportation system would occur only if no additional freeways were built, constraints are placed on the attainment of minimum cost. In other words, it is impossible to achieve our objective of providing better transportation and also meet an objective of zero cost. Similarly, there are constraints on attaining the objective of moving the maximum number of people in a minimum amount of time. It is possible to maximize only one value function at a time. Whenever we have more than one value function we would like to attain, a constraint is established and a compromise is necessary.

Alternatives

Referring back to our model of the decision-making process, the first part of the model represents a definition stage, and the balance of the model represents analysis and action. First, the decision maker prepares a list of alternatives. Each alternative must be a plan that is capable of achieving the objective or that offers a realistic solution to the problem. In other words, alternatives must be **feasible** and not violate the constraints discussed above.

In solving traffic problems in large cities, decision makers may consider typical alternatives such as building more and larger freeways, adding a rapid-transit system, and staggering working hours to level traffic usage peaks. Other alternatives may be more creative, including such possibilities as: (1) a videophone system that would decrease the need for business travel, (2) a tube or conveyor system that would reduce the amount of materials and goods to be transported in and out of the city by highway, (3) a limitation on the width of automobiles so that three lanes of traffic would be possible for each two lanes now in existence, or (4) designing work-recreation-shopping-living areas as integrated units to reduce the need for travel. These choices are only a few of the possible alternatives from which the creative decision maker will select the best decision or plan.

Another possible alternative is to postpone making a decision. Postponement may not suggest procrastination but rather a conscious recognition that now is not the best time to make a choice, that the objectives and goals may change, or that the constraints that have limited potential alternatives may be removed in the near future. In addition, anticipated technological breakthroughs may add new, more attractive alternatives than those now available. Of course, the risk associated with the decision is an important factor in deciding whether or not to delay. For example, if the decision involves extensive capital outlays, it may be prudent to wait for an expected technological development. On the other hand, if the decision involves selecting an alternative that is planned as a short-run solution, it may be worthwhile to make this choice now, knowing that a subsequent decision can be made later.

In analyzing alternatives, the decision maker gathers and analyzes data, asking such questions as: What are the expected outcomes? What will be the costs and benefits? What are the probabilities that these outcomes will occur? What is the relative importance of the outcomes? The decision maker attempts to predict the outcome of each alternative—a process that requires a great deal of information and analysis. Much of the work in management science and decision theory focuses on finding ways to facilitate the prediction of outcomes.

The decision maker relies on the expertise of staff members to furnish technical knowledge on the problem at hand. For example, a specialist may be used to develop a model for describing the system. A technical committee may be formed to discuss the implications of the alternatives. In addition, experts may be consulted for information about the organizational climate and the effect a decision might have on employees.

Selection Criteria

One of the more difficult problems in the decision-making process is to transform objectives and goals into criteria for evaluating the alternatives (**selection criteria**). It is also true that establishing appropriate criteria for evaluating the alternatives is another important key to intelligent decision making (in addition to building realistic models), because if the selection criteria is followed precisely, it will determine the logical choice from among the alternatives. To illustrate this point, let us suppose that our objective is to design a system to manufacture television sets at a rate of ten thousand units per month. In selecting the criteria for evaluating the various alternatives of production, we might list: (1) quality of manufacture, (2) cost of manufacture, and (3) employee acceptance of the system. Now, if we value the criterion of quality above those of cost and employee acceptance, alternatives related to automated manufacture assembly would be most attractive. Whereas, if we value employee satisfaction as the most important criterion, the alternatives involving hand work and employee skill might rank as better choices.

In most instances it is impossible to achieve all of the stated criteria; thus, it is necessary to assign weights, or values, to the criteria that will be acceptable to those who will be affected by the decision.

In some situations, the decision maker may arrive at a satisfactory, but not necessarily the best, solution. In other situations, the manager may reach an optimum solution, or one that maximizes the possibility of attaining an objective. An optimum decision requires that the objective be expressed as a value function—that is, the expected gain or loss that can be maximized or minimized. If an objective is expressed, for example, as the number of people who can be transported in a given period of time, then the system that would provide maximum movement would be the optimum solution. When an objective is expressed in terms of cost, the maximum attainment of the objective would be the solution that provided the minimum cost.

Tradeoff Analysis

An analysis of each alternative—its advantages and disadvantages—relative to the selection criteria results in tradeoffs. The process is complicated because each alternative may have several possible outcomes, depending on the occurrence of conditions over which the decision maker has little control. On the one hand, the manager must consider the organizational objectives, including such items as economic efficiency, return on investment, product improvement, social responsibility, market development, personnel development, customer satisfaction, diversification, growth, and profit. On the other hand, he must consider such factors as employee participation, decentralization, achievement, affiliation, job content, self-respect, communication, performance-reward, use of ability, recognition, need satisfaction, leadership style, goal congruence, group success, and justice. The decision maker must weigh alternatives to achieve a balance or tradeoff between the weight of the organizational objectives and those factors comprising the **organizational climate**. Any decision made without regard for the organizational climate may result in loss of planned efficiency. Although organizational objectives could be optimized in the short run by less consideration of the organizational climate, in the long run, a balance between organizational objectives and organizational climate would be more desirable. If this balance is achieved, the organization will meet its objectives while developing an internal climate that will help assure the long-term goals of employee development and satisfaction, and may actually improve the efficiency of the organization.

The decision maker operates within a personal framework of level of aspiration, a value system, perceptive ability, experience, education, time constraints, goal conflicts, peer group considerations, technical knowledge, self-awareness, concrete or abstract orientation, expectations, and responsibility. The alternatives that the decision maker recommends will be influenced by these personal values, abilities, expectations, and attributes. In addition, the decision maker operates in an organizational structure that includes decision models, computers, staff analysis, and committee participation. To the extent that the decision maker is oriented toward technical efficiency, he might emphasize short-run goal achievement without regard for the organizational climate, as he conducts his tradeoff analysis of the situation. Figure 5.3 illustrates the delicate balance between organizational goals and organizational climate.

Figure 5.3 The Balance or Tradeoff between Organizational Objectives and Climate

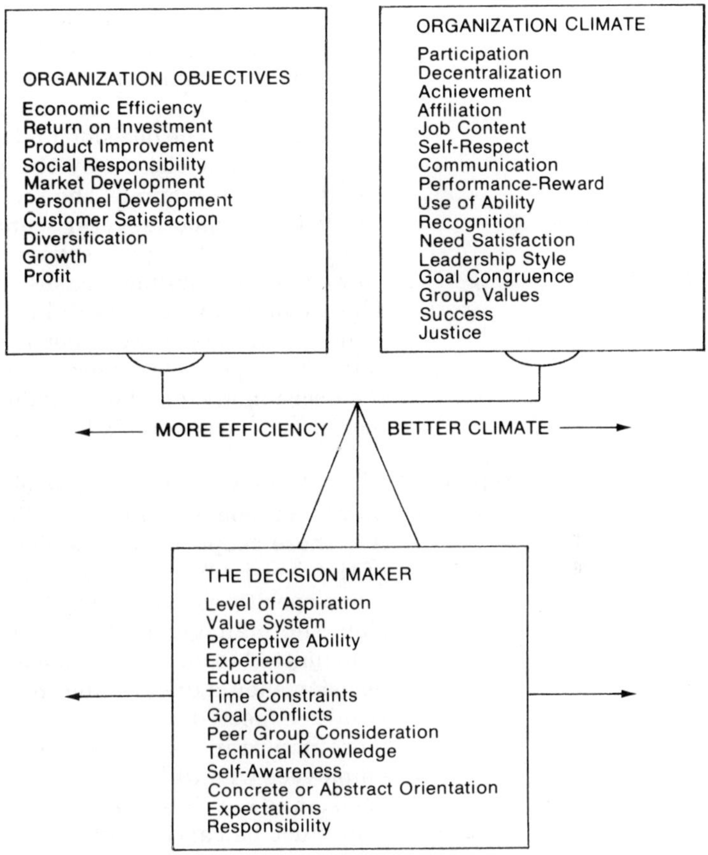

A good decision model (as in Figure 5.2) helps the decision maker recognize and weigh all of the factors in each alternative properly. In making a tradeoff analysis the more objective, impartial, and knowledgeable the decision maker can be, the better the chances are that the right alternatives will be chosen and an effective balance maintained between technical efficiency and organizational climate. Models, tools, techniques, and participation help overcome a decision maker's analytical limitations.

Synthesis

The best alternative is set forth as an action plan or decision (**synthesis**). Referring back to Figure 5.2, one of the decision model's distinct features is the emphasis on **iteration**—the process of continually refining the objectives, investigating the restrictiveness of certain constraints, developing the information and analysis, investigating new alternatives, and refining and synthesizing the actual plan or decision. In most instances, no single alternative immediately stands out as the best possibility. Often, one alternative may appear to be somewhat better than another, but even in these cases the decision maker may realize that not all of the factors pertaining to the various alternatives can be quantified, and that there are uncertainties that cannot be resolved. It is the manager's responsibility to use his best judgment and intuition to select among alternatives that may be almost equal in ranking.

At times the decision maker may decide to choose an alternative that is not chosen by the participating staff and committee members. For example, the story is told of a movie producer who decided to reshoot several scenes at a cost in excess of a million dollars, even though a consumer marketing panel that had reviewed the preliminary rushes scored the scenes as outstanding. The movie producer maintained that it was his "gut feeling" that some of the scenes were not right and had to be filmed again, in spite of the positive report by the panel. In the final analysis, it is the decision maker who must decide; it is he or she who is responsible for the success or failure of the decisions.

Implementation

After a manager has selected a course of action, he must plan how to implement his choice. He must decide: Who is to take action? Are these people capable of doing the task? Who must be informed of the decision? A decision is not complete until someone has been given specific responsibility to carry it out. In implementing a decision, the manager must be certain that planning information is transmitted to all parties concerned. Anyone who is affected by the decision should understand the rationale behind the choice and the results expected when the decision is carried out. Of course, even more important is the need to communicate information to those employees who will be involved in implementing the decision. For a decision to have a successful outcome, all phases of its implementation must be carried out according to schedule. Using our traffic problem as an example again, imagine the chaos that would occur if the people whose homes were to be purchased and demolished to make room for a new freeway were not notified, and if the real estate agents did not apprize and negotiate with these people long before the highway construction were to start.

Information Feedback

Management should try to maintain a continuous feedback of information to report whether a decision has been carried out, whether it is achieving its desired purpose, and whether the forecast and assumptions on which the decision was based are still valid. Changing conditions necessitate new decisions, and the process begins again.

We have suggested before that the decision process is a continuous cycle, each step in the sequence, a decision-making process in itself. The continual iteration through the different steps of the model refines and develops the decision. The evaluation process compares the actual results of the decision against the objective and/or problem to see if it has been successful, from the standpoints of both effectiveness and efficiency. The results of this evaluation will provide background information for making subsequent decisions.

THE HUMAN ASPECTS OF DECISION MAKING

Our discussion so far has been based on an assumption of rationality of human behavior and of the data. As we know, this is not always an accurate assumption. Much human behavior is irrational and is based on individual perceptions, assumptions, and feelings. Similarly, much of the data used in making decisions is qualitative rather than quantitative. That is, we can easily determine whether a Ford or a Volkswagen is less expensive to buy (quantitative data), but we cannot measure the intangible values (qualitative data) placed on these cars by individual buyers who believe that one car is more "prestigious" than the other, or more "aesthetically appealing." Similarly, we cannot accurately weigh intangible matters of judgment such as those involved in deciding which common stock or which real estate investment will increase most in value over a period of time.

Group Problem Solving

Many decisions are made by problem-solving committees or groups. We use executive committees to decide matters of policy, technical committees to evaluate alternatives for solving operating problems, and teams of experts to help us judge the effects of decisions on such things as productivity, morale, organizational climate, and so on. This method of decision making is called **group problem solving**. Research has shown that for decisions requiring a judgment or the pooling of knowledge, a **consensus** process is superior to other group decision methods, such as majority vote, averaging, minority control, or unilateral decisions forced on groups by an individual. Consensus is a process that involves both participation in, and acceptance of, a group decision. Everyone in the group must be involved, must adopt, and must support what the group decides. Hall and O'Leary have summarized the research on group problem solving, as shown on the graph in Figure 5.4.

Figure 5.4 Relationship between Methods of Utilization of Group Resources and Probable Adequacy of Decision

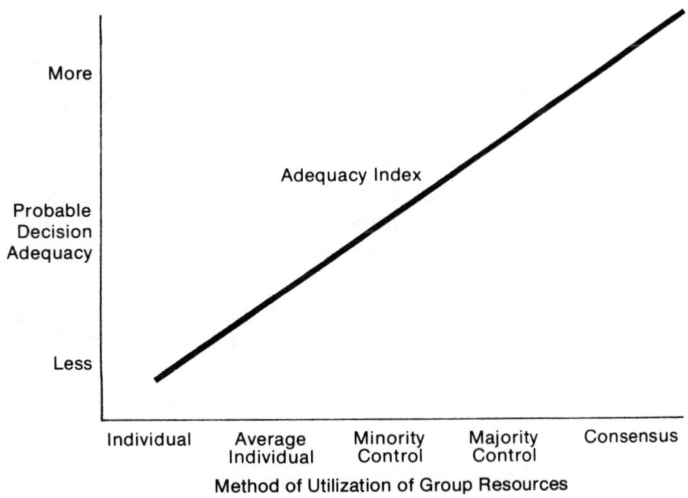

Source: Jay Hall and Vincent O'Leary, "The Utilization of Group Resources in Decision Making," National Training Laboratories, PSOTD, 1967, p. 4.

As we move along the graph's horizontal axis from "individual" to "consensus," the "adequacy index" increases. Note also that the amount of interaction among members of the group increases, since each successive method involves a higher level of mutual influence, or acting upon each other, by group members. In addition, a statistical effect (or averaging process) occurs, which cancels individual errors and reduces the influence of extreme judgments on the final result.

Thus, there is the *probability* that as a group moves toward consensus in a complex situation, the quality of a decision that requires judgment and the pooling of knowledge increases. At the same time, evidence shows that participation in arriving at the decision increases the membership's acceptance and willingness to abide by the decision. Having helped to make the decision, the group members have, in a sense, become "ego involved" with it, feeling that they "own it" and will therefore support it.

Problems with Group Decisions

Group decision processes, however, take much time and energy, in addition to suffering from other ills. Many times, they result in what Irving Janis calls "groupthink,"[2] meaning that decisions do not represent a combination of group views but rather, because of social pressure, are the product of a few, unopposed popular members. According to Janis, some notable governmental decisions—for example, United States' involvement in Vietnam and in the Bay of Pigs—were made in this manner. In other words, so-called group decisions may in fact be unilateral decisions made by a single important member who is strong enough to impose his will on the group. In addition, group meetings may sometimes stretch on endlessly without reaching a decision, while the members fret and fume and finally leave with feelings of relief mixed with frustration at wasting time and failing to act.

Collins and Guetzkow suggest several interpersonal obstacles that interfere with group effectiveness in reaching decisions.[3] First, there are obstacles that result from the self-oriented behavior of some group member. We have all attended committee meetings in which one member seemed to delight in baiting another member, succeeding in diverting the group's attention away from the problem. Psychological needs for status, dominance, and affiliation are examples of other obstacles that prevent a group from giving full attention to its task.

A second problem in groups is so-called personality conflicts. An example of this problem would be a clash between a person who believes the group can make progress only if it follows rules of order, with another who believes that progress depends on people's

[2]Irving Janis, "Groupthink," *Psychology Today* (November 1971), p. 44.

[3]Barry E. Collins and Harold Guetzkow, *A Social Psychology of Group Processes for Decision Making* (New York: John Wiley & Sons, 1964), pp. 88–119.

liking each other and being "good guys." We have also experienced the "pulling and hauling" that results when people have different prejudices on some subject or different ideas about such managerial traits as taking the initiative and being forceful.

A third problem in groups results from status inconsistencies. In our society, we place a high value on status differences. For example, if the age, educational level, or official rank of a designated or official leader of a group is inconsistent with that of other members of the group—for example, if he is younger or less well educated—some members may have difficulty accepting him as leader. As a consequence, not only their interpersonal relationships with him may be strained but also their effectiveness as a decision-making body will be affected.

Self-oriented behavior in groups is nearly always detrimental to group performance. However, incompatible personalities and other individual differences can often be constructive and stimulating to the group simply because they introduce new or contrasting ideas and thus help to improve the final product. Unfortunately, such differences are often "swept under the carpet" by group leaders. This is done either in an effort to avoid conflict, in the belief that such problems have no place in group processes, or simply because the leader (and the group) do not know how to cope with differences and conflict. The wide use of majority rule in group decision making is symptomatic of this attempt to avoid conflict.[4]

Obtaining Consensus

If the advantages of consensus (such as better decisions and higher acceptance) are desirable for the particular decision involved, then the group membership must possess the essential skills required to move the group toward consensus—that is, to get people involved in the decision process and gain their commitment, while at the same time keeping order and directing the group's attention to the problem. One way to accomplish this task is to follow basic leadership functions, classified by Benne and Sheats under the general headings of **task functions** and **maintenance functions.**[5]

Task Functions

Task functions focus on getting the group's task done and keeping the group on the track.

[4]We recognize, of course, that lack of time may force the use of voting procedures or even of unilateral decisions by an authority figure. Our point is that when these procedures are used simply to avoid dealing with differences, as they often are, they may have the effect of *imposing* a rule of the majority, resulting in poorer decisions, suppression of the opposition, and a lower level of acceptance among those who may have to act on the decisions. This can lead to subtle forms of sabotage and to postponement of time loss from the present to the future.

[5]K.D. Benne and P. Sheats, "Functional Roles of Group Members," *Journal of Social Issues,* vol. 2 (1948), pp. 42–47. See also, W.G. Bennis, K.D. Benne, and R. Chin, *The Planning of Change* (New York: Holt, Rinehart and Winston, 1961).

Initiating. Introducing ideas, proposing rational methods for solving problems, suggesting alternative approaches, contributing agenda items, setting times.

Seeking information or opinions. Asking questions, soliciting opinions or feelings about ideas, methods, alternatives, asking for facts and data.

Giving information or opinions. Responding to questions and requests for facts and data, giving feelings and opinions.

Clarifying. Making ideas more readily understandable.

Elaborating. Filling in essential details, perfecting and enlarging on ideas and opinions.

Summarizing. Putting together related ideas, restating important points at appropriate intervals, checking where the group is from time to time.

Consensus testing. Finding out how much agreement there is in the group at various points and at the end of the meeting.

Maintenance Functions

These functions have to do with keeping the group in good working order by ensuring that the legitimate social and emotional needs of the membership are being met. Behavior of the leadership and members that is appropriate to maintaining freedom to communicate, the opportunity to participate, and a safe psychological climate is essential for these purposes.[6]

Harmonizing. Helping people explore differences, adjusting and settling disagreements, striving to control tension levels in the group.

Gatekeeping. Attending to others, protecting the rights of all to participate and communicate.

Encouraging. Accepting others, conveying warmth and empathy.

Compromising. Being willing to concede based on mutual consent, blending ideas.

Standard setting and testing. Periodic review of group procedures and processes, weighing ideas against established group norms and social objectives.

To obtain consensus from group decision processes, both the group leader and the group members must perform the functions described above. In addition, performing these functions will ensure that the objectives of the task and the emotional needs of group members are kept in equilibrium. Bales classified types of group behavior into four broad categories: positive reactions, negative reactions, questions, and attempted answers.[7] By observing many task groups, Bales determined that successful problem solving tends to demonstrate fairly consistent patterns of behavior. In a typical profile, twenty-five percent of the activity showed tension

[6]T. Gordon, "Group-Centered Leadership and Administration," in Carl R. Rogers, *Client-Centered Therapy* (Boston: Houghton Mifflin, 1951).

[7]R.F. Bales, "The Equilibrium Problem in Groups," in M. Haire et al., *Small Groups* (New York: Alfred A. Knopf, 1961), pp. 424–56.

release, solidarity, and agreement; about fifty-seven percent was spent in giving suggestions, opinions, and orientation; about seven percent consisted of asking for information and opinions; and eleven percent consisted of negative reactions. The problem-solving performance of groups that departed significantly from this mix of activity tended to be less effective.

Group Decisions in Formal Organizations

The typical group behavior described by Bales was derived from *peer groups*—that is, groups consisting of individuals of approximately equal status. Many decision-making groups in formal organizations, however, are not peer groups but contain members of different ranks and members who represent other groups. For example, the executive committee of a corporation might consist of the president, the executive vice-president, and four or five vice-presidents representing departments such as manufacturing, sales, engineering, finance, and personnel. These variations in members' roles and status introduce difficulties that interfere with the consensus process. For example, members tend to defer to the president and give perhaps undue weight to his ideas and opinions. Members who represent other groups may be compelled to remain loyal to their own groups. These reactions may result in intergroup hostility, divergence of interest, competition, and limited cooperation.

The leader in formal organizational settings, therefore, is faced with problems not encountered in peer groups. Awareness of these problems and their causes and an adaptive leadership style are important prerequisites to solving conflicts. Intergroup conflicts that interfere with the consensus process, social pressures that bias members' contributions, "me-tooism" of the type that produces followers rather than creative thinkers, and other unconstructive forces generated out of the specialization and the hierarchical structures of formal organizations must be controlled to achieve effective group performance. Methods of coping with such problems are discussed further in the chapters of this book dealing with communication, leadership, and motivation.

SUMMARY

A decision is a choice among alternative courses of action. Prior to making the actual choice, there is a need for definition, investigation, and analysis. When decision making is considered as a process beginning with the choice among alternatives and leading to a program of action, it is identical to the planning process.

A model is an abstract representation of a system. Being able to abstract from the real world the dimensions and behavior of a system makes it possible for a decision maker to analyze, test, and evaluate alternatives. A decision (or planning) model describes the decision process as a pattern of thinking that produces a closed loop for continuous feedback. The first sequence includes developing general objectives or defining the problem, translating objectives or detailing the problem, identifying constraints, and analyzing basic data. The next stage includes developing alternatives, establishing criteria for selection, doing a tradeoff analysis, and synthesizing alternatives into a course of action. Finally, the decision or plan is activated and the resultant action is evaluated in comparison to objectives.

There are many human problems associated with group decision making. Self-oriented behavior, personality conflicts, and status inconsistencies are factors leading to ineffective group performance. On the other hand, individual differences that are "worked through" are potentially instructive influences. Processes directed toward obtaining consensus allow this working through and can be supported by an appropriate mix of task and maintenance behavior by leaders and group members.

DISCUSSION QUESTIONS

5.1. What is the difference between a decision and a plan?

5.2. Why is it important to detail a problem or spell out objectives in as much detail as possible?

5.3. How might constraints become a negative influence in decision making? What approach should a creative decision maker take?

5.4. How do objectives relate to the selection criteria?

5.5. How will the decision maker's own frame of reference influence the decision-making process in an organization?

5.6. What is the difference between organizational objectives and organizational climate?

5.7. Collins and Guetzkow list the following conditions associated with high consensus in groups experiencing conflicts over facts and *feelings* (Collins and Guetzkow, *Group Processes*, p. 119)
   a. When there is little expression of personal, self-oriented needs
   b. When whatever self-needs are expressed tend to be satisfied during the course of the meeting
   c. When there is a generally pleasant atmosphere and the participants recognized the need for unified action
   d. When the group's problem-solving activity is understandable, orderly, and focused on one issue at a time

   Discuss how you would translate these conditions into the different task and maintenance functions outlined in the text.

5.8. Discuss some important socioemotional problems encountered in problem-solving groups.

5.9. What is consensus? Do you think it is attainable in committees in formal organizations?

APPENDIX
A Tradeoff Analysis

John and his wife decided to buy a house to live in while they attended the university. They spent several days visiting different houses for sale and finally narrowed their choice to five alternatives. There were features they liked about each property, which made their final decision difficult. John suggested that they list the characteristics that appealed to them the most, and then rate each property from 1 to 5, with 5 being the most desirable.

Characteristic	Property Alternative A	B	C	D	E
Location relative to university	5	4	5	2	3
Condition of property	3	4	4	5	4
Quality of the neighborhood	3	4	3	5	4
Size and kind of study area	4	3	5	3	4
Quality of appliances and heating	4	5	3	4	5

The prices, taxes, and other financial arrangements for the five homes are summarized in the following table.

	Sale Price	Taxes	Down Payment	Closing Costs	Length of* Mortgage	Mortgage Payment†
A	$20,000.	$350.	$2,000.	$560.	30 years	$183.40
B	19,000.	225.	3,000.	380.	25 years	150.65
C	22,000.	400.	2,000.	580.	30 years	204.40
D	23,000.	340.	5,000.	380.	16 years	191.95
E	21,000.	200.	1,000.	580.	30 years	187.35

*In the case of houses number B and D, it was possible to assume a mortgage.
†The mortgage payment includes taxes and insurance.

1. Do you agree with the decision criteria John and his wife used?
2. Should all of the criteria have the same weight?
3. On the basis of the information you have, how would you advise John and his wife to make a decision?
4. What kinds of additional information would you want if this were your decision to make?

SELECTED READINGS

Collins, Barry E., and Guetzkow, Harold. *A Social Psychology of Group Processes for Decision Making*. New York: John Wiley & Sons, 1964.

Johnson, Richard A., Kast, Fremont E., and Rosenzweig, James E. *Theory and Management of Systems*, 3rd ed. New York: McGraw-Hill, 1973.

Johnson, Richard A., Newell, William T., and Vergin, Roger C. *Production and Operations Management*. Boston: Houghton Mifflin, 1974.

Knowles, Henry P., and Saxberg, Borje O. *Personality and Leadership Behavior*. Reading, Mass.: Addison-Wesley, 1971.

LeBreton, Preston P., and Henning, Dale A. *Planning Theory*. Englewood Cliffs, N.J.: Prentice-Hall, 1961.

Steiner, George A. *Top Management Planning*. New York: Macmillan, 1929.

BUT I THOUGHT THAT'S WHAT YOU SAID...
NO! THAT'S NOT WHAT I SAID!
I AM NOT AN ANIMAL... AND I WON'T BE TREATED LIKE ONE!
If I wanted it YESTERDAY I WOULD HAVE ASKED FOR IT YESTERDAY!
I AM LATE, I AM
RUSH

Chapter 6 Design in Management and Organization

"To design" means to fashion according to a plan. It involves creating the means to accomplish the stated mission or objective. The design establishes the relationships among the various parts of the organization (or the system), linking them together and outlining the composite whole. The design includes arranging physical facilities and technological processes to carry out production and auxiliary activities. It also determines the nature of the task, the job, or the work to be performed by the people in the organization. Therefore, the design must consider the network of communication that will unite technology, structure, and people into a coordinated effort, a total system.

The design of a new organization is only the first phase of the design function. The scope of system design also covers the function of "redesign," which implies assessing the existing system to improve its effectiveness and efficiency.

CREATIVITY AND DESIGN

Most of man's behavior can be attributed to learning, whereas other forms of animal life are more apt to behave according to instinct. Learning, or the experiences gained from the cycle of **stimulus**-response-reward, becomes a reservoir of knowledge that tends to pattern future behavior. The accumulation of knowledge and understanding is associated with mental capacity, or intelligence. "**Creativity**, on the other hand, refers to the ability of a person to develop unique and effective solutions to problems, and to create new knowledge from that which is known" [boldface added].[1]

There are many kinds of creativity. Albert Einstein's innovative work in physics represents a different kind of creativity from a prisoner's achievement in escaping from a maximum security prison, having few, if any, tools at his disposal. However, the creative process may be the same in both instances. Although most people have a potential to be creative, unfortunately only a few make use of their talent.

The Creative Process

Relatively little is known about the brain and how it works. To generate new ideas or to solve problems, individuals call into their conscious mind from their memory, information that is rele-

[1]Richard A. Johnson, William T. Newell, and Roger C. Vergin, *Production and Operations Management–A Systems Concept* (Boston: Houghton Mifflin, 1974), p. 111.

Figure 6.1 The Information Search Process.

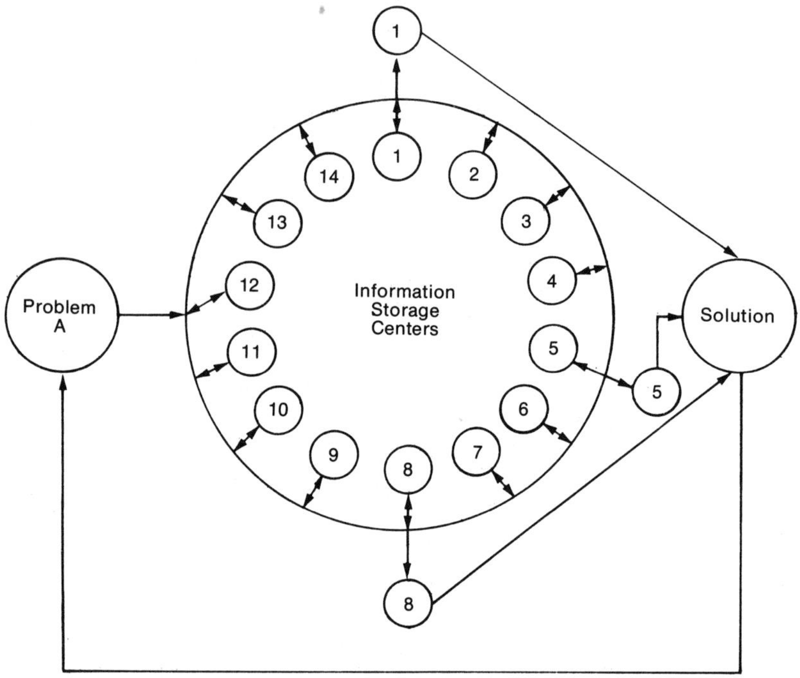

vant to the situation at hand. For example, let us assume that in the diagram in Figure 6.1, many units of information have been recalled, all of which relate to a specific problem (A). The second stage in the information search process concerns systematically testing this information, first each unit individually and then in various combinations. Perhaps the use of information units (1), (5), and (8) will suggest a new alternative or even solve the problem, thereby creating something new. Note that this example is oversimplified to illustrate the process.

Some people are more creative than others. It is possible that the creative person has a greater capability or perhaps simply makes more effective use of his or her creative potential. The model of the creative process shown in Figure 6.2 would suggest that it may be a combination of both—that is, greater capability linked with more effective use of potential. Everyone possesses a certain amount of native intelligence, which can be assessed according to: (1) the amount of information accumulated from past experiences, (2) the ability to understand and organize this information, (3) the ability to recall pertinent information as needed, and (4) the capacity to create a mental picture of actual or hypothetical situations (that is, to visualize).

People who make excellent use of their intelligence may be motivated by the hope to gain social recognition from their superiors, peers, or subordinates. Monetary gain is both a motivation for, and an indication of, such recognition. Others may gain incentive from fear of losing their job, a need to compete, curiosity

Figure 6.2 A Model of the Creative Process

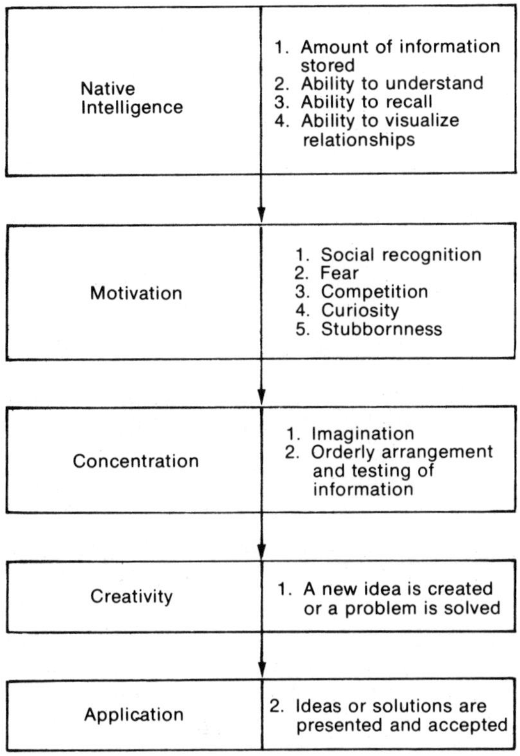

about things they don't understand, or from simply a stubborn refusal to admit that they can't do something. The greater the motivation, the more attention—**concentration**—is directed toward solving the problem. During this concentration stage of the process, the imaginative person tends to experiment with new and untried combinations of knowledge, but it is always vital to use a logical, or systematic, approach to test the various combinations in the most direct and efficient way.

When a new solution has been created, it is important to present such ideas in a way that will verify their validity. It is extremely unfortunate when innovative ideas are not recognized for their value, possibly because they were not presented effectively or convincingly.

Stimulating Creativity

An organization can stimulate latent creativity among its employees by encouraging them to try new things (expecting occasional failures) through a program of rewards and recognition, and by increasing each person's potential by providing additional education and special training. Many techniques have been developed to stimulate the process of idea generation, three examples are **brainstorming**, the **checklist**, and **value analysis**.

The technique of *brainstorming* is a group approach to problem solving. Members are given a problem to solve and then encouraged to suggest any idea that occurs to them—no matter how far-fetched it may seem—and also to build on the ideas suggested by other members of the group. The suggestions are reviewed and evaluated later.

The *checklist* is a method for developing idea clues by comparing items from a prepared list against the problem. This technique tends to stimulate ideas by association and also logically explores all facets of a problem.

Value analysis is a technique for testing the true value that each part contributes to a product. The objective is to increase the total value of the product or system. Each part is carefully reviewed according to (1) what it is, (2) what it does, (3) what it costs, (4) possible substitutes, and (5) the cost of substitutes.

THE DESIGN PROCESS

An organization develops its broad objectives first and then creates the organizational and technological systems and subsystems necessary to achieve its general objectives and more specific goals. This often involves designing complex networks of systems in precise hierarchies. One basic goal of systems management is to optimize the total system. If a change in the performance of a subsystem reduces the effectiveness or efficiency of the total system, even though it may improve the operation of the subsystem, it should not be introduced. Thus, the design of subsystems is limited by the nature of the output that the next level in the system's hierarchy requires.

Each subsystem's objectives serve those of the next higher system in the hierarchy. Therefore, in view of such major subsystem relationships, the total mission must be subdivided into segments having distinct and measurable units of input and output developed on a scale where operations are economical. Determining the total combination and nature of subsystems necessary to complete the task, although possibly not as glamorous as outlining total systems objectives, is the key to achieving effective and efficient operations or organizational goals.

Constraints on the Design

Constraints place limits on the design and operation of systems. The external environment tends to influence any open, or flexible, system. And this external influence places constraints on the design of many of the organization's subsystems. For example, ecologists can influence management's plans in the location of a new production plant and also affect the design of the manufacturing process as it relates to the disposal of wastes. Federal and state regulations may dictate the nature of cost reporting, which could determine the nature of the internal management information system. In effect, each system is a subset of a higher-level system, which may dictate or influence design and operating decisions.

The organization's general design and operating policies will also determine the limits within which each subsystem must function. Examples of such policies may include top management's standards of social and ethical conduct, the amount of capital and/or resources available for each subsystem, the kind of input furnished by external systems in addition to other subsystems of the organization, and the nature of output outlined for each subsystem as its contribution to organizational goals.

The production process also may restrict details of the design—that is, the technology available to process the input or the layout prescribed by the use of a particular process. For example, the transformation or decomposition of nitrogen pentoxide (N_2O_5) in a carbon tetrachloride (CCl_4) solution requires a specific temperature in order to maintain a predictable reaction. Such processing may require computer control, which in turn will determine the character of plant layout, and air conditioning facilities.

The constraints mentioned above are only a few illustrations of potential limitations on design. It is important to recognize such constraints early in the design process and either work around them or attempt to eliminate them.

Characteristics of Effective Design

Some systems are effective and efficient whereas others are not. Successful systems may be attributable to the skill exercised in designing the system or to the quality of management practiced during operations, or both. Successful systems are characterized by their simplicity, **flexibility**, **reliability**, economy, and acceptability.[2] Simplicity, flexibility, and reliability tend to be a function of design, whereas economy and acceptability pertain to both design and operations. Numerous relationships exist among these characteristics; for example, simplicity will affect economy and possibly reliability. Moreover, management must reach a compromise between economy and reliability, and between technical efficiency and organizational climate. The balance reached will determine whether short- or long-run objectives are optimized.

Simplicity

An effective organizational system need not be complex. On the contrary, simplicity in design is an extremely desirable quality. Consider the task of communicating information about the operation of a system and the allocation of its inputs. The task is not difficult when components are few and the relationships among them are straightforward. However, the problems of communication multiply with each successive stage of complexity.

The proper method for maintaining simplicity is to use precise definitions and to outline the specific task for each subsystem. Total systems often become complex because of the sheer size and nature of operations, but effectiveness and efficiency may still be achieved if each subsystem maintains its simplicity.

[2]Richard A. Johnson, Fremont E. Kast, and James E. Rosenzweig, *The Theory and Management of Systems*, 3rd ed. (New York: McGraw-Hill, 1973), pp. 144–46.

Flexibility

Conditions change and managers should be prepared to adjust operations accordingly. There are two ways to adjust to a changing operating environment: to design new systems or to modify operating systems. An existing system should not be modified to accommodate a change in objectives, but every system should be sufficiently flexible to integrate changes that may occur either in the environment or in the nature of the inputs. For example, a company should not use the same system to build missiles as it uses to build airplanes, nor the same system to sell insurance as the one originally designed to sell magazines. However, it should be possible to modify an existing system to produce different sizes, varieties, or types of the same product or service.

A practical system must be well designed but it cannot be entirely rigid. There will always be minor variations from the general plan, and a system should be able to adapt to such changes without excessive confusion. The advantages associated with having a flexible system will become more apparent when we consider the difficulty of administering change.

Reliability

System reliability is an important factor in organizations. Reliability is the consistency with which operations are maintained, and may vary from zero output (a complete breakdown or work stoppage) to a constant or predictable output. The typical system operates somewhere between these two extremes. The characteristics of reliability can be designed into the system by carefully selecting and arranging the operating components; the system is no more reliable than its weakest segment. When the requirements for a particular component—such as an operator having unique skills—are critical, it may be worthwhile to maintain a standby operator. In all situations, provisions should be made for quick repair or replacement when failure occurs. One valid approach to the reliability-maintenance relationship is to use a form of construction that permits repair by replacing a complete unit. In some television sets, for example, it is common practice to replace an entire section of the network rather than try to find the faulty component. Reliability is not as critical an issue when prompt repair and recovery can be instituted.

Economy

An effective system is not necessarily an economical (efficient) system. For example, the postal service may keep on schedule with mail deliveries but only by hiring a large number of additional workers. In this case, the efficiency of the postal system would be reduced. In another example, inventories may be controlled by using a comprehensive system of storekeeping. However, if the cost of the storekeeping were greater than the potential savings from this degree of control, the system would not be efficient. It is often dysfunctional and expensive to develop much greater capacity for one segment of a system than for some other part. Building in redundancy or providing for every contingency usually neutralizes the operating efficiency of the system. When a system's objectives include achieving a particular task at the lowest possible cost, there must be some degree of tradeoff between effectiveness and efficiency. When a system's objective is to perform a certain mission regardless of cost, there can be no tradeoff.

Acceptability

Any system, no matter how well designed, will not function properly unless it is accepted by the people who operate it. If the participants do not believe it will benefit them, are opposed to it, are pressured into using it, or think it is not a good system, it will not work properly. If a system is not accepted, two things can happen: (1) the system will be modified gradually by the people who are using it, or (2) the system will be used ineffectively and ultimately fail. Unplanned alterations in an elaborate system can nullify advantages associated with using the system.

Differentiation and Integration

A basic consideration in the design of organizations is dividing work into reasonable tasks (**differentiation**) while giving simultaneous attention to coordinating these activities and unifying their results into a meaningful whole (integration). Two guidelines may be followed in grouping activities:[3]

1. Units that have similar orientations and tasks should be grouped together. (They can reinforce each other's common concern and the arrangement will simplify the coordinating task of a common manager.)
2. Units required to integrate their activities closely should be grouped together. (The common manager can coordinate them through the formal hierarchy.)

When units neither have similar orientations nor share their activities, the task of grouping becomes more difficult. For example, when units are similar in nature and function but are also relatively independent, the manager must base his decision on the most appropriate way to group activities according to his past experience.

A difficult task associated with system-subsystem determination is to establish proper boundaries of operations. The more specific and distinct the goals of the operation, the easier it is to set boundaries. Other factors such as the influence of the environment, the availability of men and machines, the time schedule for design and operation, the cost of alternative designs, and the particular biases of the designers must be considered when establishing boundaries.

[3]P.R. Lawrence and J.W. Lorsch, *Organization and Environment: Managing Differentiation and Integration* (Boston: Division of Research, Harvard Graduate School of Business Administration, 1967), pp. 213–18.

Planning the Study

In this chapter, we emphasize the importance of creativity in design. Designers with imagination have the best chance to group people and machines into workable combinations having the greatest efficiency and effectiveness within the recognized constraints. We also suggest that there are certain characteristics that *should* be designed into an effective and efficient system—simplicity, flexibility, reliability, economy, and acceptability.

At this point, the designer must determine what has to be done to achieve the stated objective(s) and how the total task can be divided into meaningful units. Of the many possible combinations, one must be selected as that which satisfies the decision criteria better than the other alternatives. Of course, the balance between technical efficiency and the human factors that determine organizational climate should be included in making this decision. The eventual success or failure of the project is somewhat predetermined by management's attitude and the relationship between the designers and those who must implement the process.

The systems approach in design stresses the need for creating an orderly process. General Electric's plan of activities for developing an information system provides a good illustration (see Figure 6.3). Note how this project was organized to follow the definition, design, and operations stages. Note also how iterations (repeating the same logical process in order to improve or refine it) were included at different steps to strengthen the plan. This plan illustrates the necessity of describing completely and systematically everything that had to be accomplished to carry out a project in an orderly manner.

THE ROLE OF MANAGEMENT

The systems approach suggests a new role for management. In the traditional view, the manager operated in a highly structured, rigid system having well-defined goals, clear-cut relationships, tight controls, and hierarchical information flows. In the flexible (or open) systems view, the organization is not static but is continually evolving to meet both external and internal changes. The manager's role is to develop a viable organization, cope with change, and help participants establish a dynamic equilibrium. Leonard Sayles has expressed the manager's problem as follows:

> *The one enduring objective is the effort to build and maintain a predictable, reciprocating system of relationships, the behavioral patterns of which stay within reasonable physical limits. But this is seeking a moving equilibrium, since the parameters of the system (the division of labor and the controls) are evolving and changing. Thus, the manager endeavors to introduce regularity in a world that will never allow him to achieve the ideal.*[4]

The systems approach does not offer a prescription for making a manager's difficult and complex job easier. Rather it helps him understand and operate more effectively within the reality of

[4]Leonard R. Sayles, *Managerial Behavior* (New York: McGraw-Hill, 1964), pp. 258–59.

Figure 6.3 Plan of Activities for Development of Information System

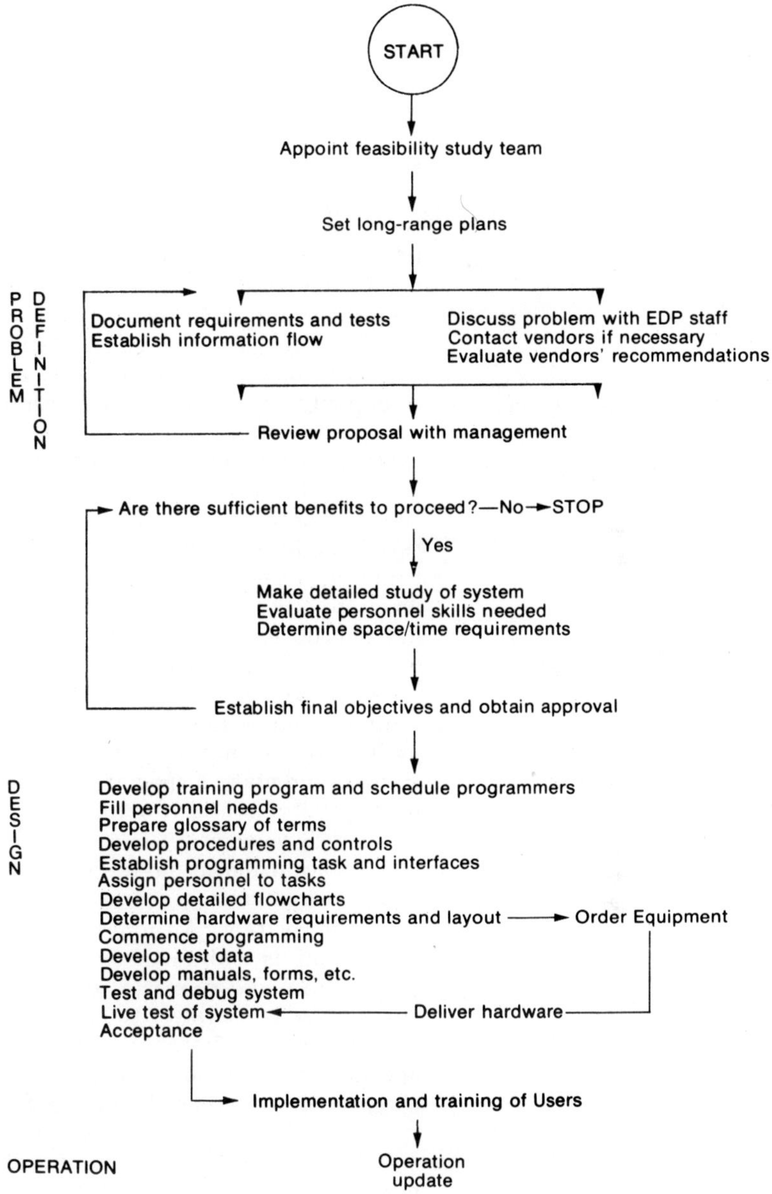

Source: R.E. Breen et al, *Management Information Systems, A Subcommittee Report on Definitions* (Schenectady, N.Y.: General Electric, 1969), p. 21.

complex systems. The systems approach suggests that operations cannot be neatly departmentalized but must be viewed as overlapping subsystems. In addition, it suggests that leadership patterns must be modified, particularly when dealing with professionals and highly trained specialists, and motivation must take the form of active, willing participation rather than forceful subjugation.

Systems design involves establishing project and facilitating subsystems to accomplish certain tasks or programs. In this approach, the network of human independence required to accomplish a given task is based on the shared responsibility of all members of the subsystem. In contrast, the traditional organization is geared to functional performance and the integrating force is authority. Instead of gearing participant activities to obedience to rules and closely structured behavior, the systems approach provides a basis for active cooperation in meeting task requirements. The manager is looked upon as a resource person who can help the group meet its goals and also as a source of authority and control. Thus, systems theory lends a structure by which the concepts of motivation, leadership, and participation can be applied effectively within the organization.

Implementation is, of course, implicit in the connotation of systems design; otherwise it would be nothing more than an empty exercise. It follows that the **interface** between managers and systems designers is critical, and mutual understanding must be fostered to maximize returns from design efforts. The system must be tailored to the needs of the organization and adapted continually as circumstances change. In a general sense, managers engage in systems design on a day-to-day basis when they plan activities and organize systems to accomplish objectives. Specialized staff groups have evolved to perform tasks such as long-range planning, organizational studies, and systems design. However, since managers are ultimately responsible for organizational endeavors, they should make a special effort to help ensure the development of useful systems and to make design activities an extension of the manager's role rather than a separate function.

Operating managers need to understand the organizational decision-making requirements and the information needed to support the system. Although the probability of success in implementation is enhanced considerably if management is vitally interested in the project, technical expertise and motivation for change are more likely to be found in staff groups. The solution to the apparent dichotomy would seem to be a team approach, with specialists supporting operating managers who are responsible for the project's success. A manager might devote either part-time to such an effort or full time temporarily, if the task requires it.

A project involving an integrated system for the entire company might well require years to complete. If operating people are delegated responsibility and authority for such a project, particularly if they are delegated the authority to outline specifications, they should also maintain sufficient contact with the day-to-day operations and its attendent information flow to retain their expertise for decision making. If the environment is dynamic or internal capabilities undergo change, it might be wise to rotate people from operations to systems design periodically, so that operating expertise is updated continually. Similarly, programmers and data processing specialists must maintain their expertise on systems design as well as operations so that an optimal mix of skill and up-to-date information can be brought to bear on systems design problems at all times.

The Process of Change

The idea of progress is symbolized by the introduction of new processes, new ways to organize, new knowledge, and new product ideas. The rate at which such change occurs depends in large part on the difference between the countervailing forces of inertia and initiative. The same assertion applies to the management of systems, for unless there is a constant review of operations through a sound critical analysis, any system, once designed, will tend to continue operating with a minimum of modification.

Few systems operate at either optimum effectiveness or efficiency. Problems relating to layout or arrangement may exist; innovations in equipment may be added to improve operating efficiency; and sometimes more appropriate techniques of analysis can be applied to optimize resource inputs. Adjusting and improving operating systems is a necessary and continual process.

Every system should be reviewed in relation to its objectives. If the basic objectives change, the system should be phased out of operation. When a project has been completed, a summary evaluation of performance provides useful information for designing better systems in the future. Let us stress, however, that change merely for the sake of change is foolish and wasteful and is not advocated in this text.

People and Change

To say that man resists change would be an oversimplification of the problem. On the contrary, in most activities people welcome and even demand change. Consumers continually look for new means and products to express their desires for change: witness the constant tendency to give formerly standardized products (such as men's underwear) higher style and fashion differentiation. Perhaps people's reception to change depends on the nature of the change. For example, let us compare two situations involving change. In the first, a consumer has some degree of control over whether he accepts or rejects change in a product; in the second, employees in an organization are met with changes that are beyond their influence or control. In other words, people tend to resist change in interpersonal and job relations because such changes threaten their sense of security and the way they have become accustomed to doing things. People may use indifference or else outright opposition and rebellion to resist change, because most changes disturb the interpersonal equilibrium with the environment in which individuals and social groups operate. It is generally not the technological change that is resisted; resistance is generated mainly by changing sociological relations and threatened economic well-being.

Management is frequently unaware of the full impact of technological change. Managers look primarily at changes in the physical setup or work arrangement in the physiological requirements of employees relating to fatigue and productivity on the job. Such changes may result in differences that would suggest new time-and-motion studies and new job specifications to compensate for physical changes in the work. Even more important, however, are changes in social equilibrium. Generally, any change will alter the informal or formal social hierarchical relationships that exist among people functioning in groups. For some people, social changes may increase status, whereas for others, their positions may be apparently diminished. Resistance to change will be discussed further in subsequent chapters.

SUMMARY

People are able to solve design problems by developing and using new and better ideas, but many people fail to develop their potential creativity. The creative process occurs when people are properly motivated and focus their abilities on a particular project or problem. Orderly arrangement and testing of information linked with imagination can improve design.

The design of subsystems begins with determining the mission or objective to be accomplished by segments of the total system. The next step is to outline carefully the constraints that may limit design alternatives. Alternative proposals are described, analyzed, and revised according to the criteria listed for selection. Finally, the best alternative is selected, tested, and implemented.

Designers wish to create systems that are simple, flexible, reliable, economical, and acceptable to the people who will operate them. In other words, the ideal is a system that is no more complex than necessary, that is able to adjust to changing environmental influences, that will tend to perform consistently in a dependable and predictable manner, and that will maximize output relative to input. These characteristics of good design apply to any man-to-man, man-to-machine, or machine-to-machine system.

People do not resist change as such. Rather, the average human being seeks satisfactory systems of interpersonal relationships that guide his activities. The idea in systems design in the organization is not to require an individual to blindly change a total pattern of living and to adapt. Instead, the objective of systems design is to change employees from old systems of work and interpersonal relationships to new situations for the advantage of both the workers and the organization.

DISCUSSION QUESTIONS

6.1. How does intelligence contribute to the creative process?

6.2. What can an organization do to stimulate creativity among its workers?

6.3. How does the technique labelled "the checklist" relate to the model of creativity in Figure 6.2?

6.4. List some of the constraints that may affect the design of a system that is to care for elderly people under the Medicare program.

6.5. How does the efficiency of operating a system relate to the reliability of operations?

6.6. Outline the steps you would follow in designing a system.
6.7. Should the design of a system be a staff function? Why or why not?
6.8. What do you believe are the most important reasons causing people to resist change?

EXERCISES

6.1. Cover the nine dots by drawing four straight lines. The lines must be connected.

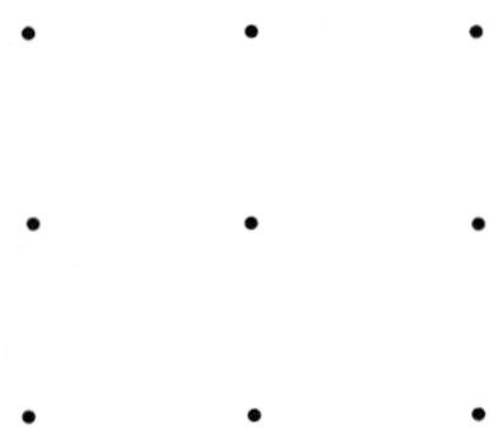

6.2. Determine the number of squares that can be outlined in the diagram below.

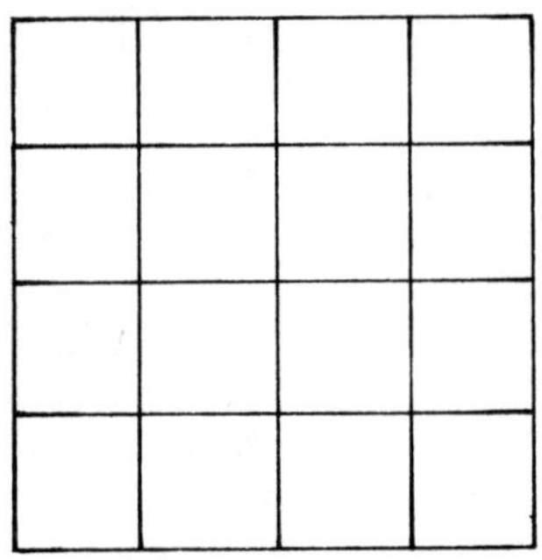

6.3. There are twelve billiard balls. All appear to be identical except that each bears a number from one to twelve. However, one ball has a different weight from the rest. Using only a balance scale, devise a system for identifying the "odd" ball in three weighings of the scale. Determine also whether the ball is heavier or lighter. It will be necessary to plan each weighing carefully: for example, if you place six balls on each side of the balance and the right-hand balance goes down, it will tell you nothing. The ball may be among those on the right-hand side, and heavier; or among those on the left-hand side, and lighter.
6.4. A city official organized a study to solve the traffic problems of that city. Proposals were made to add another freeway system, install a transit system, and to expand the bus system. Assuming that you were a member of the study group, outline:
   a. The objectives of the study.
   b. The logical constraints that your proposal included.
   c. The kinds of information necessary to describe alternative proposals.

APPENDIX
A Study in Design

A university students' organization decided to help people in need. They discovered that many people living in the university area (within five miles of the campus) did not have enough to eat. Those lacking ample food were either not eligible for government assistance or were not receiving sufficient help to satisfy their needs. Rather than investigating each individual's or family's ability to manage resources properly, the students accepted all identified need as valid. For planning purposes, they assumed that there would be no essential changes in the need for the next six months. The need gap was summarized as follows: Sixty-two families (average of five persons per family) plus thirty-two individuals could provide only seventy-five percent of their monthly needs. Thirty-eight families and eighteen individuals had the means to support only fifty percent of their monthly needs.

Before designing a system to raise money and to purchase, store, and distribute food to the designated one hundred families and fifty individuals, the students planned to outline their objectives and detail these objectives to include the amount and kind of food to supply and how often to distribute it. The system design should include the method and place of storage, distribution, office space, record keeping, and systems for controlling inventory and costs.

1. Design a system to accomplish the general objective of furnishing food for the needy. Detail this objective and outline the program in as much detail as possible. Use Figure 5.2 as your model of analysis.
2. What additional information should you have to refine your model?

SELECTED READINGS

Davis, Louis E., and Taylor, James C., eds. *Design of Jobs*. Harmondsworth, England: Penguin, 1972.

French, Wendell L., and Bell, Cecil H., Jr. *Organization Development*. Englewood Cliffs, N.J.: Prentice-Hall, 1973.

Galbraith, Jay. *Designing Complex Organizations*. Reading, Mass: Addison-Wesley, 1973.

Prehoda, Robert W. *Designing the Future*. Philadelphia: Chilton Books, 1967.

Symonds, Curtis W. *A Design for Business Intelligence*. New York: American Management Association, 1971.

Chapter 7 Information-Decision Systems

Effective information flow throughout the organization is essential for making informed decisions. In this chapter, we will examine the fundamentals of information systems—information requirements, generation, storage, and also technological advances in the field—in relation to their role and contribution to the process of management.

Information may be defined as "an established relationship among data that adds to intelligence; that which can be evaluated in terms of its relevance for decision making; and that which is subject to communication." *Communication* is defined in the next chapter as "the process of transferring information or meaning between persons or groups." The effectiveness of the transfer depends on the quality of transmission and the capability and receptiveness of the receiver. What constitutes meaningful information depends on both the problem at hand and the decision maker's ability to use the information. For example, technical data concerning manned space flights are more meaningful to someone specialized in that particular science than they would be to someone engaged in economic analysis.

Organizations are complex networks of systems that have many decision points, ranging from individual decision makers at the lowest operating levels to policy-making groups at the top of the ladder. Each decision point may be considered a subsystem with inputs of data and information, a process of transformation or analysis, and output, i.e., decisions. A manager decides issues based on the information he receives in conjunction with previously developed strategies, procedures, or rules. Depending on the level in the organization and/or the type of decision (programmed or unprogrammed), the decision maker may have several alternatives. If the question or problem is fairly commonplace and routine, the decision can be made quickly, particularly if there are rules or procedures governing the situation (programmed). If the question is more complex (unprogrammed), the decision maker may require additional inputs of data and may consult with subordinates, superiors, or peers. In every instance, it is vital that appropriate information flow be directed to the proper decision points.

THE FUNCTIONS OF INFORMATION SYSTEMS

All information systems share several functions in varying degrees of significance or intensity: (1) determination of information needs, (2) data collection, (3) data reduction, transmission, and storage, (4) data processing or information generation, (5) information reduction, transmission, and storage, and (6) information use. Figure 7.1 illustrates the sequence in which these functions take place. Note from the diagram that data (raw facts and figures) are collected and may be reduced, transmitted, or stored prior to being processed into information (data analyzed or prepared for a particular purpose). The information also may be reduced, transmitted, or stored prior to its use. The effectiveness of the information will help determine future information needs. After the information has been used, it may be returned to storage. In discussing these functions, we will sometimes combine the terms data and information to reduce redundancy.

Determination of Information Needs

The first step in the process is to outline the objectives of the system and to determine what information will be needed, when, and in what form. One method of determining information needs is

Figure 7.1 The Functions of an Information System

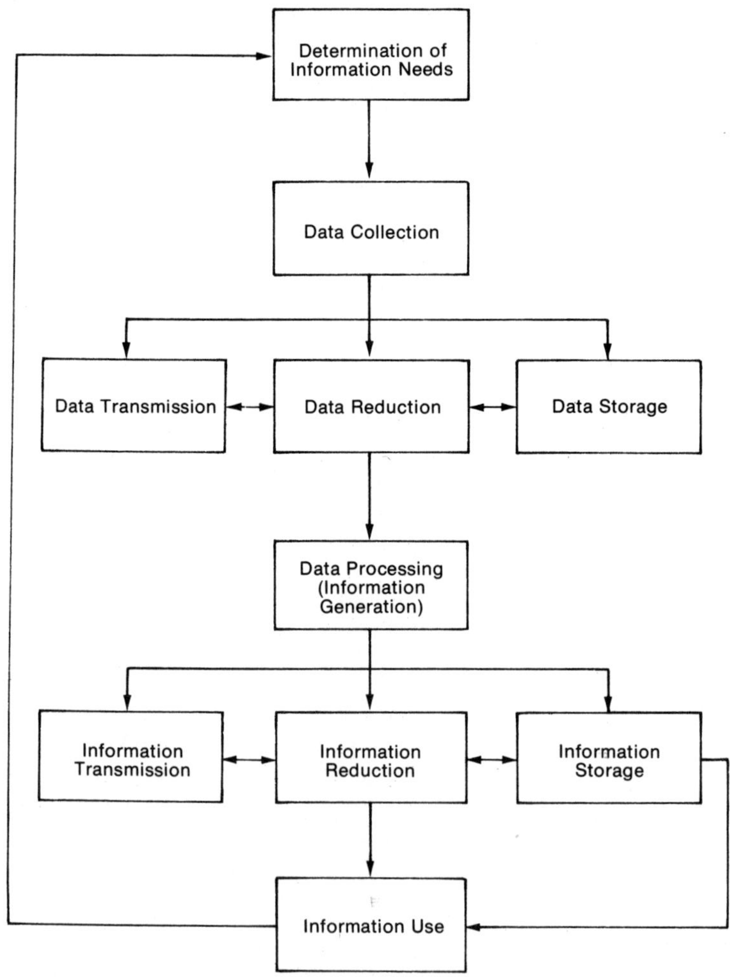

to analyze and define the information currently in use, but this approach would be unsuccessful if the current information system were not effective. A much better method is to start by asking such questions as: How much information is necessary? How will it be used? When is the information needed? In what form should it be provided? This approach begins by examining the system's output requirements, preferably in relation to the organization's long-range needs. (These long-range needs often change as the total system evolves within its environment.) An organization must consider the environment in which the system will be operating: the organization should expect to grow, change its operating objectives from time to time, and anticipate changes in competition and other outside influences.

One way to classify information requirements is to organize it according to the level of management where it will be used. What information is required to decide which new systems are needed? What information is necessary to design and allocate resources within the new systems? What information is needed to operate the systems, including controlling operations? What information is essential for evaluating the system's operating performance?

Table 7.1 lists some of the activities occurring at each of the four stages of systems management and may suggest some of the information requirements needed for these different situations. For example, the kind of information needed for establishing objectives is very different from that needed for scheduling operations. An information system should be designed to collect the right information for the appropriate user.

Table 7.1 Activities Occurring during the Four Stages of Systems Management

Systems Planning
1. Deciding on objectives
2. Establishing organizational policy
3. Developing operating strategies
4. Approving new or revised systems designs
5. Formulating budgets

Design and Creation
1. Designing new systems
2. Redesigning existing systems
3. Assigning people, machines, and technology

Operation and Control
1. Allocating inputs of material and energy
2. Controlling operations (for example, schedules, quality, inventory, and costs)

Review and Evaluation
1. Evaluating output in relation to objectives
2. Auditing efficiency

Data Collection

The data used to plan and operate an organization are collected to provide the raw material for data processing. In the modern organization, the quantity of relevant information for decision making at various levels is staggering. The number of daily activities is so great that only the most pertinent facts should be selected and reported in the normal course of operations. The information system must be designed to garner pertinent facts and screen out irrelevant or unusable data.

Collecting data and generating information are expensive processes, particularly if the decisions that follow do not optimize the operation. Substantial savings can be made by reducing the volume of data collected (through astute determination or through sampling), and by processing into information only those data relevant to the problem.

One method of reducing the amount of data gathered is to collect only that information which suggests the need for a change in operations. This method follows the "exception" principle of management—that is, data indicating that previous strategy and decisions were within an acceptable range of variation would be filtered out and not stored; indications of a changing trend in operations, however, would cause the data to be recorded and new information generated for corrective decision making. Following this approach, control data would be recorded only when a trend indicated the need for a change in input allocation.

Data or Information Reduction

Data or information can be reduced in two ways: (1) by changing the physical form in which the data and information exist (transformation), or (2) by reducing the physical volume of the data or information (compression). Although the original data may not have to be transformed to be transmitted or transported, there may be other advantages resulting from changing the form. For example, a change in form might reduce the space required for storage or permit the data to be processed, transmitted, or used more effectively.

One concern related to transformation is the possibility of losing or misinterpreting the original data. Thus, it is generally advisable to minimize the number of times data are transformed. Once data or information have been translated into machine language, for example, the systems analyst would most likely prefer to keep the information in that form for subsequent computer processing. When there is a possibility of error from misinterpretation during the transformation process, the probability and the expected consequences of such an error should be calculated. If the probability of a loss is low and the consequences not significant, it would be impractical to install extensive control measures. In contrast, if the probability is high and the possible loss is great, it would seem advisable to store the source documents even after the transformation (reduction) has been made.

Because the volume of data in most information systems is large, some compression of data and information may be essential. One method of compressing data is to reduce its physical size. Recording data on microfilm reduces a document to 1/250th of its original size, and the microfilm technology continues to advance. A second method of compression is statistical sampling, which uses a random selection of the population, for example, to represent an approximate miniature of the total. The validity of the sample as representative of the data depends on the size and design of the sample. Finally, changing the form of the data may also result in a substantial compression. A mathematical equation, for example, changes the form of data originally expressed in words and also permits a great volume of information to be condensed into a single model. As with sampling, the value of the model depends on its degree of faithfulness to the original data.

Data or Information Storage

When data and information are stored, the retention and accessibility of the documents in their original quality is important. The fundamental operations of data storage include *indexing*, *storage*, and *retrieval*. To ensure that the storage is effective and efficient, each of these three operations must be considered carefully when designing the system.

When an event is recorded and stored, it must be described in a way that will permit its retrieval within a relatively short time. Just as raw material or parts are classified and inventoried by a specific area or bin, so should data be inventoried in a manner that facilitates identification. Most often, data are classified according to some common attribute or characteristic. The information generated from these data should be categorized according to the decision requirements of the organization. Classifying the raw data according to unique characteristics provides more flexibility for the system and, even if decision requirements change, the basic data are still valid and can be processed appropriately to meet the change. For example, data may be classified according to the incremental cost of material input and transformation, and decision information can be generated periodically by totalling these costs in various ways to test alternative courses of action.

Once the data and information have been inventoried or stored, the next step is to index their locations so that they can be retrieved as needed. If data and information are not properly indexed and quickly available for decision making, the costs of their collection and generation have been wasted.

The most common method of indexing data is to use alphabetical subject headings or titles arranged in the same terminology as that used in the system. Another method is to study the information and summarize its contents according to the stage or level of the decision making to which it applies. Coding is a form of indexing in which certain words or symbols are used to file and later recover data or information. Coding makes it possible to use the computer in the search process. The computer can make a complete search very rapidly, locate the data or information, and print out the information requested while also identifying its specific location in the storage unit.

The form in which data and information are stored is a function of volume, accessibility, and cost. Storage may be temporary or permanent and range from conventional libraries with hardcover documents and index cards in card catalogs to information electronically coded and stored on magnetic tapes. One such tape can store approximately twenty million characters of data or information, which can be retrieved in seconds by a computer.

Whether data are stored in the form in which they are recorded or converted (coded) into another form that lends itself to storage and information generation depends on the compatability of the data with the processing equipment. The more available and accessible the data for information generation, the more valuable they are to the decision maker. However, coding data for computer storage may be more expensive than more traditional methods; thus, a tradeoff exists between accessibility and cost. The sheer magnitude of the data to be stored sometimes makes conversion or coding into a different language necessary. This condensation does not destroy the validity of the original input but merely converts it into another form.

Just as managers try to maintain an up-to-date inventory of raw materials and goods and occasionally purge the out-of-date inventory, so unusable or irrelevant data must be discarded. It is not always possible, though, to determine how useful or relevant data are after they have been supplemented by more current data or after they have been generated into information for decision making. One approach is to purge data and information automatically after a specified time has elapsed since their collection. Another approach is to classify data by age and regulate accessibility for information generation accordingly—that is, the more current the data, the more quickly accessible they would be to the decision maker. Older data would be removed to a less accessible storage area.

Data Processing (Information Generation)

The process of transforming inputs of data and information for decision making is outlined in Figure 7.2. Raw data are collected and stored before processing, and the program instructions are written. The output, or the part that has been generated according to programmed instructions, becomes the information that will be represented in a variety of forms, including managerial reports for decision making. This outline of a data processing system is applicable regardless of how elaborate the equipment might be at the input, processing, or output stages. To illustrate, let us suppose that this model represents a manual process for computing the payroll and determining the periodic labor costs. A predetermined sequence of operations (the *program*) is followed, and the source data (including such items as individual pay rates and hours worked) are converted into a form (such as a worksheet) that is easy to interpret and use. The processing includes multiplying the rate of pay by the hours worked for each individual, subtracting the deductions, computing the size of the individual check, and adding the totals to the departmental expense accounts. From this worksheet, the information can be reduced to other forms, includ-

Figure 7.2 Processing Data into Information

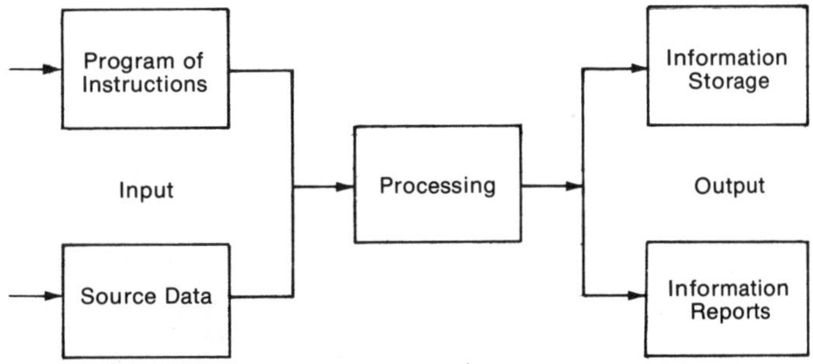

ing individual paychecks, payroll expenses, and withholding reports. The basic information such as the worker's pay grade and the sequence of operations necessary to complete the payroll procedure would be stored until the next pay period.

Processing data never ends, for the data base is continually updated by the activities or transactions that occur in operating an organization. Periodically, the data are sorted, combined, and analyzed to present meaningful information for decision making.

Data or Information Transmission

The points where information is generated and where it is needed may often be in different geographic areas. Some method must be devised to facilitate accumulating, processing, and distributing required information to decision points. Often this task is accomplished by a physical transfer. Modern communication equipment, however, can transmit coded data from their point of origin to a processing area and subsequently to wherever the results are required. Long-range transmission over hundreds of miles is possible but still costly at this time, although short-range transmission is practical and operational today.

Telephone wires can be used to transmit electronic signals between a sender and a receiver. In such cases, the data transmitted and the data received must be checked carefully to ensure they are identical. Whenever a discrepancy is discovered by the control program built into the receiver, it signals the sender that an error has been made. The sender will then back up and retransmit the correct data.

It may be necessary to change the form or coding of data to process it more efficiently; it also may be necessary to change the form again to transmit the information from one geographic point to another. In other words, it is often advisable to encode the information for effective and efficient transmission and then decode the information at the other end, for use as an information base for decision making. The ability to transmit information electronically makes it possible to establish centralized decision-making centers. In cases where such technology is unavailable, decentralized decision centers at the operations level are more appropriate.

Information Use

How information is employed depends on the quality of the information (that is, its accuracy and pertinence), its form, and its timeliness. The relative quality is contingent on the basic need outlined in the beginning. If the intended use is carefully planned and the information system is well designed, the user will receive accurate and relevant information. In addition, the form in which the information is presented is also a function of systems planning. The goal is to organize the display of information so that it can be quickly and easily understood.

The timeliness with which data and information are made available is very important to the user. Timeliness is defined by the nature of the decision to be made. For example, scientists studying the change in the depth of oceans may find that annual or semiannual reports are sufficient, whereas an air-traffic controller requires information as it changes by the second.

When a system has been designed to provide information for a particular user, its distribution is predetermined. When information is generated without predetermined distribution, the question of who should receive various information outputs arises. A filtering process can be initiated to sort the information output, but there is usually a tendency to provide more information than required rather than risk the possibility of omitting a potential user.

THE COMPUTER

The electronic computer was developed during World War II to provide a faster means for solving scientific problems. Earlier, mechanical calculators had been used to supplement human effort and improve effectiveness and efficiency, but their speed of computation was limited by the movement of mechanical parts, thus imposing a relatively low upper limit on the ultimate speed of calculation. The fundamental principle of electronic computers is that the flow of electrons, which act as signals in the circuitry of the equipment, is susceptible to direction and control. Numbers and alphabetic characters are symbolized by electronic pulses and the controlled movement, or flow, of these symbolic signals provides the basis for the computer to process data. A computer, then, may be defined as "a machine that manipulates symbols in accordance with given rules in a predetermined and self-directed manner. Speaking more technically, an automatic computer is a high-speed, automatic, electronic digital data processing machine."[1]

Speed is one of the computer's most important characteristics. The speeds at which computers are able to process data into information are fantastic and continue to increase as computers are improved. Computer technology is also working at developing larger storage capacities, or *memory*, making the electronic computer a very versatile **tool** for managerial decision making.

There are two basic types of computers: digital and analog. The **digital computer** is based on the manipulation of symbols and counting and uses the binary number system, which is based on only two digits—0 and 1. In contrast, an **analog computer** is a machine that measures some physical characteristic and creates a

[1]Ned Chapin, *An Introduction to Automatic Computers* (Princeton, N.J.: D. Van Nostrand, 1957), p. 4.

physical analog, or model (using electrical voltages), as a way to solve a specific problem. The analog computer does not use a digital representation of numbers.

The term *automatic* refers to the self-controlling characteristic of electronic computers. Internally stored programs (lists of instructions) determine the sequence of operations in a processing or computational routine. These instructions are predetermined because human effort is required to plan and set forth in detail all the steps involved in any processing job. Once the program is designed and installed, it acts as the control element in the data processing system. During processing, the program can be modified, again according to predetermined rules, and hence provide directions according to the situation represented by input data. This possibility of modifying the program parallels the feedback control concept of production automation, leading to the description of electronic data processing as "office automation." The concept of internally stored, self-adjusting programs for controlling operations sets the electronic computer completely apart from its predecessor (electromechanical equipment such as mechanical calculators). In this sense, the electronic computer is not just another stage in the evolution of mechanized equipment.

Instructions on how the processing is to take place (the **object program**), plus the source data are translated into machine language (the basic language of the computer, usually on machine cards or tape) for computer processing. The program information is stored in the computer (primary storage), while the source data are stored both in the computer and in outside storage areas such as tapes or discs (secondary storage). The computer processes the data according to the programmed instructions and the resulting output is recorded on paper or magnetic tape, which can be used as secondary storage for the information or converted into printed reports.

Because the computer is such a versatile tool and represents such an advance over its forerunners, it gives analysts the opportunity to redesign systems of information flow. Yet progress has been extremely slow in this area. There are several reasons for the lack of innovation. (1) Too much emphasis has been placed on immediate cost savings. (2) Existing procedures have been transferred to electronic data processing, resulting in the faster processing of the same old system of information flow. (3) The emphasis on the rapid introduction of new systems to demonstrate cost savings has encouraged some analysts to use a piecemeal approach with little regard for the total system. Integrated data processing is often considered a "cloud 9" concept with little practical application. (4) Finally, management has not been convinced of the merits of electronic computers in all instances. This attitude has, of course, contributed to the emphasis on immediate cost savings and the piecemeal approach.[2]

[2]Richard A. Johnson et al, *Production and Operations Management: A Systems Concept* (Boston: Houghton Mifflin, 1974), p. 268.

Hardware and Software

A computer performs the various steps of a program in essentially the same way a person would. Thus, the most trouble-free electronic data processing units seem to have developed from operations that were originally manual, rather than mechanical. A computer processes each record or item completely as a unit (in a file-maintenance approach) rather than piece by piece as in the case of punched-card systems. The similarity of electronic data processing to the manual approach encourages analysts to rethink the entire flow of paperwork and reporting for the best way to employ the high-speed computer. On the other hand, more detailed (step-by-step) instructions must be provided for the computer than for people; the machine will consider only those possibilities that are programmed before making a decision.

The task of developing the minutely detailed program of procedural steps is arduous, time-consuming, and often frustrating. The earliest programs were written directly in a special language the machine could follow to perform the arithmetic and logic operations. Each instruction had to be coded and stored in a specific location in the computer's memory. Each step included a reference to a location where the next instruction could be found. Programmers had to keep track of used and unused space allotted to the program. Simplifying this programming chore has been the focal point of much research and development over the past decade.

The programming aspect of data processing is called information **software.** Software includes all the programming systems used to facilitate effective and efficient utilization of information **hardware**—the actual physical components, or machinery (mechanical and electronic) that make up a computation system, as well as all of the supporting equipment associated with input, storage, output, and transmission. The development of improved computer hardware is moving rapidly and changes occur constantly. A summary description of such equipment is outdated before it can be published. Information system designers report their specific needs to hardware manufacturers, who then relate these requirements to the current hardware capabilities.

An early improvement in software involved the use of intermediate symbolic codes that the programmer used to develop a set of machine instructions. The computer was constructed to recognize the code and expand the written instructions by printing out the details associated with each code. For example, there might be ten steps necessary in a particular calculation but the programmer would not have to outline these ten steps each time he specified this instruction—only the code number or letter.

The next development was to combine codes or sets of instructions into a complete routine—such as regression analysis. Anytime someone wanted to instruct the computer in this type of sequence, the programmed routine could be inserted into a larger system of programmed instructions. Over the years, users have been able to accumulate libraries of subroutines. Most computers have special instructions that incorporate subroutines automatically, and programming can be accomplished by using simple reference statements, rather than full instructions, to the subroutine.

Another phase of software development came with the compiler, or **translator programs**, which are similar to assembly programs but usually more complex. The language used by the programmer in writing his general instructions has been made as simple as possible, thus requiring a more complex compiler, or translator program, to develop the machine language instructions (object program) that the computer follows.

Figure 7.3 illustrates the procedural flow for developing machine language instruction. First a programmer develops the general written instructions. Next an operator processes these instructions into machine readable information. Then the computer prepares the object program by interpreting the general instructions as outlined by the translator program.

The development of computer programming languages has progressed rapidly over the last decade. FORTRAN (Formula Translator), COBOL (Common Business-Oriented Language), and ALGOL (Algorithmic Language) are widely used languages. There are some current attempts at standardizing computer language so that program packages may be used on a variety of computers.

Integrated Data Processing (IDP)

An integrated, or total, information system is a frequent goal of information systems designers. **Integrated data processing (IDP)**

Figure 7.3 Developing Machine Language Instructions

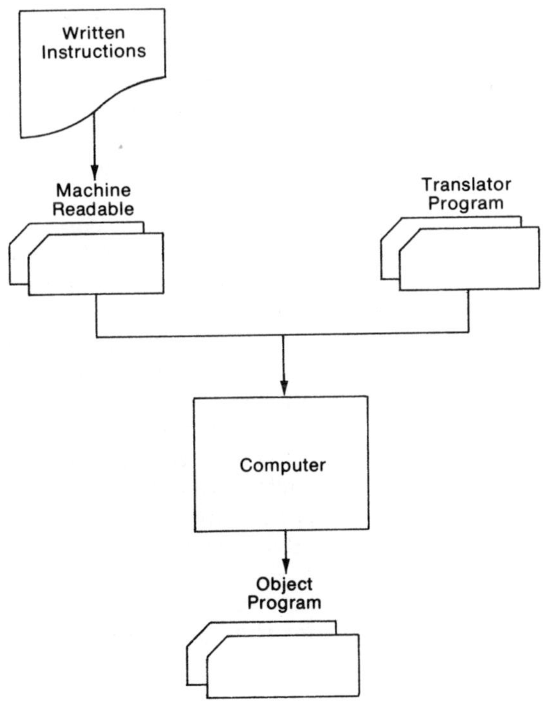

suggests data and information in a continuous flow from input to use with a minimum of human intervention in between. In addition, IDP implies a common information base or language so that the flow among various subsystems of the organization can take place without difficulty.

A problem in many information systems involves the task of converting visible source material into machine sensible language prior to processing. Normally, the justification for electronic data processing equipment is the elimination of much clerical work and electrical accounting machines. Ironically, however, the numbers of supporting electrical accounting machines (EAM) have increased with the addition of electronic data processing machines (EDPM). The job of creating and checking the cards or tape for input into the computer has caused this increasing pressure on supporting equipment. And often, human intervention occurs between the initial source document and input into the computer. Since impure input will compound errors (garbage in—garbage out), it is essential to strive for error-free input—thus making a case for integrated data processing with its goal of reducing errors by minimizing human intervention.

With integrated data processing, both visible and machine sensible source materials are prepared simultaneously. A variety of machines have been developed to provide both a hard copy and a perforated tape. More often, a punched card is generated first and used to produce subsequent documents of information with a minimum of human intervention, thus cutting time, effort, and errors. At present, the perforated tape is the most widely used common language element of an integrated, data processing system. A perforated tape can be produced by cash registers, typewriters, bookkeeping machines, desk calculators, and teletype. The tape can actuate other pieces of equipment, including card-punching machines and computers.

To eliminate the human handling of cards and paper tapes between initial recording and input to the computer system, some type of direct data transmission system is required. The feasibility of electronic data transmission among various subsystems of an organization provides the key for the broader use of integrated data processing; particularly by multiplant companies with central data processing, companies with geographically widespread sales offices, and companies with scattered, in-plant recording stations.

Many kinds of input data for computer processing originate outside of the system and, therefore, cannot be obtained through such source data by-product devices as typewriters that produce paper tape as a by-product. The logical solution, rather than the laborious, costly, and sometimes inaccurate card-punch process, is to develop machines that "talk" to the computer without human intervention. The banking industry adopted a common language for the encoding of checks early in the 1960s, and other organizations are now adopting various scanning devices, or readers, to integrate their data processing.[3] Eventually, man will be able to communicate directly with the computer through speech.

The advantages of integrated data processing or integrated information flow are probably more significant from an organizational point of view than from the efficiencies gained within the subsystem. To create a common information base for the entire organization, it is necessary to outline the proper role of each subsystem relative to the objectives of the total system. The advantages of clarifying these roles and objectives in relation to the system should be apparent.

Real-Time Data Processing

Most data processing applications are examples of *batch processing*—that is, data are collected manually or electronically over a period of time and then merged with a master file in a processing run. Such processing may take place biweekly, weekly, daily, or even at more frequent intervals. Batch processing implies, however, that the system is never completely up-to-date at any given moment. **Real-time data processing**, in contrast, updates the master file or the description of the current situation with every transaction, regardless of the frequency.[4] For example, a wholesaler may record his entire inventory in the magnetic storage of a file computer. As orders are received and processed, the inventory status is updated immediately. Thus, the inventory record is perpetually current in relation to transactions at any given time. Information on out-of-stock items is obtained as an exception printout whenever this condition occurs, and a purchase order is printed the minute the reorder point is reached. Contrast this with a system in which transactions are collected for a week and then processed, resulting in possibly serious gaps in the information available for managerial decision making.

Airline reservation systems are examples of real-time information systems. These systems are designed to match customer demand with an inventory of seat schedules for future flights. The airlines must process inquiries about availability of space in a matter of seconds to make the system practical from a customer's point of view. The system accepts inquiries from numerous points: travel agents' offices, airport ticket counters, airport check-in positions, airport standby control, and telephone requests. A request initiates a sequence of processing steps that checks the availability of seats, reports status, and updates the inventory if a sale is made. However, the real-time reservation system for an airline is only a small part of the total information system of the company. In addition, one airline's reservation system is often interconnected with those of other airlines, since a passenger may need to transfer to

[3]Carl Heyel, *Computers, Office Machines, and the New Information Technology* (New York: Business Equipment Manufacturers' Association, 1969), pp. 126–33.

[4]Even in a real-time system, some momentary delay in updating may occur subject to the capabilities of input and processing devices. A unique characteristic of real-time systems, however, is that they are always "on-line," or available.

another carrier to complete his journey. Such a supersystem of multicompany coordination, or integration among systems, achieves the ultimate benefit for the traveler.

Time Sharing

Computer information systems are sometimes organized on a time-sharing basis. This permits a number of users to employ the same computer simultaneously. The users may or may not be part of the same organization. Each user has the ability to inquire and to add or alter information in the central storage of the computer.

The original meaning of **time sharing** pertained to the kind of electrical circuitry that permitted more than one instruction or operation to be executed simultaneously by the computer, thus allowing greater utilization of the internal processing capability while inputting or outputting occurred. "Second generation," or more advanced, computers permitted several programs with large quantities of input and output to be loaded at the same time. In such *multiprogramming*, an executive program directs the utilization of input, processing, and output capability according to established priorities.[5] The next advance in time sharing occurred when remote locations were tied into a central system. The computers in these systems were the "third generation" of equipment and possessed program-interrupt features. These "third-generation" computers (such as those used in airline reservation systems) incorporate real-time processing.

Two arguments are used to justify the installation and operation of a general purpose, time-sharing system. The first is a "performance requirements or objectives" argument, which states that the need for accomplishment is so important that this type of facility is justified. Certain military applications would fall into this category. The second rationale is more common and is based on a cost-benefit analysis, which considers the advantages, the disadvantages, and the cost of time sharing. However, the major advantage of time sharing is the immediate availability of information for decision making.[6] There are clearly certain kinds of control operations in which having continuous and immediate knowledge of changes is extremely important; in many other operations, however, a decrease in the response time of information is not nearly as crucial. In still other applications, the ability to use or interact with the computer as a dynamic problem-solving tool is a real advantage, when current information and substantially reduced time required for problem solving or planning are essential.

The principal disadvantage of time-sharing systems is the high cost, particularly because of the cost of installation. A time-sharing system requires a larger central processing unit, memory, and auxiliary memory storage. In addition, the cost of high-speed auxiliary storage, terminals, communication controllers, and

[5]Heyel, *Computers, Office Machines, and the New Information Technology*, pp. 158–63.

[6]The assumption here is that the subsystem could not support its own computerized information system.

communication facilities must be included. Then, too, a comprehensive executive program to establish priorities of use must be developed and maintained. Furthermore, as the number of users on the on-line service increases and the unit cost decreases, overloading the time-sharing facilities and making the system less effective for each user becomes a potential difficulty.

There are other special problems that may influence an organization's decision to install time-sharing facilities. For example, the security and privacy of the user's data come into question. Even though there are various mechanical safeguards as well as programmed controls built into the software systems, the danger always exists that data and information stored by one organization in the computer might be used or distorted by other users. There are also possibilities of breakdowns and transmission difficulties, in addition to overloading.

THE ECONOMICS OF INFORMATION

Information is valuable to the extent that it improves decision making. The value of an increment of new information could be determined if the marginal improvement resulting from the decision-making process that followed were known. New information that changes the model representing the real world is important if the change is significant enough to alter a subsequent decision. For example, new information about how students learn could bring about changes in the design of educational systems, or additional information concerning increasing inventory cost could cause a change in the allocation of resource inputs in an operating system. Similarly, control and evaluation information may cause a change in operations or systems design.

A frequent criticism is that much data and some information are generated for no real purpose and should consequently be eliminated. There seems to be a tendency to generate large quantities of information on the assumption that a direct correlation exists between the volume of information and the quality of decision making. This assumption is reasonable only if the information is correct and relevant and properly presented to the right parties at the right time. Perhaps this lack of knowledge about what information, when and for whom it should be generated causes the excessive volume. The tendency seems to be to compensate for these uncertainties by furnishing large quantities of information to administrators at all levels and at all times, never stopping to analyze whether managers are using the information that has been generated. One possible solution would be to advocate a free-inquiry system, in which decentralized decision centers receive only information that they request from the data processing center and do not receive voluminous, and possibly irrelevant, periodic reports.

Ultimately, each user of information for decision making should determine the cost of information relative to its utility for decision making. In such a case, the cost of complete information can be weighed against the expected return from good decisions as opposed to the consequences and probability of less desirable action. To make this analysis, we must know the incremental cost of information and the incremental gain from improving the quality of the decision. Creating a model that would faithfully represent reality is difficult, since information for decision making is dynamic; casting this process as a static model is highly theoretical and of limited practical value.

It is worth noting, however, that the limitations of outlining a theoretical model to evaluate the economics of information should not cause the reader to discount this approach entirely, for a theoretical discussion may still be worthwhile. When the cost of information is formalized, particularly in making structured decisions, the value of improved information can be weighed against the cost of acquiring this information. Analyzing the cost against the value of improved information can alert managers and systems designers on the need to provide only the most appropriate information to the decision makers.

Cost of Information

Many organizations summarize the cost of information by the cost of operating the data processing department. Although this is clearly a real and determinable cost, it does not include the total cost of data collection, storage, processing, transmission, and use. For example, a decision maker tends to overlook the time spent scanning irrelevant reports or trying to extract meaningful information from data that may or may not pertain to the problem. Similarly, this method of determining the cost of information does not always consider the costs of storing outdated information. To make a meaningful analysis of the costs of information, the information flow through all of the functions must be included.

An organization may assume that the costs of information are fixed; that is, once the information equipment has been purchased or leased, the costs of this equipment will continue unchanged regardless of volume. This attitude, therefore, assumes that additional information can be generated at a small incremental cost inasmuch as the computer may be idle otherwise. In practice, this approach may result in generating information of dubious value and loading the system to the point where costs increase rapidly. Overloading the system also reduces its ability to respond to special information needs and requirements.

Costs are directly related to the response time required to identify, analyze, and present stored data into processed information. At one end of the spectrum is the real-time operation, in which the response time is instantaneous (or at least very short). At the other extreme is the periodic sequential processing, in which information is generated during specified intervals. Real-time information systems usually are more expensive because of the higher cost of storing data and information on-line and because of the higher transmission costs in connecting the remote-controlled input and output stations.

The accuracy and reliability of a system also affect costs. The more accurate and reliable the system, the greater the costs. In a given system, the increase in these costs will be moderate to a point, beyond which the cost will increase rapidly. The most feasible approach is to specify the desired accuracy and the reliability of a system within practical limits, and then add control checks that will identify errors as they occur.

The decision maker must determine the value of information. Again, he must analyze the sensitivity of the decision relative to the value of additional or more current information, as activities change the data base. In addition, he must consider the importance of timeliness and the need to receive only that information pertaining directly to the decision at hand. The marginal value of each of these conditions measured against cost will determine the optimum total system.

SUMMARY

Information can be defined as "an established relationship among events that adds to knowledge and can be evaluated in terms of its relevance for decision making." In some organizations, decisions may be preprogrammed at lower levels subject to anticipated inputs, whereas in other organizations or at higher managerial levels, information is generated to make unprogrammed decisions.

The functions of the information system are to: (1) determine the information need, (2) collect data, (3) reduce data or information, (4) store data or information, (5) process data or generate information, (6) transmit data or information, and (7) use information.

An electronic computer may be defined as "a machine that manipulates symbols in a predetermined and self-directed manner." Because of the computer's speed, versatility, accuracy, and extreme efficiency, it has added new and dramatic design possibilities for information systems. Using such equipment only to replace electromechanical equipment, without rethinking the entire flow of information in a system, does not utilize its real potential.

Software includes all of the programming systems that facilitate the utilization of computer hardware. Software is the procedural instructions to the computer telling it what to do as well as the control of what is done. Important developments in computer technology are (1) subroutines, which are instructions for performing typical calculations that can be inserted as needed in the general program, and (2) compilers, or translator programs, which permit the programmer to use simple language in writing instructions. Many special language programs have been created and may someday be standardized.

Integrated data processing, or IDP, suggests a close coupling of all the components in the information system so that human intervention between input and output is minimized. The greatest gain from IDP is the potential elimination of errors that creep into the system when data are duplicated or converted into another form by human action. The essential requirement of such a system is the use of a common language, with machine inputs generated or created from the original input.

Real-time data processing means an on-line, or constantly available, operation involving the continuous updating of the master information file whenever transactions occur. Airline reservation systems are an example of this kind of an operation. Another innovation, time sharing, originally meant the computer's carrying out more than one instruction at one time. Later on, several programs were processed simultaneously, and then remote stations were included when the interrupt feature was added to computer capability. Now when we talk of time sharing, we mean on-line, real-time, concurrent access; autonomy of subscribers; and latitude of functional capabilities.

The value of a unit of information is equal to the incremental improvement it makes in decision making. If we could measure this, we would know how much, and what kind of, information to provide. The tendency is to oversupply information. This is particularly true if we consider the cost of the information system fixed and believe that we might as well operate it at full capcity. This raises the further question of the size, accuracy, and reliability of information systems. The best solution seems to be to select equipment capable of handling the organization's specific information requirements.

DISCUSSION QUESTIONS

7.1. Outline the kinds of information required to make rational decisions in:
  a. Buying a home
  b. Operating a bowling alley
  c. Creating the Peace Corps

7.2. Does planning information differ from control information? How?

7.3. What is the difference between data and information? Give examples of each.

7.4. Is it advisable to classify data by the function they are to perform or by some characteristic unique to the data? Why? Contrast this with the classification of information.

7.5. How can the volume of data collected and stored be maintained within reasonable limits?

7.6. What effect does electronic transmission of data have on information systems design?

7.7. Discuss the significance of computers relative to the design of information systems.

7.8. A phase of software development relates to compilers, or translators. What is gained by this development?

7.9. Outline the prerequisite of an integrated data processing system. How does this differ from real-time data processing?

7.10. What kinds of organizations would be apt to use time-sharing installations? Why?

7.11. Discuss the relationship between system cost and response time. What cost factors are significant and how will they vary as response time increases?

APPENDIX
The Admissions System

A graduate school of business administration admits 210 full-time student equivalents to its MBA program each year. One class of 35 full-time equivalents is admitted for summer quarter, and five classes of 35 full-time equivalents are admitted to start the six-quarter program in September. Students are admitted according to the following guidelines.

Admission Policy

1. Admit no more than ten percent part-time students. (Two part-time students are equivalent to one full-time student.)
2. Admit no more than ten percent foreign students.
3. Admit no more than three foreign students from any one country.
4. Select a student body with diverse educational backgrounds and with a distribution from all parts of the country.
5. Encourage minority students.
6. Select students with the best potential for graduate study.

Decision Criteria for Admission

The school's objective is to look at the whole person and his or her potential. Some of the characteristics examined include:

1. Past performance as evidenced by junior-senior grade point average in consideration of school and program of study.
2. Potential as evidenced by scores from the Admission Test for Graduate Study in Business.
3. Relevant business experience.
4. Motivation as indicated through interviews with applicants, letters outlining personal objectives, and letters of recommendation.

Admission Procedure

The files of applicants are accumulated until March 1. At that time:

1. The summer class is admitted.
2. Fifty percent of the autumn class is admitted.
3. Some applications are placed on "hold" for future consideration.
4. Other applicants are denied.

On May 1, a second screening of applications is conducted. This screening includes all applications that have been completed since March 1 and those placed on "hold" on March 1. In addition:

1. The balance of the autumn class is admitted.
2. Some applications are placed on "hold."
3. Other applicants are denied.

About seventy-five percent of the applicants admitted for summer quarter actually enroll, whereas between sixty and seventy percent of those admitted for autumn can be expected to enroll. Those not enrolled may elect to go to another school or decide not to attend graduate school at this time. Each student is asked to respond within thirty days if he or she plans to attend, but because there is no deposit required until July 1, most all indicate "yes" at that time.

1. List the kinds of information needed to make admission decisions that will achieve the school's objectives.
2. Propose a system of information flow that will be effective and efficient.

SELECTED READINGS

Dearden, John, McFarlan, F. Warren, and Zani, William M. *Managing Computer-Based Information Systems*. Homewood, Ill.: Richard D. Irwin, 1971.

Hertz, David B. *New Power for Management*. New York: McGraw-Hill, 1969.

LeBreton, Preston P. *Administrative Intelligence-Information Systems*. Boston: Houghton Mifflin, 1969.

Sanders, Donald H. *Computers and Management*. New York: McGraw-Hill, 1970.

Sharpe, William F. *The Economics of Computers*. New York: Columbia University Press, 1969.

ORGANIZATION DEVELOPMENT
OVERDOSE?
WE'RE HAVING AN O.D. MEETING AT TEN

Chapter 8 Communication in Organizations

People can relate to each other only through some form of communication. The survival of an organization depends on individuals and groups who are able to maintain among themselves effective and continuing relationships. If we can understand organizational communication, we will understand the organization itself. For our purposes, we will define *communication* as "the transfer of meanings between persons and groups." The purpose of communication may range from completing a task or mission to creating and maintaining satisfying human relationships. The word *transfer* means more than the simple process of "packaging" an idea as conceived by a sender and transporting it to the mind of a receiver, where it is "unpackaged." It implies the *creation* of meaning in the mind of a sender followed by a *re-creation* of the same meaning in the mind of a receiver. If something occurs along the way to change the sender's original meaning, the communication has failed in its intent.

FUNCTIONAL ORGANIZATIONAL COMMUNICATIONS

Communication may be considered a functional part of an organizational system, and it may also be considered in an interpersonal context. First we will explore the communication function in an organization, and later in the chapter we will review the interpersonal aspects. We will also discuss communication in terms of form or method, content, direction, meaning, channels or networks, and the barriers and gateways that impede or improve its effectiveness.

Formal Communication

The structure of an organization is determined in part by the network of channels or paths along which information must flow between members or subunits. This structure is the formal organization, which we discussed in Chapter 2. The messages that move along its prescribed and regulated pathways are known as **formal communications**. The content of these messages concerns the work and the other related activities of the organization. Formal communications may be verbal, nonverbal, or written and may take the form of letters, telephone calls, teletype or radio messages, computer printouts, or memos. In some cases, a gesture may communicate as effectively as writing or talking to someone. Messages originate with people or machines who, by their position or activities, act as authorized senders. From the sender, the messages move along designated routes to responsible action takers and to other people or machines who need to know the content of the message.

As a general rule, all formal communications are recorded and filed and become a part of the organization's records. File copies are retained by the sender, the receiver, and all other offices that need to stay informed. If a manager has spoken to someone about something, the message may also become the subject of a "memorandum for the file," prepared by the manager, the receiver, or both. Some examples of formal communication include work orders, accounting records and reports, inventory and sales reports, policy statements, job descriptions, and work-method protocols.

Figure 8.1 illustrates how formal communication follows prescribed lines of authority. Messages *A* and *B* illustrate memoranda moving from superiors to subordinates in the formal hierarchy; message *A* might be a request for information on sales in a certain district, while message *B* might be a notice of a meeting to be held at some future date. Message *C*, originated by the vice-president of manufacturing and received by the vice-president of engineering, might be a routine request for technical information on a particular product being manufactured.

Formal communication may sometimes move horizontally across parallel lines of authority. Whether or not a copy of messages that cross organizational lines or functions (as message *C* does) should be sent to the common superior (in this case, the president) would be determined by how the president has defined the authority held by sender and receiver. If, in the judgment of both vice-presidents, the content of the message were entirely within

Figure 8.1 Formal Communications

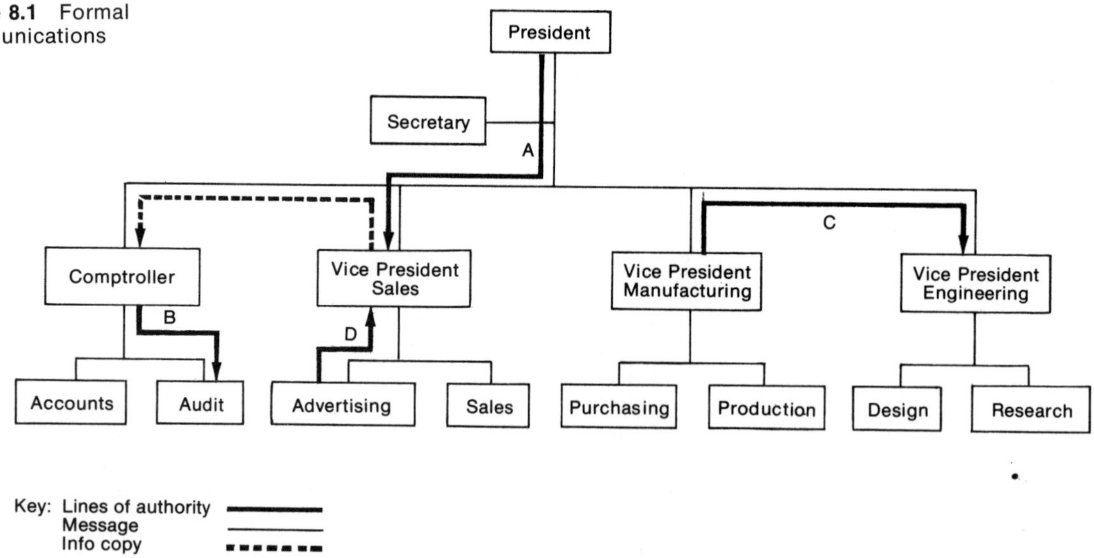

their area of decision making, they would probably not send an information copy to the president. If the boundaries of their authority are not entirely clear-cut, they might send a copy to the president or they might route the message via the president for his approval before he sends it on to the intended receiver. In either case, the objective would be to inform or to request information, and also to ensure that *prescribed lines of communication* or lines of authority are followed. In the case of message D (in Figure 8.1), which represents a progress report from a departmental subordinate to his V.P., (heavy line) an information copy (dashed line) has moved between two different functional managers who share a responsibility delegated to them by the head of the organization. In this case, the primary receiver of the message (one divisional vice-president) has decided that another divisional vice-president should have an information copy and has sent one.

The network of formal communication in an organization with accompanying filing places serves several purposes. First, it defines the routes over which all important messages will travel. Second, it provides a road map for senders and receivers of information to follow. Third, it indicates the route to those who will act and also to those who must be kept informed of planned action, work in progress, and work results. Fourth, it provides a place for the storage of information that may be required for planning, operations, and control. Finally, it delineates an orderly system for superiors and subordinates to keep each other informed.

Informal Communication

The formal communication channels and procedures provide structural guidelines for sending, receiving, and recording messages required in operations and other organizational business, but formal communications fall short of meeting all the needs of organizational communication. The gap between the needs of the formal system and those of the total system is bridged by an *informal communication* system. This consists of a large number of messages, bits of information, opinions, and expressions of feeling among people and groups over paths and by means that are chosen for convenience and necessity rather than for propriety or in conformity with formal communication patterns.

It is impossible to describe all of the types of informal messages that are sent and received, or the paths over which they move in the organization. It is safe to say, only, that without informal messages, work would not be accomplished, and that the organization itself, made up of psychological, social, economic, and technological needs, could not continue to function. Leon Festinger, in reporting the results of a number of studies on communication, found that there are three general sources of pressure that generate informal communication in human groups.[1]

1. People need to share with each other and agree on important

[1]Leon Festinger, "Informal Social Communication," in Dorwin Cartwright and Alvin Zander, *Group Dynamics* (New York: Harper & Row, 1968), pp. 182–91.

opinions and attitudes in order to feel that they *belong* together in the group. Hence, much information is exchanged unofficially in the organization as a whole and among members of subgroups, which satisfies needs to conform and to share positions on issues.

2. People need to share with superiors and others their hopes and ambitions in satisfaction of needs for achievement, affiliation, and power. In Festinger's terms, these communications arise from forces to "locomote" within a group.
3. People need to express emotions such as joy, anger, hostility, and the like as a means of "blowing off steam."

A fourth source of informal communication could be added and is related to the rigidity of the formal bureaucratic structure:

4. The need to bypass "official channels" for the sake of expediency in getting and giving information about performance on the job.

Figure 8.2 illustrates several pathways, from origins to destinations, over which informal messages may flow. Such message movements follow any convenient and expeditious route. For example, message *A* might be a cheerful "Hello, Tom," coming from a vice-president to an employee in the advertising department as they pass each other in the parking lot. Message *B* could be an unofficial verbal follow-up on a job from Audit to persons in Accounts and V.P. Sales offices. Message *C* could be a call from the president's secretary to her husband in Sales. Message *D* might be an attempt to organize a group of bridge players for a lunchtime bridge game. Message *E*, originating in the design department,

Figure 8.2 An Example of Informal Communication Flow in an Organization

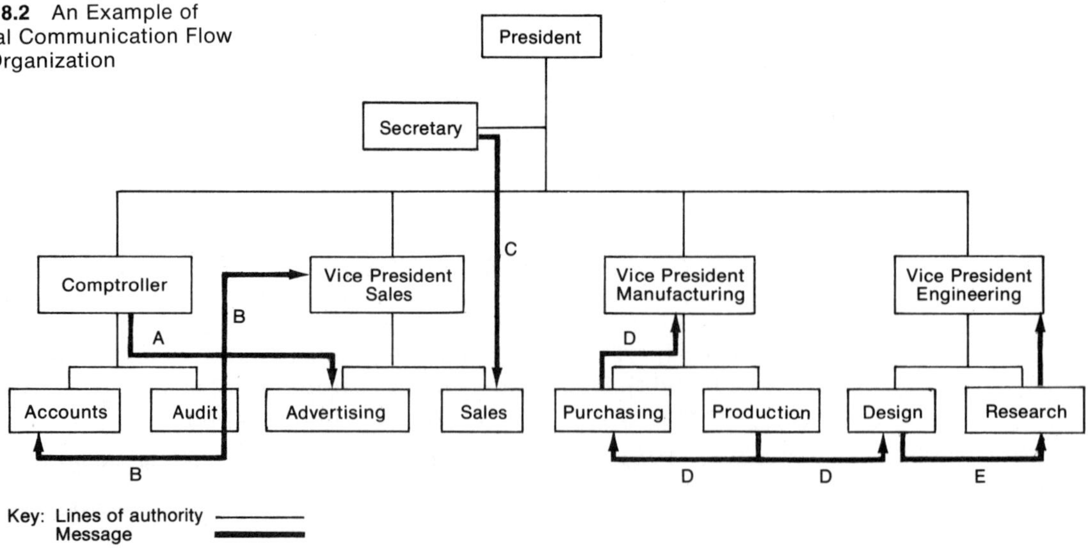

could be a rumor concerning what kind of wage agreement was coming up for discussion in current labor negotiations. (Rumors, by the way, seem to flow in the greatest volume when uncertainties affecting a number of people in the organization rise above normal levels of concern). Each of these communications could be classified as meeting social or emotional needs, as expressions of hopes, fears, and feelings about events or about other employees, or as communications that "short-circuit" the formal communication network for the sake of expediency.

Direction of Communication

If we consider formal communications as they occur in traditional military organizations, messages have a "one-way" directional characteristic. In the military organization, the formal communication proceeds from superior to subordinate, and its content is presumably clear because it originates at a higher level of expertise and experience. Military communications also carry the additional assumption that the superior is responsible for making his communication clear and understandable to his subordinates. This type of organization assumes that there is little need for two-way *exchanges* between organizational levels except as they are initiated by a higher level. Because messages from superiors are considered to be more important than those from subordinates, the implicit rule is that communication channels, except for prescribed information flows, should not be cluttered by messages from subordinates but should remain open and free for messages moving down the chain of command. "Juniors should be seen and not heard," is still an unwritten, if not explicit, law of military protocol.

Vestiges of one-way flows of communication still exist in many formal organizations outside the military, and for many of the same reasons as described above. Although management recognizes that prescribed information must flow both downward and upward, managers may not always be convinced that two-wayness should be encouraged. For example, to what extent is a subordinate free to communicate to his superior that he understands or does not understand a message? Is it possible for him to question the superior, ask for clarification, suggest modifications to instructions he has received, or transmit unsolicited messages to his superior, which are not prescribed by the rules? To what extent does the one-way rule of direction affect the efficiency of communication in the organization, in addition to the morale and motivation of subordinates?

These are not merely procedural matters but include questions about the organizational climate, or psychological atmosphere in which communication takes place. Harold Leavitt has suggested a simple experiment that helps answer some of these questions.[2] As shown in Figure 8.3, a group is assigned the task of re-creating on paper a set of rectangular figures, first as they are

[2]Harold Leavitt, *Managerial Psychology* (Chicago: University of Chicago Press, 1958), pp. 118–28.

Figure 8.3 Leavitt's Rectangles

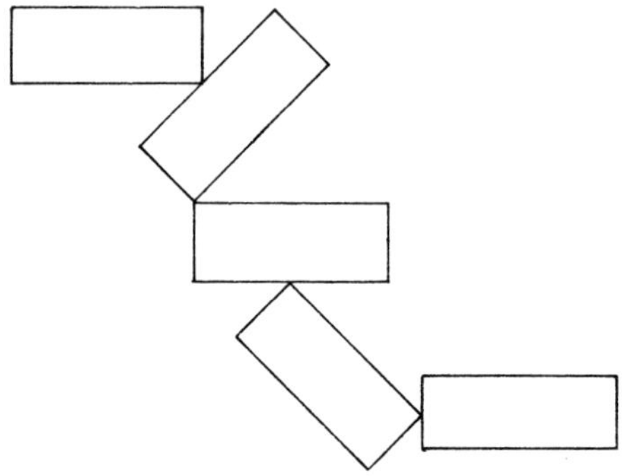

Source: Harold J. Leavitt, *Managerial Psychology* (Chicago: University of Chicago Press, 1958), p. 121.

described by the leader under *one-way* conditions, and second as they are described by the leader under *two-way* conditions. (A different configuration of rectangles is used in the second trial.) In the one-way trial, the leader's back is turned to the group. He describes the rectangles as he sees them. No one in the group is allowed to ask questions and no one may indicate by any audible or visible sign his understanding or his frustration as he attempts to follow the leader's directions. In the two-way trial, the leader faces the group. In this case, the group may ask for clarifications on his description of the rectangles and he can not only see but also can feel and respond to the emotional reactions of group members as they try to re-create his instructions on paper.

On the basis of a number of experimental trials similar to the one described above, Leavitt formed these conclusions:

1. One-way communication is *faster* than two-way communication.
2. Two-way communication is more *accurate* than one-way communication.
3. Receivers are more sure of themselves and make more correct judgments of how right or wrong they are in the two-way system.
4. The sender feels psychologically under attack in the two-way system, because his receivers pick up his mistakes and oversights and *point them out* to him.
5. The two-way method is relatively noisier and looks more disorderly. The one-way method, on the other hand, *appears* neat and efficient to an outside observer.

Thus, if *speed* is necessary, if a businesslike *appearance* is important, if a manager does not want his *mistakes* recognized, and if he wants to *protect his power*, then one-way communication seems preferable. In contrast, if the manager wants to get his message across, or if he is concerned about his receivers' feeling that they are participating and are making a contribution, the two-way system is better.

Communication Networks

Networks are another aspect of direction and flow of communication. Bavelas has shown that communication patterns, or networks, influence groups in several important ways.[3] Communication networks may affect the group's completion of the assigned task on time, the position of the de facto leader in the group, or they may affect the group members' satisfaction from occupying certain positions in the network. Although these findings are based on laboratory experiments, they have important implications for the dynamics of communication in formal organizations.

Figure 8.4 illustrates different patterns of communication. Each dot represents an individual in a communication network. The lines represent communication linkages, or channels between individuals. Pattern *A* is known as the "Chain," *B* as the "Wheel," *C* as the "Star," *D* as the "All-Channel" network, and *E* as the "Circle." The Chain can readily be seen to represent the hierarchical pattern that characterizes strictly formal information flow, "from the top down," in military and some types of business organizations. The Wheel can be compared with a typical **autocratic** organization, meaning one-man rule and limited employee participation. The Star is similar to the basic formal structure of many organizations. The All-Channel network, which is an elaboration of Bavelas's Circle used by Guetzkow, is analogous to the free-flow of communication in a group that encourages all of its members to become involved in group decision processes.[4] The All-Channel network may also be compared to some of the informal communication networks that were discussed earlier in this chapter.

If we assume that messages may move in both directions between stations in the networks, it is easy to see that some individuals occupy key positions with regard to the number of messages they handle and the degree to which they exercise control over the flow of information. For example, the person represented by the central dot in the "Star" handles all messages in the group. In contrast, individuals who occupy stations at the edges of the pattern handle fewer messages and have little or no control over the flow of information. These "peripheral" individuals can communicate with only one or two other persons and must depend entirely on others to relay their messages if they wish to extend their range.

In reporting the results of experiments involving the Circle, Wheel, and Star configurations, Bavelas came to the following tentative conclusions. In patterns with positions located centrally, such as the Wheel and the Star, an organization quickly develops around the people occupying these central positions. In such pat-

[3]Summarized from concepts developed by Alex Bavelas, "Communication Patterns in Task-Oriented Groups," pp. 503–11; Harold Guetzkow, "Differentiation of Roles in Task-Oriented Groups," pp. 512–26, in Cartwright and Zander, *Group Dynamics*; H.J. Leavitt, "Some Effects of Certain Communication Patterns on Group Performance," *Journal of Abnormal and Social Psychology* vol. 46, (January 1951), pp. 38–50.

[4]Refer to discussion of group decision making in Chapter 5.

Figure 8.4 Communication Patterns

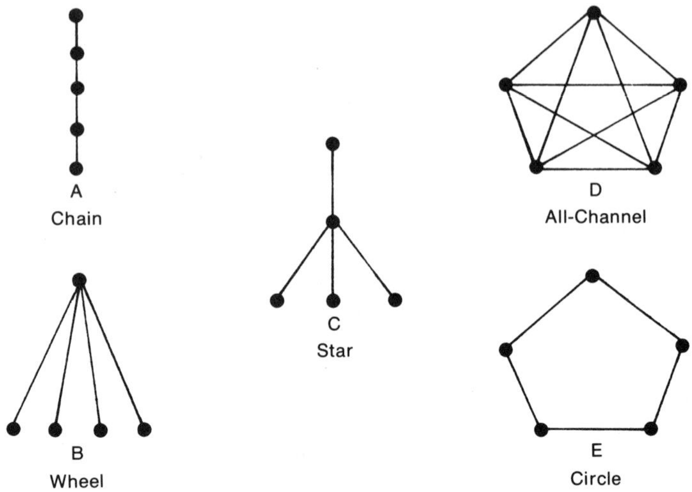

terns, the organization is more stable and errors in performance are lower than in patterns having a lower degree of centrality, such as the Circle. However, he also found that the morale of members in high centrality patterns is relatively low. Bavelas speculated that this lower morale could, in the long run, lower the accuracy and speed of such networks.

In problem solving requiring the pooling of data and judgments, or "insight," Bavelas suggested that the ability to evaluate partial results, to look at alternatives, and to restructure problems fell off rapidly when one person was able to assume a more central (that is, more controlling) position in the information flow. For example, insight into a problem requiring change would be less in the Wheel and the Star than in the Circle or the Chain because of the "bottlenecking" effect of data control by central members.

We may conclude from these laboratory results that the structure of communications within an organization will have a significant influence on the accuracy of decisions, the speed with which they can be reached, and the satisfaction of the people involved. Consequently, in networks in which the responsibility for initiating and passing along messages is shared more evenly among the members, the better the group's morale in the long run.

INTERPERSONAL COMMUNICATION

Another facet of communication in the organization is the process of face-to-face, **interpersonal communications**, between individuals. Such communication may take several forms. Messages may be verbal (that is, expressed in words), or they may not involve words at all but consist of gestures, facial expressions, and certain postures ("body language"). Nonverbal messages may even stem from silence—Out of sight, out of mind! When we crook a finger at someone, we are sending a message without using words. When we frown, we are announcing that we are displeased about something. When a diplomat walks out of a meeting, he may be saying, "I do not approve and wish to emphasize how far apart we are by my absence." If the boss is silent about that "important" memo you sent him a month ago, his silence may fill your mind with uncomfortable thoughts.

Ideally, the meanings sent are the meanings received. This is most often the case when the messages concern something that can be verified objectively. For example, "This piece of pipe fits the threads on the coupling." In this case, the receiver of the message can check the sender's words by actual trial, if necessary. However, when the sender's words describe a feeling or an opinion about something that cannot be checked objectively, meanings can be very unclear. "This work is too hard" or "Watergate was politically justified" are examples of opinions or feelings that cannot be verified. Thus they are subject to interpretation and hence to distorted meanings. The receiver's background of experience and learning may differ enough from that of the sender to cause significantly different perceptions and evaluations of the topic under discussion. As we shall see later, such differences form a basic barrier to communication.

Nonverbal content always accompanies the verbal content of messages. This is reasonably clear in the case of face-to-face communication. As Virginia Satir has pointed out, people cannot help but communicate symbolically (for example, through their clothing or possessions) or through some form of body language.[5] In messages that are conveyed by the telephone, a messenger, or a letter, the *situation* or *context* in which the message is sent becomes part of its nonverbal content. For example, if the company has been losing money, and in a letter to the production division, the front office orders a reorganization of the shipping and receiving departments, this could be construed to mean that some people were going to lose their jobs—unless it were made explicitly clear that this would not occur.

A number of variables influence the effectiveness of communication. Some are found in the environment in which communication takes place, some in the personalities of the sender and the receiver, and some in the relationship that exists between sender and receiver. Figure 8.5 depicts these variables and suggests some of the difficulties of communicating with understanding from one person to another. The sender wants to formulate an idea and communicate it to the receiver. This desire to communicate may arise from his thoughts or feelings or it may have been triggered by something in the environment. The communication may also be influenced or distorted by the relationship between the sender and the receiver, such as status differences, a staff-line relationship, or a learner-teacher relationship.

Whatever its origin, information travels through a series of filters, both in the sender and in the receiver, before the idea can be transmitted and re-created in the receiver's mind. Physical

[5]Virginia Satir, *Conjoint Family Therapy* (Palo Alto, Calif.: Science and Behavior Books, 1967), p. 76.

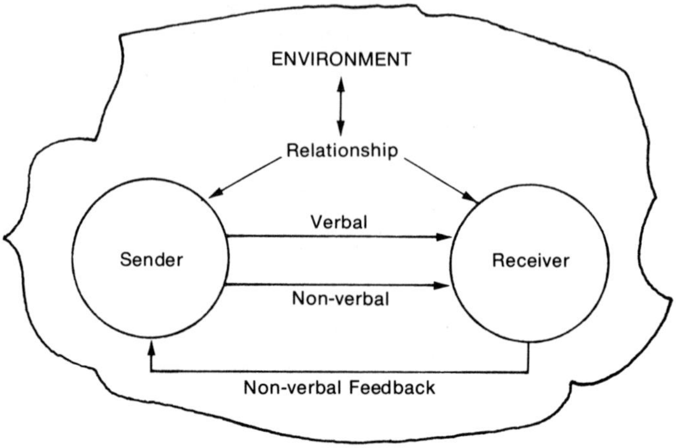

Figure 8.5 Factors Affecting Meaning in Communication

capacities to see, hear, smell, taste, and touch vary between people, so that the image of reality may be distorted even before the mind goes to work. A ship's master once shouted to a man on the shore, "What bay is this?" The man on the shore heard him say, "What *day* is this?" and responded, "Friday!" The bay was duly entered on the ship's chart as Friday Harbor and goes by that name to this day.

In addition to physical or sense filters, cognitive filters, or the way in which an individual's mind interprets the world around him, will influence his assumptions and feelings. These filters will determine what the sender of a message says, how he says it, and with what purpose. Filters are present also in the receiver, creating a double complexity that once led Robert Louis Stevenson to say that human communication is "doubly relative." It takes one person to say something and another to decide what he said.

Physical and cognitive, including semantic filters (which decide the meaning of words) combine to form a part of our memory system that helps us respond to reality. In this sense, March and Simon compare a person to a data processing system.[6] Behavior results from an interaction between a person's internal state and environmental stimuli. What we have learned through past experience becomes an inventory, or data bank, consisting of values or goals, sets of expectations and preconceptions about the consequences of acting one way or another, and a variety of possible ways of responding to the situation. This memory system determines what things we will notice and respond to in the environment. At the same time, stimuli in the environment help to determine what parts of the memory system will be activated. Hence, the memory and the environment form an interactive system that causes our behavior. As this interactive system responds to new experiences, new learnings occur which feed back into memory and gradually change its content. This process is how people adapt to a changing world.

[6]James G. March and Herbert A. Simon, *Organizations* (New York: John Wiley & Sons, 1958), pp. 9–11.

Differences in Background

Communication between persons brings individual personalities and individual views of the environment into contact. People can agree on many things if they are products of the same experiences. But the fact that they have had different experiences may lead to disagreement. Extremely different backgrounds can cause serious communication problems. In other words, if you and I are trying to communicate with each other but do not see the same world, we are simply not talking about the same things. There are several possible consequences.[7]

1. *I can assume that I know what I am talking about and you don't.* This can cause inattention and create an emotional impression in such basic reactions as: "You are wrong, I am right," and even "You are evil, I am good." The struggle over differences may thus intensify.

2. *I can assume that since I am right, my objective must be to get you to agree with my point of view.* At first, this may lead me into trying to be logical. I assume that you will be convinced once the facts are set straight. If I fail in this, I may resort to strategies of winning at any cost. I will dominate the discussion, talk instead of listening, and generally demean your ideas. All of these things would tend to heighten emotions and increase frustration, leading to an impasse in which we both would lose.

3. *I will interpret what you say according to my understanding of the situation.* In many cases, this would be about as appropriate as trying to find a city in Russia using a map of France. Carl Rogers has described this "tendency to judge, to evaluate, to approve (or disapprove) the statement of the other person" from the listener's point of view as the major barrier to interpersonal communication.[8] Rogers' remedy to this problem is what he calls "listening with understanding"—that is, trying to understand through a deliberate effort to see the other person's point of view, to see the world as he sees and experiences it. In other words, we must abandon the "I-know-what-I'm-talking-about-you-don't" attitude and, instead, open our minds and our ears to the other person's viewpoint. This means that we must admit to ourselves that there may be ideas that, though they are different from ours, are just as valid and just as worthy as our own. We may, in fact, learn something if we listen. This attitude is clearly difficult to achieve, since few people like to admit they may be wrong. But opening our minds to others' opinions is the only way we can gain the advantage of perceiving another side of the problem.

Really listening (and not just "hearing") has another important advantage. If I listen attentively to another person, I am expressing to him a form of respect, and in a very substantial way contributing to his feeling of self-worth. This strengthens his ego and at the same time evokes in him a feeling of respect for me.

[7]We are indebted to Carl Rogers and Fritz Roethlisberger for many of the ideas in this section. See, for example, Carl Rogers and Fritz Roethlisberger, "Barriers and Gateways to Communication," *Harvard Business Review* (July–August 1952), pp. 28–34.

[8]Rogers and Roethlisberger, "Barriers and Gateways to Communication," pp. 28–34.

These mutually supportive feelings help to chase out antagonisms, fears, and defensive tactics. A supportive attitude can lead the way to cooperative problem solving, in which both of us emerge winners.

Levels of Communication

Differences in perception are not the only sources of misunderstanding and difficulty in interpersonal communication. Communication is also complicated by the fact that it takes place at different *levels* simultaneously. As we send verbal messages by word and voice, we also send nonverbal messages by our gestures, expressions, posture, status, and even by the way we dress and comb our hair. We cannot avoid these silent comments on what our words are meant to say. Sometimes we may deliberately twist or distort messages to achieve our purposes, and sometimes we send distorted messages without being aware that they are distorted. Occasionally, we may be able to transmit what we mean so that it means the same thing to the person on the receiving end.

One way to think about the different levels from which messages emerge is the **Johari Window.** Imagine that the human personality could be divided into four parts according to the level or degree to which each part is "open" and "known" to both the sender and the receiver of a communication. This structure could then be represented in matrix form, as shown in Figure 8.6. Each of the areas in the figure can now be defined. The "Open" area contains motivations and behavior whose meanings are shared by the individual and others with whom he is in contact. The individual's feelings and his understanding of these feelings and what he communicates (verbally and nonverbally) are consistent, and they are received and understood by others in the same sense as they are understood and sent by the sender. There is no cover-up and no confusion between his words and his gestures, his expression, and how others interpret his meaning. This is free, honest, and relevant behavior, unburdened by cynicism, distrust, naïveté, or any other hidden attitude or feeling. The meanings experienced and sent are the same as those that are experienced and received.

The "Hidden" area includes concealed motivations that are known to the sending individual but unknown to others. In this

Figure 8.6 Johari Window

	Known to Others	Unknown to Others
Known to Self	Open	Hidden
Unknown to Self	Blind	Unknown Potential

Source: Joseph Luft, *Group Processes: An Introduction to Group Dynamics* (Palo Alto, Calif.: Mayfield Publishing Company, formerly National Press Books 1970), p. 11. Reprinted by permission.

category are all the "little white lies" in which we indulge, including the bigger deceptions we sometimes use in communicating with others. One illustration is the use of ingratiating behavior for our own gain, aimed at a manager whom we do not like or with whom we privately disagree. Or suppose a friend goes by and calls out, "Hello, how are you!" It is a social convention in our culture that such greetings are not to be taken at face value. They are most often merely a form of recognition. But we cannot always be sure. Is our friend merely using her greeting as a form of recognition, a signal of acknowledgment, an automatic pleasantry to which we automatically respond, "Fine! How are you?" Or is she genuinely concerned with the state of our health and should we tell her that we feel lousy and just lost money in the stock market? In this case, the greeting may be influenced by some hidden agenda or concealed motive that is not clear to us. And to that extent, our communication is not on the same wavelength.

The "Blind" area in Figure 8.6 includes motivations and behavior that are known to others but to which the individual is blind. This is sometimes indelicately called the "bad breath area." We have all known people who have feelings that they do not verbalize, but that manifest themselves in little mannerisms, nervous tics, habitual gestures, grimaces, and the like, which actually change the implications of things they say to us. As an example, think of the person who says, "I'm not scared!" while his face whitens and tightens with visible nervous tension.

Consider the spectacle of high government officials arriving at the White House to discuss the nation's energy crisis in their big limousines on a wintry day, keeping the motors running and the heaters on during the meeting, and later driving off as news cameras clicked and passersby stared.[9] These public servants seemed blissfully unaware of the inconsistency between what they were saying and how they were behaving. In other words, their "blind" side was showing in a way that would have been comical if it had not been so serious.

The "Unknown Potential" area in Figure 8.6 is unknown to both the individual and to others. It is the area that Freud describes as the "unconscious." This area probably contains aspects of ourselves that, if available to us, could increase our general effectiveness as persons. By definition, however, this area is available only through a process of self-discovery, sometimes requiring deep and prolonged **psychoanalysis.** For our purposes here, we need say only that this unconscious part of our personality influences in unknown ways our communications with other people, as well as affecting our internal communication with ourselves.

To increase our effectiveness in interpersonal communication, it would appear helpful to enlarge the "Open" area of our personality, while at the same time reducing the "Hidden" and "Blind" areas. This may be accomplished through the two inter-

[9]From a front-page story in the *Seattle Times*, (November 8, 1973), p. 1.

Figure 8.7 Effect of Exposure and Feedback on the Johari Window

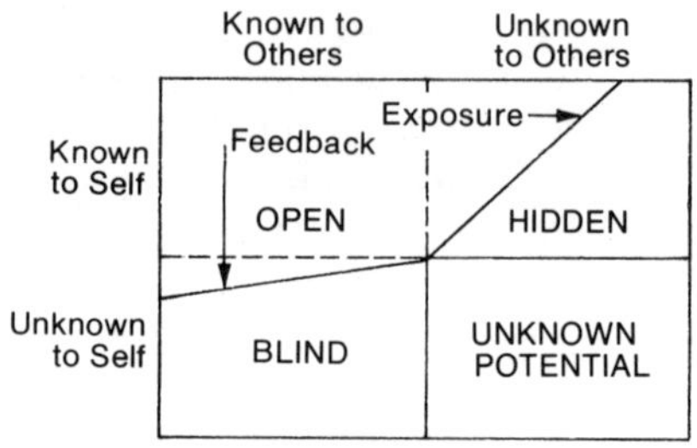

dependent processes of *exposure* and *feedback*, as illustrated in Figure 8.7. If we trust others in a relationship, we may be more willing to reveal some of the motives that we would otherwise keep hidden out of fear of consequences, should our motives become known. At the same time, by giving us information about those nonverbal messages that originate in the "Blind" area, others can help us become aware of the effect that such messages have on the meanings we are trying to convey. This feedback, however, must be in a form that helps to create a supportive, nonthreatening psychological climate. Lacking this emotional support, we would probably continue to be defensive.

Sometimes we complicate the clarity and reliability of verbal messages by either unknowingly or perhaps purposely expressing something that is inconsistent with what we think or feel. Such behavior is not always dysfunctional. For example, we may occasionally feel the need to defend our self-esteem against threats, for example, a seemingly (to us) unfair reprimand by a teacher or a friend that might reduce our personal effectiveness in a particular situation. In addition, there are social conventions that require that we mask our true feelings to avoid hurting someone unneccessarily. These social conventions often help us to maintain stability in relationships with other people.

On the other hand, hidden agendas and blind spots can be dysfunctional if they hide information that could improve rather than hinder our ability to solve the problems we share with others. If someone appears to feel one way about something when in fact he does not, and we base our plan for dealing with him on a mistaken impression, we may miss an opportunity to solve a mutual difficulty. Or, if we are unknowingly doing something that garbles our messages but, if brought to our attention, could be corrected, we would increase our potential for effective communication and for effective action.

SUMMARY

In this chapter, several dimensions of communication and information flows in organizations have been described. Communication is an integral part of the interaction and functioning of the parts of organizational systems. It also controls the design and technological performance of the organization, as well as the quality of the social and psychological climate.

Formal communications are messages that tell members of the organization what to do, how to do it, and when to do it. They also keep track of when and how well objectives and goals are achieved. Formal communications flow through prescribed channels, or lines of authority, in accordance with protocol drawn from the organization's policies and procedures. They relate primarily to technological, job, and legal matters of particular concern to the organization's management.

In contrast, informal communications relate to the social and emotional needs of organizational members and arise out of a motivation to avoid formal bureaucratic red tape in achieving organizational or personal objectives. These communications follow any paths that are convenient and may cross lines of authority and jump organizational levels.

The speed and accuracy with which groups complete tasks requiring cooperative behavior are influenced by the pattern of communication. Patterns of centralized control over the flow and collection of information, like the Chain, Wheel, and Star, tend to speed up task performance and eliminate errors. Emergent leadership is accelerated in such networks. However, morale and perhaps creativity seem to increase in networks of decentralized control, like the Circle and the All-Channel patterns.

Differences in our perceptual filters lead to differences in our interpretation of people and situations in organizations. These filters create communication barriers that make straightforward exchanges of meaning difficult. At times, when another's view seems forced upon us or when we believe that we are being judged, we tend to become defensive. Defensive behavior can inhibit cooperation and exchange of ideas, thus decreasing the possibility for effective action.

DISCUSSION QUESTIONS

8.1. Distinguish between formal and informal communications. What purposes does each serve in the organization? Your class is a formal organization within the school's total structure, your instructor is the formal leader. How would you distinguish between formal and informal communications in the class and in the school? What purposes and functions does each serve?

8.2. Using Leavitt's Rectangles, try out one-way and two-way communications. Discuss each mode of communication in relation to speed, accuracy, and the feelings of both sender and receivers. (Speed and accuracy can be measured by timing each trial and by gathering data on the receivers' errors in each mode.)

8.3. Is the Circle likely to be a better problem-solving network than the Star? For what kinds of problems? Why?

8.4. Describe the various perceptual filters that influence interpersonal communications. How can their effects be reduced?

8.5. Discuss some emotional barriers to communication. What causes them? How can they be avoided or minimized?

8.6. Could the Johari Window be used to analyze the communications between students and the administration or faculty at your school? How? Discuss.

APPENDIX
The Case of the Tardy Secretary

Mrs. Carter paused before knocking on Mr. Jackson's door and thought to herself, "I wonder how I'm going to explain being late again this morning?" Several times recently she had been held up at home getting the kids off to school, and she knew Mr. Jackson had noticed. With four kids and a husband in the family, meeting morning schedules was not always easy.

Behind his office door, Mr. Jackson stared thoughtfully out the window. He had left word for Mrs. Carter to come in to see him as soon as she reported for work. He liked Mrs. Carter. She was a good secretary and was well liked by everyone in the office. In fact, in many ways he thought of her as indispensable. If she just didn't come in late all the time! It was setting a bad example, he thought. He knew he had to handle this problem carefully if he were to avoid harming their friendly relationship, but he felt he must say something.

Just then he heard a knock on his door. "Come in!" he called. Mrs. Carter opened the door and entered his office.

1. What perceptions and assumptions does Mr. Jackson seem to be bringing to his problem with Mrs. Carter? How would you evaluate her perceptions and assumptions as a basis for his meeting with her?
2. What perceptions and assumptions do you think Mrs. Carter is bringing to this meeting with her boss? How would you evaluate her perceptions and assumptions in relation to the meeting with Mr. Jackson?
3. Take the role of Mr. Jackson and ask someone else to play the role of Mrs. Carter. Enact the meeting between them in a way that you think would have useful and constructive results.
4. Discuss the role play with the members of the class.

SELECTED READINGS

Haney, William V. *Communication and Organizational Behavior*, rev. ed. Homewood, Ill.: Richard D. Irwin, 1967.

Hayakawa, S. I. *Language in Thought and Action*, 2nd ed. New York: Harcourt, Brace and World, 1964.

Leavitt, Harold. *Managerial Psychology*. Chicago: University of Chicago Press, 1958.

Rogers, Carl. *On Becoming a Person*. Boston: Houghton Mifflin, 1961.

Turner, Arthur N., and Lombard, George F. F. *Interpersonal Behavior and Administration*. New York: The Free Press, 1969.

HOW TO PUT A STAMP ON A LETTER. BUSINESS PERSON LOWERS STRING, DANGLING FLY (A) IN FRONT OF FROG (B). FROG JUMPS AFTER FLY, LANDS ON SCALE STICKING STAMP ON DOGS TONGUE (C), WHILE AT THE SAME TIME LOWERING BANNANA ELICITING MONKEY (D) TO PEDDLE BICYCLE ACTIVATING CHAIN (E). CHAIN TURNS WHEEL (F) INVERTING DOG WITH STAMP ON TONGUE TO LETTER (G) THUS FINISHING THE PROCESS.

Chapter 9
Motivation

During a routine rearrangement of desk assignments in the typing pool, the office manager moved one of her clerk-typists from a desk near the central aisle to one near the window. Since the office manager considered seats near the window to be the most desirable (and a bonus for good work) because of the availability of good natural light and fresh air, she made this change as a routine move without consulting her young employee. During the next two weeks, she noted a decrease in the typist's output. At the end of that time, the clerk-typist asked the office manager if she could move back to a place near the central aisle, complaining about noise and drafts at the window. The office manager granted her request, but she was puzzled. In her view, the noise level was no higher than elsewhere in the room and no one else, not even the older women, had ever complained about draftiness by the windows. Later the clerk-typist confided to a friend, "I missed being by the aisle. There was always someone coming by whom I knew, and I used to enjoy kidding with the messengers and getting all the news. It was pretty lonely by the window."

Each of us, either as a manager or as an employee, has been involved in a similar situation. Stories like these raise many questions about what motivates and directs human activity. Much activity is motivated by the desire to achieve individual goals. Let us examine some general theories of human motivation. Figure 9.1 is a general model of behavior. Beginning with the assumption that an individual experiences *tension* as a result of unfulfilled needs, this diagram shows that an individual behaves in such a way as to reach a goal which will reduce the amount of tension. For example, if a man experiences hunger (stomach contractions), he will try to find food. When he eats the food, his symptoms will stop and the uncomfortable sensation in his stomach will go away. In the language of motivation, an inner state of tension (hunger, in this case) motivated him to reach a goal (food) which, when achieved or realized, will reduce the tension. When the tension leaves him, that is, when his stomach is full, he no longer feels an urge to find something to eat. In its simplest terms, this is the general model of behavior—it is goal directed and powered (motivated) by an internally felt need to act. A *motive*, therefore, may be defined as

Figure 9.1 General Model of Behavior

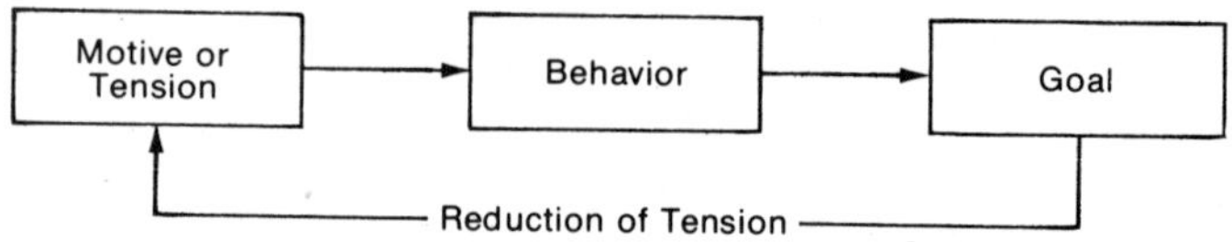

"energy generated by inner tension which moves a person to behave in a manner that will reduce this tension."

THEORIES OF MOTIVATION

One comprehensive study of twenty distinct psychological theories of motivation[1] (a study in progress reviews ten additional theories!) illustrates the superabundance of possibilities facing the manager in the formidable task of attempting to sort out answers to the questions, What makes people work and what can I do to make them *want* to work? Is motivation rooted in primitive biological needs and drives several times removed from conscious thought, or is it acquired in the process of socialization? Or are both true? Is all human behavior traceable to a few, simple human needs? Or do differences between individuals make any attempt to identify general motivational patterns fruitless? Can motivation be inferred from observed behavior or is the connection so oblique and obscure as to defy analysis? Can motivation be measured? Can human behavior be controlled in any objective sense? These are a few of the questions still debated by behavioral scientists. How, then, should the nonspecialist—the practitioner of individual and collective supervision—proceed in his attempts at motivating people on the job?

Our purpose here is to try to draw from this abundance of conceptual schemes some practical ways to think about what makes people tick. We will begin with an overview of current motivational theory, which includes three broad classifications: (1) psychological hedonism, (2) drive theory, and (3) personality theory.

Psychological Hedonism

The idea that people behave in ways that will increase pleasure and reduce pain has had a long history in the social sciences as well as a practical application in the workaday world. The philosophy, called **hedonism**, is believed to have originated around 270 B.C. with the Greek epicurean philosophers: "Whatever we do," said Aristippus, "is done through hope of pleasure or fear of pain—even when we impoverish ourselves for our friends, or give our lives for our generals. . . . The art of life lies in plucking pleasures as they pass, and making the most of what the moment gives."[2]

In classical economics, "economic man" weighed the anticipated effect of actions to be taken in the market place and, accord-

[1] K.B. Madsen, *Theories of Motivation* (Kent, Ohio: Kent State University Press, 1968).

[2] Will Durant, *The Life of Greece* (New York: Simon & Schuster, 1939), p. 504.

ing to theory, always acted to maximize his gains and minimize his losses. Adam Smith outlined this psychology: "It is not from the benevolence of the butcher, the brewer, or the baker that we expect our dinner, but from their regard of their own interest. We address ourselves not to their humanity, but to their self-love, and never talk to them of our own necessities but of their advantage."[3]

In psychology, Sigmund Freud theorized that people are motivated by what he called "the pleasure principle."

> *It seems that our entire physical activity is bent upon* procuring pleasure *and* avoiding pain, *that it is automatically regulated by the Pleasure Principle. Now of all things in the world we should like to know what are the conditions giving rise to pleasure and pain, but that is just where we fall short. We may only venture to say that pleasure is* in some way *connected with lessening, lowering, or extinguishing the amount of stimulation present in the mental apparatus; and that pain involves a heightening of the latter.*[4]

In psychology, the pleasure-pain concept became generally known as the "law of effect": Behavior that seems to lead to reward is repeated; behavior that seems to lead to punishment is not repeated. Conceptually, this is perhaps the simplest of the theories of motivation because it forms a part of all other such theories and easily matches our commonsensical notions of what makes people act as they do. However, we encounter difficulty with this theory *because* of its simplicity or, more precisely, because it is simply inadequate to explain so complex a phenomenon as human motivation. Freud commented on the difficulty of defining pleasure and pain for different individuals, having in mind, of course, his own struggles to understand his patients. The phenomena of masochism and sadism, for example, are well-known puzzlers as pleasure-pain choices. The masochist takes pleasure in being abused or dominated by another; the sadist takes delight in being cruel to those he loves. It is also known that adult interests and preferences are typically multimotivated. This means that a simple activity such as eating may satisfy many tensions—for example, hunger, association, and the need for status.

Drive Theories

In their most general form, **drive theories** are based on the hypothesis that the primary source of motivating energy is found in man's biological nature: hunger, sex, and other physiological needs. Secondary, or learned, motives are then derived from this physiological base during the socialization process, during which an individual develops preferences for certain foods over others and learns to behave in acceptable ways through adopting values, attitudes, and habits taken over from society's teachers.

[3]Adam Smith, *An Inquiry into the Nature and Causes of the Wealth of Nations* (New York: Random House, Modern Library edition, 1937), p. 14.

[4]Sigmund Freud, *A General Introduction to Psychoanalysis* (New York: Washington Square Press, 1960), p. 365. (Orig. Boni & Liveright, 1924.)

Hence, drive theory includes not only the idea of physiological urges but also learned ways of translating these urges into tension-relieving goal achievement. Edward Tolman is representative of this school of thought. His "purposive behaviorism" provides us with a systematic theory designed to explain and predict human behavior.[5] Behavior, according to Tolman, is the result of influences or stimuli (environmental events or objects) evoked by an "initiating physiological state" (motive) acting through "intervening variables" (other knowledge elements) in the human mind.

B.F. Skinner, one of the most influential American psychologists, uses a somewhat different approach from that of the drive theorists, because he excludes from his theoretical formulations "inner causes" such as needs and drives.[6] Skinner's strategy is to view behavior as though both stimulus and response exist only outside a person—that is, in observable objects and events. A rat can be taught to press a lever that releases food which the rat then eats. In this case, the stimulus (immediate prod to activity) is the lever and the rat's response (pressing the lever) is reinforced (rewarded, in this case) by the food. The rat will then learn to continue to press the lever until his stomach is full.[7]

Skinner extended this theory of learning to humans. In his view, human behavior can be explained, predicted, and controlled through the association of stimuli and responses that are "public"—for example, we shy away from the club raised over our head or we relax after a long hike. "Public" events or objects are those that are observable and open to analysis. Skinner forecasts that what are now "private" events—thoughts, ideas, and other inner processes—will gradually become "public" with advances in technology and behavioral science. Skinner has become the center of a philosophical debate over his advocacy of "behavior modification" through the manipulation of external variables. In other words, objects and events manipulated by an "expert" can govern human behavior. His views on determinism and free will, or choice, are illustrated in the following quote: "A world in which people are wise and good without trying, without "having to be," without "choosing to be" could conceivably be a far better world for everyone."[8]

[5]Edward Tolman, *Purposive Behavior in Animals and Men* (Los Angeles, Calif.: University of California Press 1932, 1951).

[6]B.F. Skinner, *Science and Human Behavior* (New York: Macmillan, 1953).

[7]This is a description of *instrumental*, or *operant* conditioning: pressing the lever is instrumental in the rat's obtaining food. *Classical* conditioning, investigated by Pavlov in relation to salivation in dogs, is a process in which a dog salivates when given food and later salivates to the sound of a bell, after a period when the food and bell have been repeatedly presented simultaneously, and the dog has learned to associate one with the other. Pavlov called the food the *unconditioned* stimulus and the salivation the *unconditioned* response; the bell was named the *conditioned* stimulus and the salivation to the bell the *conditioned* response. For a good discussion of instrumental and classical conditioning and other aspects of stimulus–response theory, see A.L. Baldwin, *Theories of Child Development* (New York: John Wiley & Sons, 1967).

[8]Carl R. Rogers and B.F. Skinner, "Some Issues Concerning the Control of Human Behavior," *Science* 124 (November 30, 1956), pp. 1057–66. See also B.F. Skinner, *Beyond Freedom and Dignity* (New York: Alfred A. Knopf, 1971).

Drive theories serve the needs of scientists seeking to achieve a sound theoretical explanation of behavior. Elements of drive theory also have an appeal to practical men who use a "carrot-and-stick" approach in controlling behavior. Reward and punishment have long been used by the family to socialize children and by business organizations in the belief that man's primary motive is economic gain. The "carrot" has usually meant something tangible like a piece of candy in the case of a child and a raise in pay or a new title in the case of a worker. The "stick" has usually been a spanking for the child and an economic penalty such as a wage reduction or a firing in the case of the worker.

Drive theorists maintain a historical point of view with respect to the development of human motives. Present behavior toward objects or events and the motives that energize behavior reflect *learned* responses to similar influences encountered in the individual's past. For example, if an individual learned in earlier years that hard work leads to financial success—a goal which he still values—he can normally be counted on to work hard to increase his income.

This emphasis on the robotlike, or mechanical, nature of man as a way of explaining behavior and its underlying motivations worries many psychologists. Is man not a thinking, feeling being who possesses the power of choice? Is he not a pilot, planning and charting his own course on the sea of existence? Doesn't individual choice in the present influence his behavior? Aren't his choices based on motives that relate to his expectations as well as his experience? Attempts to adapt drive theory to these concepts of human nature have not been altogether successful, and have led some behavioral scientists down different paths in their search for answers to what motivates human behavior.

Personality Theories

Whereas drive theory has generally remained in the realm of experimental psychology, personality theories are the product mainly of clinical psychologists and psychotherapists. Drive theories are more easily adaptable to systematic experimentation; personality theories thus far are not. Derived in large part from classical Freudian ideas about the human mind, personality theories tend to be subjective and based on case-by-case observations of clinicians who are more concerned with their patients' mental health than with documenting the proof.

Personality theories of motivation can be subdivided into "self" and "cognitive" theories, though both are based on the idea that motives are the product of free choice, of the *thinking* process. This contrasts with the assumptions underlying drive theories, according to which energy can ultimately be traced to biological needs. "Self" theories of motivation emphasize the influence of the *self-concept* or *self-image* on behavior. Motives are related to the protection and development or growth of the individual's idea of the kind of person he is, should be, or hopes to become. How a person responds to a situation, to a person, or to change depends on his interpretation of it and his attempts to make his behavior compatible with his self-image in that context.

For example, the student who thinks of himself as a scholar will make an effort to obtain a high grade on an examination, because getting a high grade is in harmony with his self-concept. Or the woman who considers herself a good hostess will strive in every possible way to cater to the needs of her guests to ensure that they will recognize her qualities of hospitality and skill in making them comfortable and happy. In these cases, the motive is an inner state that energizes and channels behavior in the direction of maintaining, enhancing, and actualizing the self. Carl Rogers, one of the foremost proponents of this theory of motivation, suggested that, "Rather than many needs and motives, it seems entirely possible that all organic and psychological needs may be described as partial aspects of this one fundamental need."[9]

Thus, self-confirmation and development is man's constant goal and every encounter with the world around him is directed toward realizing this goal. All behavior is an attempt to use personal resources in conjunction with environmental forces to achieve self-actualization. This approach places human motivation in a current, rather than a historical context: "Behavior is not 'caused' by something which occurred in the past. Present tensions and present needs are the only ones which the organism endeavors to reduce or satisfy. While it is true that past experience has certainly served to modify the meaning which will be perceived in present experiences, yet there is no behavior except to meet a present need."[10]

By shifting the focus from past events or remote and hard-to-define biological drives to present sources of behavior, personality theories of motivation seem to be more in accord with common sense and experience. For this reason, they have achieved popularity among administrators and students of organizational behavior.[11]

SOME SPECIFIC APPLICATIONS OF MOTIVATION THEORY

Much motivational theory is esoteric and technical and seems to serve the purposes of academicians attempting to develop laws that will explain and predict the *general behavior* of man. We do not disparage scientific efforts to understand human nature. On the contrary, more study is needed here than in any of the physical sciences, where progress has outrun man's capacity to control his inventions—as witness war, destruction, and pollution. But practitioners of the art of administration are generalists, not

[9]Carl R. Rogers, *Client-Centered Therapy* (Boston: Houghton Mifflin, 1951), pp. 487–88.

[10]Ibid., p. 492.

[11]The two theories of motivation that have dominated research and practice in the field of organizational behavior are those of Abraham Maslow and Frederick Herzberg, discussed later in this chapter. Both of these theories contain derivatives of personality and drive theory. In fact, almost all psychologists today are "theoretical mixtures" in relation to our classification scheme.

specialists, and they must, therefore, have rules of thumb in language they can understand to provide them with a useful way of thinking about the everyday motivational problems they encounter in the organization. These problems run the gamut from union negotiations, wage incentives, profit-sharing plans, and work rules to interpersonal conflicts, helping or coaching relationships, and influencing or changing specific individuals.

Maslow's Hierarchy of Needs

One theory that seems to meet this practical requirement is Maslow's hierarchy of needs.[12] Combining elements of both drive and personality theories, Maslow's need hierarchy rests on three fundamental propositions: First, there are five broad categories of human needs or motives that form a hierarchy of prepotency, or priority. In order of prepotency, the need categories are (1) *physiological,* (2) *safety,* (3) *belongingness and love,* (4) *esteem,* and (5) *self-actualization.* This means that physiological needs have priority over safety needs, safety needs have priority over belongingness and love, and so on (see Figure 9.2). Second, needs are activated and become motivators of behavior progressively. As each level of

Figure 9.2 Maslow's Hierarchy of Needs

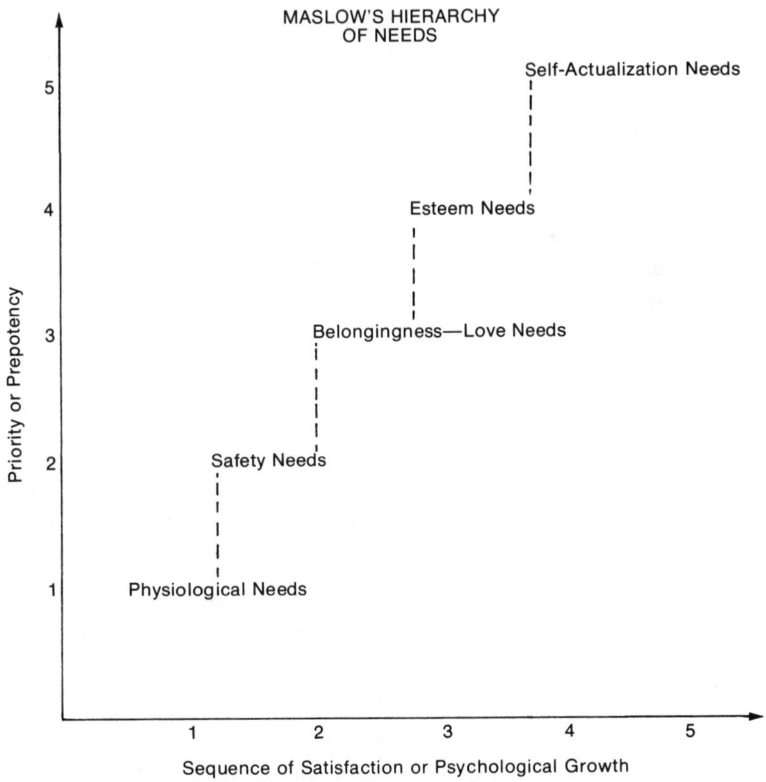

[12]Abraham Maslow, *Motivation and Personality* (New York: Harper & Row, 1959). This section on Maslow is based especially on ch. 5 of his book.

needs is "reasonably well" satisfied (dotted lines, Figure 9.2), the next level emerges and dominates behavior. And third, as a corollary to the second proposition, a satisfied need becomes dormant and no longer controls behavior. For example, if a person's need for food or safety has been satisfied, he no longer is energized by a need for something to eat or for conditions that will make him feel safe.

Physiological Needs

These are the basic animal needs for such things as food, warmth, shelter, sex, water, and other body needs. If a person is hungry or thirsty or his body is chemically unbalanced, all of his energies turn toward remedying these deficiencies, and other needs remain inactive. Maslow explains that, "Anyone who attempts to make an emergency picture into a typical one, and who will measure all of man's goals and desires by his behavior during extreme physiological deprivation, is certainly blind to many things. It is quite true that man lives by bread alone—when there is no bread."[13]

Safety Needs

With his physical needs relatively satisfied, the individual's safety needs take over and dominate his behavior. These needs have to do with man's yearning for a predictable, orderly world in which injustice and inconsistency are under control, the familiar frequent, and the unfamiliar rare. In the world of work, these safety needs manifest themselves in such things as a preference for job security, grievance procedures for protecting the individual from unilateral authority, savings accounts, insurance policies, and the like. Religious beliefs are, in part, a response to safety needs. For many people, religion provides a reasonable explanation of the universe and the relationship between life and death.

For the most part; physiological and safety needs are reasonably well satisfied in our affluent and relatively lawful society. The obvious exceptions, of course, are people outside the mainstream—the poor, the disadvantaged, and members of minority groups. If frustration has not led to apathy and weakness, such people still struggle to satisfy the basic physiological and safety needs. They are primarily concerned with survival: obtaining adequate food, clothing, shelter, and seeking justice from the dominant societal groups.

Belongingness and Love Needs

Those for whom physiological and safety needs are satisfied find themselves at the next level of the need hierarchy. At this point, all their yearnings for love and acceptance surface and engage their attention. People at this level seek the love and affection of family and friends and affiliation with meaningful groups, both in the community and at work. For most people, being deprived of

[13]Ibid., p. 83.

love (affection) is a severe psychological blow, since it leads to feelings of unworthiness and inadequacy. Unless and until these yearnings for affiliation and association and the giving and receiving of affection are reasonably well satisfied, people will struggle to fulfill these needs. Having been successful, they are ready to progress to the next level of needs.

Esteem Needs

These are the needs for self-respect and for recognition—that is, for feelings of accomplishment and autonomy, and feelings of status, prestige, and the respect of others based on real competence: "Satisfaction of the self-esteem need leads to feelings of self-confidence, worth, strength, capability, and adequacy, of being useful and necessary in the world. But thwarting of these needs produces feelings of inferiority, of weakness, and of helplessness."[14]

Self-Actualization Needs

In Maslow's scheme, the final stage of psychological development comes when the individual feels assured that his physiological, security, affiliation and affection, self-respect, and recognition needs have been satisfied. As these become dormant, he becomes filled with a desire to realize all of his potential for being an effective, creative, mature human being. "What a man *can* be, he *must* be," is the way Maslow expresses it. The opportunity to engage this need for self-actualization in our bureaucratic society comes to only a select few, and then only under special conditions conducive to bringing out the best in man.

Figure 9.3 shows the sequencing, or priority, of needs as the individual advances toward psychological growth, that is, the essential satisfaction of all needs.[15] The height of each curve indicates the relative importance of each need category in the process of growth. Thus, satisfaction of the esteem needs are shown as more important to growth than the belongingness and love, safety, and physiological needs. In addition, the graph suggests that a person is multimotivated at any moment, although clearly one category of needs is predominant.

In Maslow's terms, *growth* and *health* of the human personality are similar phenomena. He postulates that the perfectly healthy, normal person has been able to satisfy all of his needs "reasonably well," except self-actualization, a need that in essence is insatiable. In terms of psychological growth or mental health, the normal person has achieved the condition illustrated by the line *CD* on the graph. Note that the state of psychological growth or health in the human adult has nothing to do with chronological age. If, for example, an individual in the motivating condition

[14]Ibid., p. 91.

[15]David K. Krech, Richard S.C. Crutchfield, and Egerton L.B. Ballachuy, *Individual in Society* (New York: McGraw-Hill, 1962), p. 77.

Figure 9.3 Relative Importance and Sequence of Needs in Psychological Growth

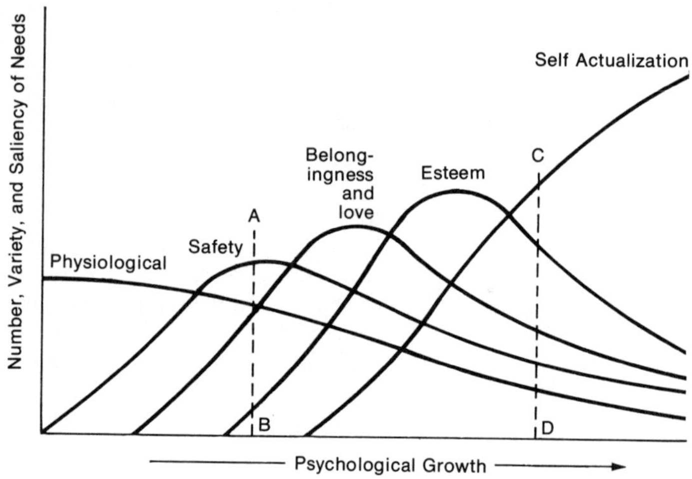

Source: Adapted from Krech, et al, *Individual in Society* (New York: McGraw-Hill, 1962), p. 91.

shown at line AB should suddenly be deprived of food and water, all of his energies would return to finding the means of satisfying these more basic and priority needs. A person who is still struggling to satisfy some needs is, according to Maslow, deprived, and could be in some such condition as illustrated by line AB. His safety needs are nearly satisfied but still motivating, and his belongingness and love and esteem needs are just emerging and will begin to influence his behavior when he is ready to attend to them.

Maslow's need hierarchy is set forth as a general proposition and does not imply that everyone's needs follow the same rigid pattern. For some people, self-esteem seems to be a stronger motivation than love. Mussolini, for example, alienated his closest friends by undertaking reckless military adventures to achieve status as a conqueror. (This example can also be used to illustrate the means-to-an-end dilemma of human motivation. That is, Mussolini may have reached for status as a means to gaining the affection of Adolf Hitler. More will be said about this problem later.) For some people, the need to create is often a stronger motivation than the need for food and safety. Thus, the artist living in poverty is a classic example of reversing the standard hierarchy of needs. Similarly, persons who have suffered hunger or some other deprivation for protracted periods may live happily for the rest of their lives if only they can get enough of what they lacked. In this case, the level of aspiration may have become permanently lowered and the higher-order, less prepotent needs may never become active. There are also cases of people's martyring themselves for causes and suffering all kinds of deprivations, particularly in the physiological, safety, and sometimes social categories, to achieve their goals.

To further confound the problem of understanding motivation, Maslow points out that motives are not always conscious. In the average person, he believes, they are more often unconscious than conscious—showing the influence on his thinking of Freudian psychologists who have long been concerned with the hidden causes of human behavior.

In Maslow's theory, then, human needs are arranged in a hierarchy of importance. Needs emerge only when higher-priority needs have been satisfied. By the same token, satisfied needs no longer influence behavior. This point seems worth stressing to managers and administrators, who often mistakenly assume that money and other tangible incentives are the only cures for morale and productivity problems. It may be, however, that the need to participate, to be recognized, to be creative, and to experience a sense of worth are better motivators in an affluent society, where many have already achieved an acceptable measure of freedom from hunger and threats to security and personal safety, and are now driven by higher-order psychological needs.

The Motivation-Hygiene Theory

Frederick Herzberg's studies of job attitudes and their connection with industrial mental health are related to Maslow's theory of motivation. His findings have had a considerable theoretical, as well as a practical, influence on attitudes toward administration.[16] According to Herzberg, man is not content with the satisfaction of lower-order needs at work—for example, those associated with minimum salary levels or safe and pleasant working conditions. Rather, he looks for the gratification of higher-level psychological needs having to do with achievement, recognition, responsibility, advancement, and the nature of the work itself. So far, this appears to parallel Maslow's theory of a need hierarchy. However, Herzberg added a new dimension to this theory by proposing a two-factor model of motivation, based on the notion that the presence of one set of job characteristics or incentives leads to worker *satisfaction* at work, while another and separate set of job characteristics leads to *dissatisfaction* at work. Thus, satisfaction and dissatisfaction are not on a continuum with one increasing as the other diminishes, but are independent phenomena. This theory suggests that to improve job attitudes and productivity, administrators must recognize and attend to both sets of characteristics and not assume that an increase in satisfaction leads to a commensurate decrease in dissatisfaction.

The two-factor, or *motivation-hygiene theory*, developed from data collected by Herzberg from interviews with a large number of engineers and accountants in the Pittsburgh area. From analyzing these interviews, he found that job characteristics related to what a man *does*—that is, to the nature of the work he performs—apparently have the capacity to gratify such needs as achievement, competency, status, personal worth, and self-realization, thus making him happy and satisfied. However, the *absence* of such gratifying job characteristics does not appear to lead to unhappi-

[16]Frederick Herzberg, *Work and the Nature of Man* (Cleveland: World Publishing, 1966); F. Herzberg et al, *The Motivation to Work*, 2nd ed. (New York: John Wiley & Sons, 1959).

ness and dissatisfaction. Instead, dissatisfaction results from unfavorable assessments of such job-related factors as company policies, supervision, technical problems, salary, interpersonal relations on the job, and working conditions. Thus, if management wishes to increase satisfaction on the job, it should be concerned with the nature of the work itself—the opportunities it presents for gaining status, assuming responsibility, and for achieving self-realization. If, on the other hand, management wishes to reduce dissatisfaction, then it must focus on the job environment—policies, procedures, supervision, and working conditions.[17] If management is equally concerned with both (as is usually the case), then managers must give attention to both sets of job factors. Figure 9.4 summarizes the results of Herzberg's studies.

The data shown in Figure 9.4 were derived from an interview analysis:

> *Briefly, we asked our respondents to describe periods in their lives when they were exceedingly happy and unhappy with their jobs. Each respondent gave as many "sequences of events" as he could which met certain criteria including a marked change in feeling, a beginning and an end, and contained some substantive description other than feelings and interpretations. . . .*
>
> *The proposed hypothesis appears verified. The factors on the right that led to satisfaction (achievement, intrinsic interest in the work, responsibility, and advancement) are mostly unipolar; that is, they contribute very little to job dissatisfaction. Conversely, the dissatisfiers (company policy and administrative practices, supervision, interpersonal relationships, working conditions, and salary) contribute very little to job satisfaction.*[18]

Unlike Maslow, who offered little data to support his ideas, Herzberg and others have presented considerable empirical evidence to confirm the motivation-hygiene theory. Their work, however, has been criticized on methodological grounds. Nevertheless, Herzberg and his associates have rendered a valuable service to science and to management through their efforts to apply scientific methods to understanding complex motivational problems at work and have stimulated others to continue the search.

The Path-Goal Hypothesis and Expectancy Theory

A number of behavioral scientists have pointed to inadequacies in the need hierarchy and motivation-hygiene theories. The most basic is the criticism that both of these theories contain the relatively explicit assumption that happy and satisfied workers produce more. Another problem is that these and other statistical theories are concerned with explaining "average" behavior and

[17]In *The Motivation to Work*, ch. 9, Herzberg presents some evidence of a positive relationship between morale and productivity, although he recognizes the existence of considerable evidence to the contrary. See our later discussion in this chapter of the path-goal hypothesis.

[18]Herzberg, "The Motivation-Hygiene Concept and Problems of Manpower," *Personnel Administration* (January–February 1964), pp. 3–7.

Figure 9.4 Comparison of Satisfiers and Dissatisfiers

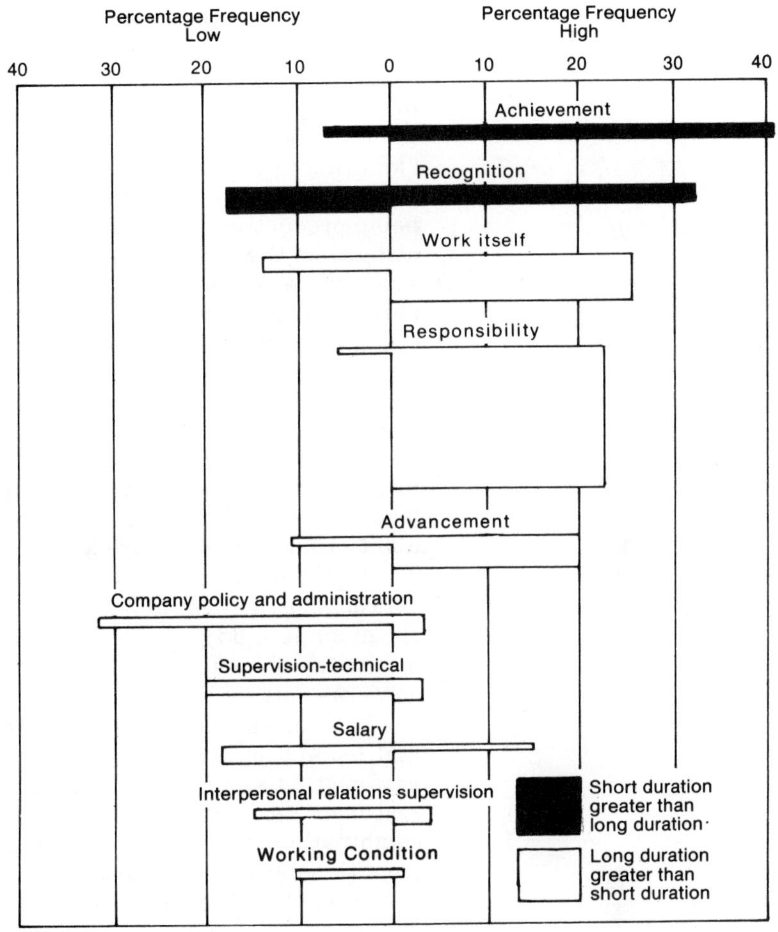

Source: Frederick Herzberg. *The Motivation to Work*, 2nd ed. (New York: John Wiley & Sons, 1959), p. 81.

cannot effectively cope with the problem of individual differences in motivation.

Do happy workers work harder than unhappy ones? There is an absence of conclusive evidence linking need satisfaction and improved performance on the job. Actually, there is considerable evidence that there is no necessary relation. The fact that a worker may not perceive job productivity as a path to achieving personal goals can confuse the relationship between motivation and the quality and/or quantity of work produced. Several writers point out that a complete analysis of the motivational situation may involve the individual's relation not only to his work but also to the organization's social system, his work group, the union, the community, his family, and other influential elements of his life in and beyond the work setting. In addition, the individual's subjective perception or interpretation of the situation confounds the problem. If recognition by management, including rewards through promotion, increased pay, and desirable assignments are viewed as correlative with productivity, it would seem to follow that an individual will work harder and perform a better job to satisfy his needs for advancement, recognition, or status. On the other hand, if "playing politics," belonging to particular committees, and "knowing the right people" are seen as the paths to need satisfaction, productivity may of necessity have to take a back seat. Or, if being productive on the job in any way interferes with satisfaction away from the job, such as would be the case if take-home work or study were required or if the extra effort invested at work were perceived as contributing to physical or mental exhaustion or other adverse health effects, then productivity on the job would have to be adjusted.

The *path-goal hypothesis* and its descendant, *expectancy theory*, are concerned specifically with the question of whether or not a particular path—for example, productivity—will be included in goal-oriented behavior. This approach to motivation includes measures of a person's perception, expectancy, or estimated probability that he will receive a desired return from a given behavior. These theories thus explicitly recognize subjective judgments concerning the desirability of alternative paths as they relate to the amount and type of reward received. Figure 9.5 illustrates this motivational model.

As this model shows, motives, behaviors, and goals comprise the parts of a subjective system. Of course, individuals are complex and usually multimotivated. For illustrative purposes, however, assume that an individual has an unsatisfied ego need—a need for status or recognition. This need causes him to look for a goal or combination of goals that will reduce the inner tension resulting from this need. Several possible goals are a higher income level, a promotion to a higher, more prestigious position, and an improved golf game. If this person finds himself valuing income or position or some combination of these goals more than the goal of improving his golf, his path choice will be to expend effort on job productivity or some other organizational objective—that is, if these, rather than golfing ability, are perceived as the route to success. On the

Figure 9.5 Illustration of Expectancy Theory

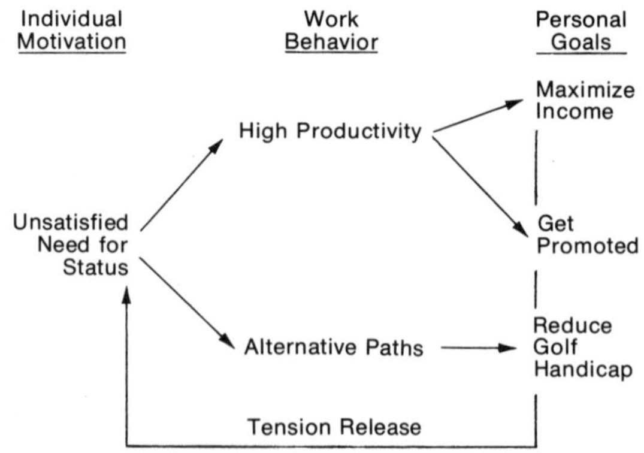

other hand, if playing a better game of golf is the means he chooses to satisfy his need for recognition, then he will find ways to play and think about golf more often, perhaps resulting in an accompanying lower output on the job. Finally, in his pursuit of status he might take a balanced view and strive to pursue several behavioral paths in an effort to achieve a combination of personal status objectives.

In other words, this individual's expectation or estimated probability that a given behavior will bring a valued outcome determines his choice of means and the effort he will devote to these means. In effect, this diagram of expectancy depicts an employee asking himself the question posed by one investigator, "How much payoff is there for me toward attaining a personal goal while expending so much effort toward the achievement of an assigned organizational objective?"[19]

This approach to the study and understanding of motivation would appear to have certain conceptual advantages over other theories: First, unlike Maslow's and Herzberg's theories, it is capable of handling individual differences. Second, its focus is toward the present and the future, in contrast to drive theory, which emphasizes past learning. Third, it specifically relates behavior to a goal and thus eliminates the problem of assumed relationships, such as between motivation and performance. Fourth, it relates motivation to ability: Performance = Motivation × Ability.

The Achievement Motive

To conclude our discussion on motivation, David C. McClelland's and his associates' investigations of achievement motivation have particular relevance to the emergence of leadership. McClelland was interested in the possibility of deliberately arousing a motive to achieve in an attempt to explain how individuals express their preferences for particular outcomes—a general problem of motivation. In this connection, the *need for achievement* refers to an individual's preference for success under conditions of competition. The vehicle McClelland employed to establish the presence of an achievement motive was the type of fantasy a person expressed on the **Thematic Apperception Test (TAT)**, a series of pictures that subjects were asked to interpret by writing stories about them.

The procedure in McClelland's initial investigation was to arouse in the test audience a concern with their achievement. A control group was used in which arousal was omitted. In the course of this experiment, McClelland discovered through analyzing the stories on the TAT that initial arousal was not necessary. Instead, members of the control group—individuals who had had no prior arousal—demonstrated significant differences in their stories, some writing stories with a high achievement content and some submitting stories with a low achievement content. Using results based on the Thematic Apperception Test, McClelland demon-

[19]Basil S. Georgopolous, Gerald M. Mahoney, and Nyle W. Jones, Jr., "A Path-Goal Approach to Productivity," *Journal of Applied Psychology* 41 (December 1957), p. 346.

strated that individuals in a society can be grouped into high achievers and low achievers based on their scores on what he called "nAch" (need for achievement).[20]

McClelland and his associates have since extended their work in fantasy analysis to include different age groups, occupational groups, and nationalities in their investigations of the strength of need for achievement. These investigations have indicated that the nAch score increases with a rise in occupational level. Invariably, businessmen, managers, and entrepreneurs are high scorers. Other investigations into the characteristics of the high achievers have revealed that accomplishment on the job represents an end in itself; monetary rewards serve as an index of this accomplishment. In addition, these other studies found that the high achievers, though identified as managers, businessmen, and entrepreneurs, are not gamblers. They will accept risk only to the degree they believe their personal contributions will make a difference in the final outcome.

These explorations into the achievement motive seem to turn naturally into the investigation of national differences based on Max Weber's thesis that the industrialization and economic development of the Western nations were related to the Protestant ethic and its corresponding values supporting work and achievement. McClelland and his associates have satisfied themselves that such a relationship, viewed historically through an index of national power consumption, indeed exists. Differences related to individual, as well as to national, accomplishments depend on the presence or absence of an achievement motive in addition to economic resources or the infusion of financial assistance. High achievers can be viewed as satisfying a need for self-actualization through accomplishments in their job assignments as a result of their particular knowledge, their particular experiences, and the particular environments in which they have lived.[21]

UNCONSCIOUS MOTIVATION

Earlier in the chapter, we noted that some psychologists believe that a significant portion of human behavior is energized and directed by unconscious motives. According to Maslow: "Psychoanalysis has often demonstrated that the relationship between a conscious desire and the ultimate unconscious aim that underlies it need not be at all direct."[22] In other words, stated motives do not always match those inferred by skilled observers. For example, it is possible that a person can be accident-prone

[20]David C. McClelland, "Methods of Measuring Human Motivation," in John W. Atkinson, ed., *Motives in Fantasy, Action and Society* (Princeton, N.J.: D. Van Nostrand, 1958), pp. 12–13; and *The Achieving Society* (Princeton, N.J.: D. Van Nostrand, 1961), pp. 41–43.

[21]David C. McClelland, "Achievement Motivation Can Be Developed," *Harvard Business Review* 43 (November–December 1965), pp. 6ff. McClelland's theory can thus be seen to contain elements of personality theory (self-actualization) and of the learning aspects of drive theory. McClelland played down the importance of biological needs.

[22]Maslow, *Motivation and Personality*, p. 66.

because he has an unconscious desire to hurt himself and not because he is careless or ignorant of the safety rules. Similarly, some overweight people are not really hungry for food but for attention and love. Eating is merely a defensive reaction to lack of attention. Some workers damage more equipment than others because they harbor unconscious feelings of aggression toward authority figures.

Psychotherapists point out that some behavior is so automatic that the reasons for it are not available in the individual's conscious mind. Compulsive cigarette smoking is an example. Sometimes maintaining self-esteem is so important and the motive for an activity is so threatening that it is simply not recognized and, in fact, may be disguised or repressed. Rationalization, or "explaining away," is one such disguise, or *defense mechanism*, as it is called. Another is projecting or attributing one's own faults to others. "I feel I am to blame," becomes "It is her fault; she is selfish." Repression of powerful but socially unacceptable motives may result in outward behavior that is the opposite of the repressed tendencies. An example of this would be the employee who hates his boss but overworks himself on the job to show that he holds him in high regard.

Unconscious motives add to the hazards of interpreting human behavior and, to the extent that they are present, complicate the life of the administrator. On the other hand, knowledge that unconscious motives exist can lead to a more careful assessment of behavioral problems. Although few contemporary psychologists deny the existence of unconscious factors, many do believe that these are activated only in times of anxiety and stress, and that in the ordinary course of events, human behavior—from the subject's point of view—is rationally purposeful.

Frustration

Now let us shift our attention to the result of blocking motivated behavior. An individual may react in several different ways. He may respond with rational problem-solving methods to overcome the barrier. Failing in this, he may become **frustrated** and behave irrationally. An example of blockage of motivational energy would be the case of the worker who wants time off to go fishing but is denied permission by his foreman. Another example would be the executive who wants a promotion but finds he lacks certain qualifications. If, in these cases, an appeal to reason does not succeed in reducing the barrier or in developing some reasonable alternative approach, the frustrated individual may resort to less adaptive methods of trying to reach his goal.

He may, for example, attack the barrier physically or verbally or both. The worker who is refused time off to go fishing may "cuss out" his supervisor to his face or behind his back. If he is sufficiently aroused, he may strike out at him with his fists or with the nearest weapon. If the supervisor is not present or the worker's fear of the consequences of direct attack is stronger than his desire to attack, he may transfer his aggression to someone or something else. Taking his frustration out on his family or on some object like his car or his equipment are typical ways of transferring aggression.

Another "solution" to frustration is *regressive behavior*—becoming childish or reverting to earlier and more primitive ways of coping with the goal barrier. Throwing a temper tantrum, bursting into tears, or sulking are examples of regression. Wearing a long face and a worried look are other signs of this method of dealing with frustration.

Stubborn refusal to respond to new conditions affecting the goal, such as removal or modification of the barrier, sometimes occurs. As pointed out by Brown, severe punishment may cause individuals to continue nonadaptive behavior blindly:

Either it may have an effect opposite to that of reward and as such, discourage the repetition of the act, or, by functioning as a frustrating agent, it may lead to fixation and the other symptoms of frustration as well. It follows that punishment is a dangerous tool, since it often has effects which are entirely the opposite of those desired.[23]

An example of nonadaptive behavior of this sort might occur in the case of the executive who feels persecuted by his failure to be promoted. Even when offered a training course to improve his chances of promotion, he turns down this opportunity and continues to sulk.

Flight, or leaving the scene, is another way people have of dealing with their frustrations. In the above example of the executive, we might find him quitting his job rather than face up to the consequences of being passed over for promotion. Or, a player quits the football squad because he is not given enough playing time or fails to win the starting berth as quarterback.

Managers must learn to recognize the symptoms of frustration to avoid responding in ways that intensify rather than ameliorate the problem. The main point to remember is that the affected person is often not in a rational, problem-solving frame of mind and is, therefore, not attuned to the "facts" or to logical procedures for dealing with his situation. Some frustrated people need to be guided back to "reality." They cannot be reasoned with in their present mental state. Listening with understanding to such a person is one effective way to reduce frustration. Talking to a sympathetic listener provides a way for him to vent his feelings and regain control of himself.[24]

SUMMARY

Motives provide energy and direction for behavior. Appropriate behavior, in turn, reduces the inner tensions that signal the motivated state. An understanding of the relationships among motives, behavior, and human goals provides the manager, administrator, or leader with a way of thinking about human activity and a framework within which to gather, sort, and analyze data related to behavioral problems.

[23]J.A.C. Brown, *The Social Psychology of Industry* (Baltimore, Md.: Penguin, 1954), pp. 253–54.

[24]See discussion of nondirective counseling in Chapter 11 of this text.

Some of the numerous theories of motivation serve the needs of scientists for systematically investigating motivational phenomena, some serve psychotherapists concerned with their patients' mental health and associated problems, and some are useful for the common man who would understand his fellowmen better. Psychological hedonism, derived from the oldest and most general philosophy of motivation, states that man acts to obtain pleasure and tries to avoid pain. Drive theory focuses on our physiological or biological demands and their translation through a process of reinforced learning into socially approved needs. Personality theories of motivation emphasize thinking and choosing processes and preserving the self-concept.

Maslow's hierarchy of needs, along with Herzberg's two-factor (motivation-hygiene) modification of this theory, serve as useful conceptual schemes for thinking about the behavior of "average" man. These theories are widely used in executive training and in motivational research by organizational psychologists, largely because of their adaptability and their practical, common-sense qualities. Expectancy theory, which attempts to relate means and ends in motivation and also to unravel the connection between personal and organizational goals (such as the relationship between personal need satisfaction and performance) is still in its developmental stages. Expectancy theory attempts to make scientific sense out of an old aphorism: Human Performance = Motivation × Ability. McClelland's achievement motive, or nAch, focuses on the relationship between cultural and family background and individual's ambition to achieve economic success.

Frustrated needs often result in irrational behavior, while unconscious motives, especially in times of stress, often obscure direct connections between behavior and motivation. These phenomena complicate the attempts of observers to understand other people. Thus, supposedly conscious and logical thought neither always creates nor always explains our behavior. Understanding complexities and some of the generally accepted hypotheses concerning human motivation, coupled with a keen eye and a sharp ear for what goes on around us, can help us measurably to a better understanding of our fellowman.

DISCUSSION QUESTIONS

9.1. What support do we find in the writings of Adam Smith and Sigmund Freud for the theory of psychological hedonism?

9.2. Compare and contrast Skinner's theory of motivation to the drive theories.

9.3. Compare the personality theories of motivation to the drive theories.

9.4. How would you classify Maslow's need hierarchy in terms of the basic theories discussed in the first part of the chapter?

9.5. Explain an example of human behavior according to Maslow's need hierarchy.

9.6. Compare Herzberg's theory of motivation with Maslow's theory.

9.7. Do you believe happy workers produce more? Why or why not?
9.8. What were McClelland's contributions to the literature on motivation?
9.9. List the different forms of behavior associated with frustration.
9.10. Do you believe that unconscious motivation plays a significant role in human behavior? Why or why not?

APPENDIX
The Benson Motor Company

The Benson Motor Company, located in the Middle West, produced and marketed a small automobile. Until recently, it had competed successfully with three other automobile makers. Its manufacturing plant used a standard assembly-line format. Normal production was about one hundred cars per day. In the last three months, however, the number of cars per day that met final inspection had dropped to about eighty. The indirect costs per car increased. The assembly lines were operating at normal speed.

After an extensive investigation, Tom James, plant superintendent, concluded that there were no technical problems contributing to the quality problem. He believed the situation to be the culmination of a long-term trend, which he described as a loss of worker motivation. Pay scales and working conditions in the Benson plant were no different from those in other plants in the region. "These people," he said, "are just doing sloppy work and I don't know what to do about it!"

In discussing this problem with Frank Horton, the personnel director, James learned about a new motivational plan that had been used successfully in a local engine plant. This plan, called "behavior modification," was based on the "law of effect" and the operant, or instrumental, conditioning ideas advocated by B. F. Skinner. Under this system, worker behavior is controlled through the application of positive reinforcement. Desirable behavior is rewarded, undesirable behavior is not. Negative reinforcement is not used. Another feature of the plan, again following the theory of operant conditioning, is that the rewards are deliberately planned to follow closely behind the desired behavior. In the local engine plant, for example, supervisors made it a point to compliment workers often for good work. Mr. James wondered if he could in some way change supervisory practices, payment schedules, and other aspects of the work environment on the Benson assembly line so that he could increase the motivation of the workers to do work that would meet quality standards.

1. How would you advise Mr. James to proceed?
2. What other motivational approaches would you suggest he consider?

SELECTED READINGS

Herzberg, Frederick et al. *The Motivation to Work*, 2nd ed. New York: John Wiley & Sons, 1959.

McClelland, D. C. *The Achieving Society*. Princeton, N.J.: D. Van Nostrand, 1961.

Maslow, Abraham H. *Motivation and Personality*. New York: Harper & Bros., 1954.

Skinner, B. F. *Science and Human Behavior*. New York: Macmillan, 1953.

Vroom, Victor H. *Work and Motivation*. New York: John Wiley & Sons, 1964.

JOEL A. NAIVE

Chapter 10 The Manager as a Leader

Whenever we think of groups or organizations, we think of leaders. Organizations are identified by their leaders. For example, when we think of the Ford Motor Company, we automatically think of the gaunt, innovative genius who gave it his name. We look to leaders for ideas, for visions of the future, for objectives and goals, and for the enthusiasm and energy that will carry the organization and its members forward through growth and development. Leadership may refer to one or several individuals in the organization who play important roles as leaders. It may refer also to the functions these individuals perform to enable the organization and its members to achieve their objectives and goals. In addition, leadership may be considered in relation to style or behavior that attracts followers and influences the nature and degree of their performance.

For our purposes, we will define leadership as the process of influencing other members of the organization or the group to cooperate in achieving the organization's purpose or in pursuing a common cause. Thus, leadership reflects (1) the social influence of leaders over followers, (2) followers' voluntary cooperation, support, and effort toward achieving the organization's aims, and (3) the demands of the particular situation in which leadership occurs.

LEADERSHIP IN FORMAL AND INFORMAL ORGANIZATIONS

An organization that is established as an instrument or means for achieving defined objectives has been referred to as a formal organization. Its design specifies how goals are subdivided and reflected in subdivisions of the organization. Divisions, departments, sections, positions, jobs, and tasks make up this work structure. You will recall from Chapter 2 Max Weber's description of this formal arrangement as a bureaucracy—characterized in its ideal form by a hierarchy of positions, where each position is governed by a job description, and where written rules and procedures govern all activities. Thus, the formal organization is expected to behave impersonally in regard to relationships with clients or with its members. According to Weber's definition, entry and subsequent advancement is by merit or seniority. Each employee receives a salary and enjoys a degree of tenure that safeguards him from the arbitrary influence of superiors or of powerful clients. The higher his position in the hierarchy, the greater his presumed expertise in adjudicating problems that may arise in the course of the work carried out at lower levels of the organization. It is this bureaucratic structure that forms the basis for the *appointment* of heads or chiefs of administrative subdivisions in the organization and endows them with the authority attached to their position.

In contrast to the appointed head or chief of an administrative unit, a leader emerges within the context of the informal organization that underlies the formal structure. The informal organization expresses the personal objectives and goals of the individual membership. Their objectives and goals may or may not coincide with those of the formal organization. The informal organization represents an extension of the social structures that generally characterize human life—the spontaneous emergence of groups and organizations as ends in themselves.

In prehistoric times, man was preoccupied with his personal security, maintenance, protection, and survival. Now man spends a major portion of his waking hours working for organizations. His need to identify with a community that provides security, protection, maintenance, and a feeling of belonging continues unchanged from prehistoric times. This need is met by the informal organization and its emergent, or unofficial, leaders.

Leaders emerge from within the structure of the informal organization. Their personal qualities, the demands of the situation, or a combination of these and other factors attract followers who accept their leadership within one or several overlay structures. Instead of the authority of position held by an appointed head or chief, the emergent leader wields influence or power. *Influence* is the ability of a person to gain cooperation from others by means of persuasion or control over rewards. *Power* is a stronger form of influence because it reflects a person's ability to enforce action through the control of a means of punishment.

An individual who is appointed to a managerial position has the right to command and enforce obedience by virtue of the authority of his position. However, he must possess adequate personal attributes to match his authority, because authority is only potentially available to him. In the absence of sufficient personal competence, a manager may be confronted by an emergent leader who can challenge his role in the organization and reduce it to that of a figurehead. However, only authority of position has the backing of formal sanctions. It follows that whoever wields personal influence and power can legitimize this only by gaining a formal position in the hierarchy, with commensurate authority.

Theories of Leadership

As we have said, a leader may occupy a formal position to which he is appointed or he may emerge as a leader in response to the needs of the informal group. Ideally, a manager appointed to a position of authority in the formal organization will possess the necessary personal qualities for leadership. The organization needs an appointed head or chief who is also a leader in every sense of the word. Like the emergent leaders of the informal organization, the formal leader will command cooperation from his followers by virtue both of his position and their acceptance of him as a leader.

"Great Man" Approach

Why are some people natural leaders? How is this potential identified? There is very little precise knowledge about the process whereby an individual gains a leadership position. One of the earliest attempts at explaining the emergence of leaders was labelled the "great man" theme. Thomas Carlyle, in *Heroes and Hero Worship*, and Ralph Waldo Emerson, in *Representative Men*, reviewed the careers of great historical figures, such as Napoleon, Alexander the Great, Plato, and Mohammed and ended by implicitly suggesting that leaders were born and not made. There is a random process of selection at work, in which some individuals appear to have been destined to greatness because of a combination of favorable circumstances, including family background, education, and such personal attributes as intelligence, drive, and ambition. These people then find themselves in situations where their particular combination of talents and skills provide a glorious match with the opportunity and accrue resultant benefits to mankind. Such a view of leadership is in the tradition of the literary artistry of a Carlyle and the worshipful ardour that some biographers have for their subject's life story.

Traits Approach

Finding little substance in the "great man" theory of leadership, later investigators began to explore leadership from the point of view of the particular personal traits that characterize leaders, and those that a person might need to qualify for a leadership role. A number of investigations have focused on compiling lists of leadership traits: intelligence, drive, perseverance, honesty, seriousness, diligence, physical stamina, ambition, and moral integrity. Some lists contained a hundred or more alleged leadership traits. It soon became evident that not much progress could be made using this approach, since the various traits appear in every individual to some degree and in various combinations. There seemed to be no way that the traits approach to leadership could be used to predict the emergence of a leader or to differentiate between leaders and nonleaders. However, the hypothesis remains that qualities reflected in the personality of an individual constitute one avenue to understanding leadership.[1]

Situational Approach

The investigations on leaders and leadership suggested that whether an individual would or would not emerge as a leader was

[1]Cecil A. Gibb, "Leadership," in Gardner Lindzey, ed., *Handbook of Social Psychology*, vol. 2 (Reading, Mass.: Addison-Wesley, 1954), pp. 884–89. Cf. also Henry P. Knowles and Borje O. Saxberg, *Personality and Leadership Behavior* (Reading, Mass.: Addison-Wesley, 1971).

closely tied to the demands of the situation. Each situation calling for leadership is in some measure unique and requires a particular response from the organization and its members. The individual who seems best prepared for the position, as judged by his background, intelligence, and knowledge, in addition to other qualities such as physical stamina, particular skills, and so on would emerge as the leader. Once the demands of the particular situation had passed, his leadership role might disappear.

For example, in putting together an expedition to climb Mount Everest, the needs of such a venture in terms of financial backing, publicity, composition of the team of climbers, transportation, and other logistics problems might well result in leadership being conferred on a film producer committed to the venture for a photographic record of the expedition. Once the expedition has begun climbing the steep walls of the mountain, perhaps leaving part of the expedition stranded on an exposed ledge, the most experienced mountain climber is likely to emerge as the leader of the expedition. If, in such a situation, the personal glory to be the first to reach the top tempts the most experienced climber to strive for this at the expense of his stranded team members, the expedition may fail for the lack of accepted leadership among the expedition's members.

Democratic group processes may be used to select a leader as a result of a voting procedure that ensures the support of the majority for the elected leader. Talcott Parsons has pointed out, however, that election of a leader fails to take into account either the specialized knowledge of the individual elected into office or the requirements of the situation, since *the future remains unknown*.[2]

Appointing an individual to a position of authority by itself confers importance on him, and thus power. He will be assumed to possess adequate personal competence to make his position in the hierarchy effective. We referred to the formal leader as one who combines authority of position with personal qualities commensurate with that authority—and thus acceptance by followers. Many writers on leadership make no distinction between a manager in the bureaucracy and a leader, but assume that leadership applies to the activities of individuals who wield power or hold a supervisory or managerial position in the organization.[3]

Max Weber referred to several sources of leadership authority. He suggested that *traditional authority* may serve as one basis for the emergence of leadership. This is illustrated in the preliterate society by kinship rules of inherited leadership, which determine the heir of a chief or the guardian of religious symbols, just as do rules of succession for royalty even in modern times. Beyond

[2]Talcott Parsons, "Components and Types of Formal Organization," in Preston LeBreton, ed., *Comparative Administrative Theory* (Seattle, Wash.: University of Washington Press, 1968); p. 11.

[3]Cf. Stanley Lieberson and James O'Connor, "Leadership and Organizational Performance," *American Sociological Review* 37 (April 1972), pp. 117–30.

traditional rules of inheritance, leadership may be based on **charisma**. Max Weber referred to charisma as personal magnetism, "by grace of God," which attracts followers who have an implicit belief in such a person. However, once the charismatic leader has achieved a position of power and authority, he must create a *legal-rational* organization, or bureaucracy, to support his rulership and to help him deliver on the promises he has made to his followers. Fidel Castro in Cuba is an example of a charismatic leader who subsequently developed a strong legal-rational organization to support his regime.

We can extend this discussion to suggest that there may appear a circular flow: Traditional leadership will be overthrown by charismatic leadership—the old guard versus the young turks—which in turn will become managerial bureaucratic leadership. This bureaucratic stage may last for a long time. Then, as the organization accumulates experience in its operations, it begins to act again as a tradition-based organization. At this point, the cyclical evolution of leadership authority will begin anew.

LEADERSHIP BEHAVIOR AND INTERACTION

What does a leader do, how does a leader behave, and how are leaders ranked within their elite grouping? These are some of the questions discussed in the following section.

Functions of Leadership

There is an additional dimension attached to leadership—the functions that a leader is expected to perform.[4] The most important function is that of creating a commitment to the objectives and goals of the organization and the needs of the group. Leadership assumes followers and the cooperation of the members to work for the organization's objectives. Men have chosen not to live as isolates but in the fellowship of natural gregariousness, which serves as a survival mechanism in human society. For such a human community to find a unifying theme, a point of identification for its members, a few must point out the path to be followed by the rest.

In military organizations, leaders are frequently characterized by their success in creating an *esprit de corps*, or spirit of community, among the soldiers, even if the odds against victory are small. During World War II, the German military commander in the African desert war, General Rommel, whose legendary leadership gained the respect of the Allies, refused to stay behind his troops with his headquarters staff but showed up in front-line positions, apparently oblivious to personal danger. Called the "Desert Fox," his reputation grew so strong that the Allied Command had to issue orders forbidding reference to Rommel by name because of the psychological effect his leadership of German forces was having on the Allied Forces in Africa.

From Talcott Parson's famous discussion of the problems that face every organization, we can differentiate between the

[4]Edgar H. Schein, *Organizational Psychology* (Englewood Cliffs, N.J.: Prentice-Hall, 1970), p. 128.

problems of instrumental or task performance and the problems of integration or maintenance of the organization—organizational efficiency versus organizational climate. Specifically, the organization strives to achieve objectives and goals. It must either acquire or have at its disposal the resources necessary to achieve this objective. In addition, the organization must preserve itself as an entity, in which the efforts of individuals and groups are integrated in a coordinated whole dedicated to performing the organization's mission. And further, the individual members of the organization must remain motivated to make their contribution to the organization. Individuals must find strength by being convinced that their personal objectives and goals can best be reached through the organization's achieving its mission.

In this context, Bales's analysis of group interactions has demonstrated that each member can be characterized by his behavior in the group.[5] As we discussed in Chapter 5, the group must develop from its membership two different types of leaders. The *task leader* fulfills the role of calling the group's attention to the work at hand and the need to carry through on the part of the other members. The *socioemotional leader*, by the nature of his personality, provides support and understanding of the feelings of group members and helps alleviate tension. When these two leaders work to complement each other's efforts to help the group realize its objectives and goals, both the task and the social needs of the members can be fulfilled.

In many organizations, the division of roles between the two types of managers becomes quite apparent. One will represent the stern task master, frequently autocratic in his leadership behavior and exclusively concerned with getting the job done. Complementing him there may well be another manager who is supportive of employee needs and easily approachable for a discussion of personal or professional problems. Sensitivity and awareness of the psychosocial needs of the organization characterize the socioemotional leader.

Ranking—The "Iron Law of Oligarchy"

In any group or organization of men, a **ranking** is bound to occur. There are several possible bases for this ranking. Members of the organization will possess different degrees of status and prestige, or ranking may reflect differences in knowledge of various members, which will thus ensure the continued survival of the organization. In other situations, a difference in ranking may be related to age. That is, wisdom may be associated automatically with old age and the maturity and experience accompanying it; this wisdom, and thus ranking, may be credited to family and kinship relationships as a reflection of divine guidance, and thus superior to what might be offered by commoners. Or, superior intelligence combined with superior knowledge may be reflected in superior performance, and thus earn superior rank for an indi-

[5]Cited in Gibb, "Leadership," in Lindzey, ed., *Handbook of Social Psychology*, pp. 881–82, 894–97.

vidual. Accumulation of desirable goods or wealth stands as mute testimony to superior performance or perhaps to inherited abilities, and thus results in high ranking.

According to the concept of the "iron law of oligarchy" first introduced by Robert Michels, any organization or group will be governed by a minority.[6] Carried to its ultimate conclusion, this suggests that the governing minority will be ruled by a minority of one. However, such a feature in organizational life need not be interpreted in derogatory terms. Rather, it may reflect particular individuals' talent, enthusiasm, energy, interest, and knowledge in relation to the affairs of the organization. Their superior commitment will help earn the governing minority the right to be involved in decision making at the top of the organizational structure. To be effective, as we have pointed out, these individuals must also possess the requisite personal capabilities to ensure performance not only of task-oriented roles but also of those roles associated with the social and emotional needs of the organization's members.

Elite as Value Carriers

The top management of an organization can be viewed as its elite. These people are the value carriers of the organization. Their decisions, their actions, and their behavior provide a pattern of leadership for the other members to emulate. Since leadership represents the focal core of values for the organization, it is only natural that leaders tend to try to conserve these values. This may, however, represent the outcome of a particular leadership that has gained the upper hand. For example, there may have been a period of transition during which different interest groups were competing to gain a commitment to their cause from the organization's members. The prevailing leadership was victorious in introducing their values into the organization. This leadership naturally proceeds to protect its position and act as a conserver and preserver of these values against usurpers and interlopers attempting a change.

Leadership for Adaptive Survival

The degree to which the elite is being constantly renewed and replaced signifies continuous change in progress. In an open society such as ours, renewal is explicitly promoted through a circulating elite providing adaptation to new conditions. In a closed society such as a caste system, adaptation to new conditions proceeds very slowly and is more difficult. Nevertheless, every organization, every society, preserves avenues for renewal of the elite. In the caste society, permissible marriage relationships frequently serve to introduce new blood, new ideas, and changing views which, crossing established caste boundaries, help to create a degree of social mobility and a basis for change. A family-owned business will provide opportunities for professional managers when family members have other interests or insufficient competence.

[6]Robert Michels, *Political Parties* (New York: Dover Publications, 1959), pp. 377–92.

The leader embodies the values of the organization and its mission. He is charged to maintain them and in fact to ensure the organization's survival. It follows that the leader is responsible for providing a vision for the future. The Old Testament warns that, "Where there is no vision, people will perish." "For if the trumpet give an uncertain sound, who shall prepare himself for the battle?" asked Saint Paul in the First Epistle to the Corinthians.

The successful leader is shouldered with a multiple responsibility. He must simultaneously conserve and protect the organization's ways of doing things and yet adapt to change to ensure survival. A manager as an adaptive leader needs sufficient sensitivity to, and awareness of, changes in the environment or in the desires of his employees to be able to introduce a new direction relying on the collective efforts of his followers.

Democratic versus Autocratic Leadership Styles

The successful leader has his own needs. These needs will be reflected in his unique leadership style that must then support the organization's values. The original investigations on leadership style were conducted by Lippitt and White at the Child Welfare Center of the University of Iowa.[7] The researchers were interested in the effect of particular leadership behavior on the behavior of members of experimental groups involved in task performance. Their classic experiment studied groups of ten-year-old boys. Each group was assigned an adult leader with a designated leadership behavior. For the first group, the style of leadership was noncommittal and uninvolved; it might be described as the abdication of leadership of the group. White and Lippitt referred to this as a **laissez-faire style of leadership** (translated as, "to let people do as they choose").

The second group was under the guidance of a leader who determined in minute detail the nature of the work that each group member was to complete. He kept each boy at work and assigned new work as soon as the prior work had been completed. The leader's total involvement with detailed instructions for each member was labelled an **autocratic** style of leadership.

The third group was assigned an adult leader who provided support and encouragement to the group and its members without becoming involved in too much detail, except when he was asked for assistance or saw a need to provide assistance. He appeared to place great confidence in the participants to perform by themselves and complete their assigned tasks. This leadership behavior was termed a *democratic* style of leadership.

Laissez-faire leadership has generally been ignored as irrelevant in the context of managerial and supervisory behavior, though it sometimes occurs in real-life situations. It does represent abdication from active leadership as in the case of a foreman or supervisor who feels himself overwhelmed by experts, and consequently refuses to take an active role as a leader.

[7]Ronald Lippitt and Ralph K. White, "An Experimental Study of Leadership and Group Life," in E.E. Maccoby, T.N. Newcomb, and E.L. Hartley, eds., *Readings in Social Psychology*, 3rd ed. (New York: Holt, Rinehart and Winston, 1958), pp. 496–511.

The results of White's and Lippitt's experiment indicated that the autocratic work group performed more work than the democratic work group, though apparently no actual productivity figures were kept. The democratic work group appeared to be characterized by a more balanced disposition on the part of the participants. They seemed to enjoy and show more interest in what they were doing and to do higher quality work. In contrast, the autocratic work group appeared generally hostile and dissatisfied and displayed actual destructive and disruptive behavior, even though these boys were generally known to be submissive and passive. This research gave rise to more questions than it answered. As a result, investigations into the effects of leadership style on employee behavior continue.

In their studies of leadership style, Schmidt and Tannenbaum have described autocratic and democratic styles of leadership as extremes on a continuum of possible alternatives characteristic of a manager, his personality, and the demands of the situation.[8] Figure 10.1 demonstrates that an extreme autocratic position, in which the manager would make decisions on life and death, does not exist. Neither is there an extreme democratic position, in which all the decisions would be made by the employees. In an organization, this might imply no structure and a state of anarchy. For these reasons, the decision allocation line is not drawn exactly from corner to corner of the diagram.

Business lives in a constant state of crisis in a market, or competitive, economy. In the same way that a nation temporarily suspends total democratic procedure during times of crisis and war (giving the president powers he does not normally have), managers tend to behave autocratically during times of crisis. An increased standard of living and standard of education, and the changed expectations and aspirations on the part of the employees, managers,

Figure 10.1 Continuum of Leadership Behavior

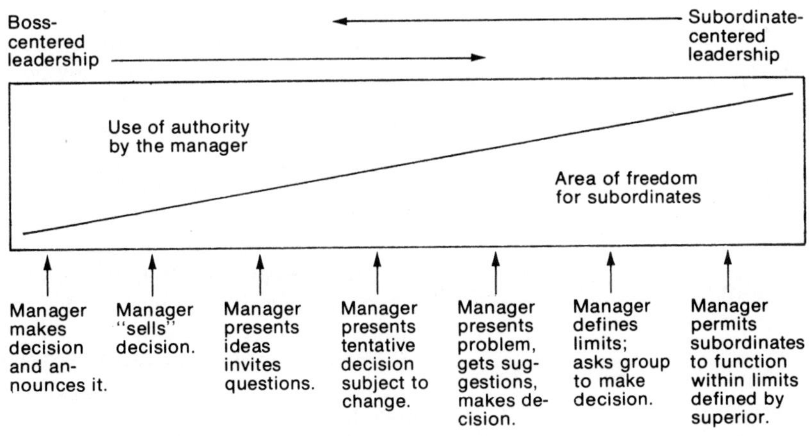

Source: Robert Tannenbaum and Warren H. Schmidt, "How to Choose a Leadership Pattern," *Harvard Business Review* 36 (March–April 1958), p. 96.

[8]R. Tannenbaum and W.H. Schmidt, "How to Choose a Leadership Pattern," *Harvard Business Review* 36 (March–April 1958), pp. 95–101.

and the public may have created some movement toward a more democratic style of leadership in the business decision climate. Many researchers and writers on management and organization have attempted to persuade management to move their behavior closer to a democratic style, but with minimal success. Perhaps this reluctance is mute testimony to the personality characteristics required of those who advance to high positions of responsibility in business and industry, as well as in other types of organizations. A manager's personal involvement, commitment, and perseverance, coupled with ambition, intelligence, foresight, and ability to carry through the organization's objectives and goals, may somewhat dull his sensitivity to the needs of the social environment, thus perhaps making him a less effective socioemotional leader, although an effective task leader.

The choice of a democratic or autocratic leadership style ultimately reflects the basic values and attitudes internalized in the personality of a manager-leader.[9] If he is able to place trust and confidence in his employees and their willingness and capacity to carry through decisions when given a position of trust and responsibility, he might operate as a democratic leader. His managerial style would be employee oriented, encouraging participation from both subordinates and managers in decisions affecting the enterprise. But if he harbors a basic mistrust and doubt about other people and their willingness to perform, he might tend toward autocratic leadership.

Each manager, as a result of his unique personality, will have a natural position on the range of leadership styles between the democratic and autocratic polar positions. Successful leadership is expressed in the manager's ability to play the whole keyboard in response to the demands of the particular situation. In some matters, perhaps in raising new capital for his firm, only the manager has sufficient information and the proper long-range perspective to make strategic decisions. In other situations, the decision will involve a short-term perspective and limited resources; that is, selection of new portable dictating machines. Note that the quality of the decision may be directly related to how much involvement there has been by those who are affected by it—for example, a tactical, short-run decision regarding distribution of overtime among eligible employees. In both cases, the nature of the work force—that is, whether transient labor or professional manpower—has much to do with the degree of involvement that can be expected. We must emphasize, however, that although leaders may consciously vary their leadership styles to meet the requirements of a given situation under normal conditions, in times of stress and tension they will tend to revert to their natural position on the range of leadership styles. During a crisis, a manager's basic attitudes take charge and his behavior follows suit.

[9]Knowles and Saxberg, *Personality and Leadership Behavior*, p. 43.

Management of scientific manpower (engineers and scientists) must rely on employees' involvement in decisions affecting them and their work. Such employees are likely to be more competent in their particular specialty than the managers under whose supervision they serve, and consequently they carry an expectation of such involvement as part of their professional ethic.[10]

The differences between the normal managerial organization, and engineers and scientists under the supervision of their own managers is vividly demonstrated in the emergence of the dual organizational structure in a technologically oriented industry. Engineers and scientists, as highly competent specialists, are encouraged to remain in their specialty while benefiting from advancement and promotion within the engineering and scientific organization. However, there inevitably exists tension in the organization, because transfer to management may be considered either a sign of success in the engineering and scientific organization or an escape before technical obsolescence occurs. The line manager of scientific manpower, despite his background as a specialist in an engineering or scientific field, will tend to lose his technical proficiency in a few years, as his administrative duties take up more and more of his time.

University of Michigan Leadership Studies

Much of the research on leadership style as it relates to satisfaction, morale, and productivity has been conducted by the Institute for Social Research, Survey Research Center, at the University of Michigan. These studies have involved various work environments, such as clerical office work, railroad-track section gangs, and factory workers in a heavy industrial equipment company.[11] The investigations were motivated by the researchers' desire to correlate supervisory behavior with employees' satisfaction, morale, and productivity as evidenced through questionnaires, figures on absenteeism, turnover, grievance rates, and production output.

In 1948, researchers at the Survey Research Center gained the cooperation of the Chesapeake and Ohio Railroad to explore in what ways productivity depended on supervisory behavior and employee morale. Employee morale was defined through workers' attitudes toward work as expressed through satisfaction with the company, with fellow employees, with promotion opportunities, and so on. The study involved a total of 72 railroad section-maintenance gangs, including 298 men and 72 foremen. These section gangs were evenly divided between high-producing and low-producing units in relation to track maintenance work.

The research findings indicated that supervisory behavior

[10]Borje O. Saxberg and John W. Slocum, Jr., "The Management of Scientific Manpower," *Management Science* 14 (April 1968), pp. 473–89.

[11]Robert L. Kahn and Daniel Katz, "Leadership Practices in Relation to Productivity and Morale," in Dorwin Cartwright and Alvin Zander, eds., *Group Dynamics*, 2nd ed. (Evanston, Ill.: Row, Peterson, 1960), pp. 554–70.

was important in determining the productivity of the work groups. Compared to the low-producing groups' foremen, the high-producing supervisor showed the following characteristics:

1. Differentiates his work from that of workers
2. Recognizes his leadership role
3. Spends time on his supervisory function
4. Defines his work in relation to his workers' needs
5. Maintains an employee orientation without abdication of leadership
6. Displays a personal interest in his employees
7. Behaves in a supportive, nonpunitive manner when workers experience problems on the job
8. Helps to train his workers for better jobs

Thus, the high-producing supervisor used a different approach to his workers and their motivation at work in contrast to the low-producing supervisor who frequently became involved in working alongside his workmen to push out more production.

One interesting finding from this research was that those employees who expressed most satisfaction with the content of their work were in low-producing work groups. High-producing sections were moderately satisfied. The explanation may lie in the generally low level of skill required for task performance and thus low-level worker aspirations. Consequently, in the high-producing group there was a higher level of aspiration that was thwarted by the low-skilled jobs available.[12]

The results of these various investigations have been inconclusive. However, the original notion that democratic, or participative, managerial behavior leads to higher productivity has been shown to be too simplistic. Neither has it been possible to conclude whether high morale creates high productivity in the work group, or whether high productivity precedes high morale. Stanley Seashore of the Survey Research Center has indicated that the research appears to offer evidence that employee-oriented behavior by supervisors leads to organizational effectiveness. Thus he concludes that there is sufficient evidence to suggest higher productivity for employee-centered supervision than for production-centered supervision. This would also hold true when supervision is general rather than close; morale would also be higher. These conclusions are tempered by the general observation that the employee's performance is dependent in part on "the degree to which he becomes a member of a work group which is characterized by group pride, cohesion, and solidarity."[13]

Later findings by the staff of the Survey Research Center led

[12]Daniel Katz, Nathan Maccoby, Gerald Gurin, and Lucretia G. Floor, *Productivity, Supervision and Morale among Railroad Workers* (Ann Arbor, Mich.: Survey Research Center, Institute for Social Research, 1951).

[13]Stanley E. Seashore, "Administrative Leadership and Organizational Effectiveness," in Rensis Likert and Samuel P. Hayes, Jr., eds., *Some Applications of Behavioural Research* (Paris: Unesco, 1957), p. 63.

its director, Rensis Likert, to modify these conclusions on effective supervisory behavior. He added an emphasis to the need for foremen to be supportive of employees by utilizing participative methods of supervision, having high performance goals, being technically informed, and receiving support from higher management.[14]

Subsequently, Floyd Mann of the Survey Research Center adopted the view that supervision and leadership can be dealt with as consisting of three separate types of competence or skill requirements:

1. Technical skill, which refers to knowledge and ability to perform the *task*.
2. Human relations skill, which refers to the ability to work with *people*—including an understanding of principles of human behavior and motivation.
3. Administrative or conceptual skill, which refers to the ability to think of the *organization* as a whole consisting of people and things working toward the realization of its objectives and goals.[15]

Ohio State University Leadership Studies

The other major leadership studies were conducted by the Bureau of Business Research of Ohio State University. These studies focused on constructing a research instrument to describe leadership. The characteristics generally identified were "consideration" and "initiating structure," which parallel in meaning "employee orientation" and "production orientation," used by the Survey Research Center of the University of Michigan, as well as the "task" and "maintenance" functions of leadership discussed earlier.[16]

Rather than classifying "consideration" and "initiating structure" as polar positions on a unidimensional scale, they were characterized as independent variables. Figure 10.2 diagrams a leader's perceived or observed behavior on a quadrant indicating different combinations of consideration and initiating structure. The Ohio State studies generally support the University of Michigan findings, suggesting that successful supervision in terms of effective organizational performance appears to involve different

[14]Rensis Likert, *New Patterns of Management* (New York: McGraw-Hill, 1961), pp. 98–102.

[15]Floyd C. Mann, "Toward an Understanding of the Leadership Role in Formal Organization," in Robert Dubin et al., eds., *Leadership and Productivity* (San Francisco: Chandler, 1965), p. 75; Robert L. Katz, "Skills of an Effective Administrator," *Harvard Business Review* 33 (January–February 1955), pp. 33–42.

[16]Edwin A. Fleishman and Edwin F. Harris, "Patterns of Leadership Behavior Related to Employee Grievances and Turnover," *Personnel Psychology* 15 (1962), p. 43. One interesting finding referred to by Fleishman and Harris indicates that supervisors who were rated high by their management were characterized by a leadership pattern high in initiating structure and low in consideration in a production environment. However this problem was also associated with high labor turnover, union grievances, worker absences and accidents, and low worker satisfaction.

Figure 10.2 Leadership Style

Consideration		
	Low Initiating Structure High Consideration	High Initiating Structure High Consideration
	Low Initiating Structure Low Consideration	High Initiating Structure Low Consideration
	Initiating Structure	

combinations of consideration and initiating structure, depending on the situation. Again, researchers have been able to show very little correlation between supervisory behavior and productivity in work groups.[17]

In reading the quadrant, we see that a supervisor may be characterized by low initiating structure and high consideration. This would suggest that he puts little emphasis on production quotas or on rules and procedures in achieving them, but he is concerned about the welfare of his subordinates and supportive of their expectations and aspirations. We might assume that the optimum behavior would show an emphasis on task achievement as well as consideration for the employees—that is, high initiating structure with high consideration.

The Managerial Grid.® The "grid" approach to managerial leadership attempts to measure managers' behavioral predispositions in organizational situations. This approach shares the concepts of consideration and initiating structure of the Ohio State leadership studies, with a midposition added.[18] As the managerial grid in Figure 10.3 indicates, the vertical axis shows concern for employees. Thus a score of 1 on the vertical axis indicates a minimal concern for people, and a 9 on the horizontal axis a maximum concern for production—a position called "task orientation." The optimum position suggested by Blake and Mouton is at 9,9, where a "team" has been created and the leader shows a maximum concern for production and a maximum concern for employees.

THE EFFECT OF ORGANIZATIONAL ENVIRONMENT ON LEADERSHIP

The leadership studies just discussed take on a different meaning if we relate them to contingency theories of management and leadership, such as the recent emphasis on the influence of the organizational environment on management, organization, and

[17]Ralph M. Stogdill, *Individual Behavior and Group Achievement* (New York: Oxford University Press, 1959), pp. 235–38. Cf. Abraham K. Korman, "Consideration, Initiating Structure, and Organizational Criteria—A Review," *Personnel Psychology* 19 (1966), p. 360.

[18]Robert R. Blake and Jane S. Mouton, *The Managerial Grid* (Houston, Texas: Gulf Publishing, 1964), p. 10.

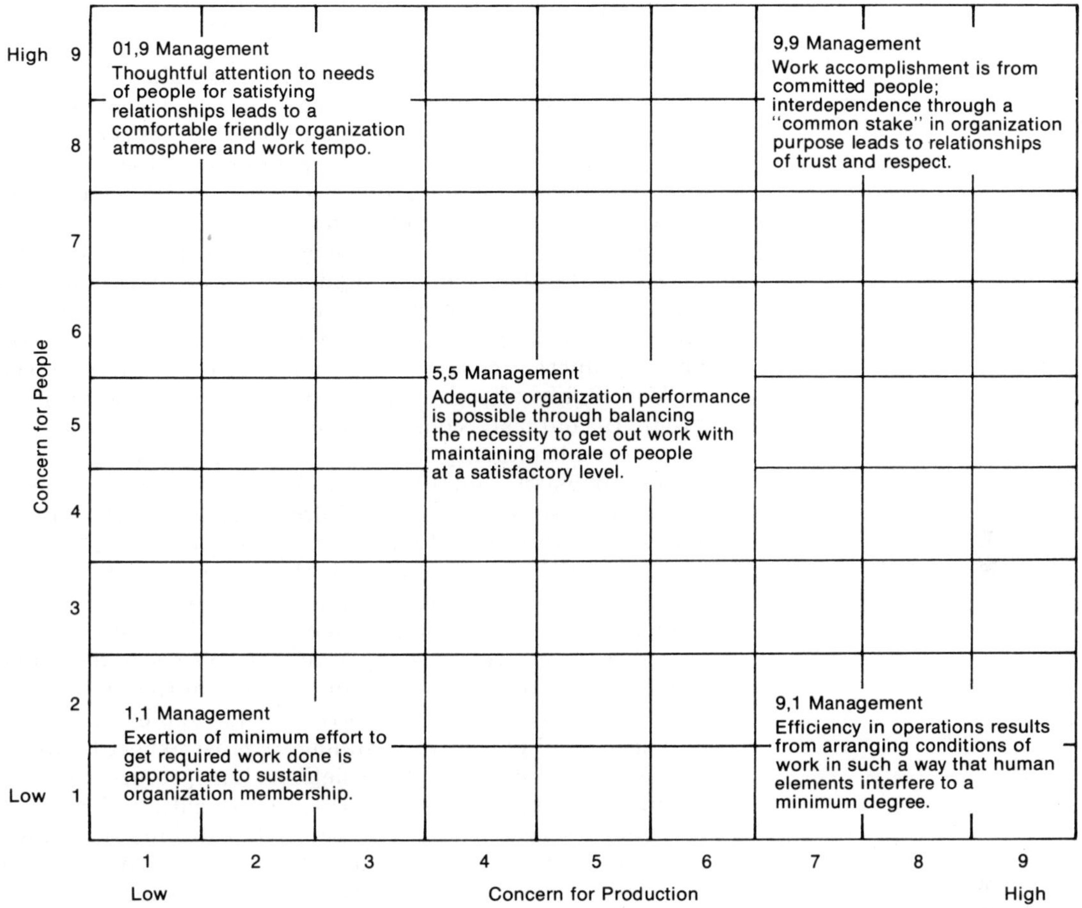

Figure 10.3 The Managerial Grid

Source: Robert R. Blake and Jane Srygley Mouton, *The Managerial Grid* (Houston, Texas: Gulf Publishing, 1964), p. 10. Reproduced by permission.

leadership effectiveness. For example, the publicly owned but closely regulated public utilities were long identified with the conditions of a stable market and relatively little innovation. Their organization, therefore, took on the features of a regimented bureaucracy. Once an electric utility was confronted with the need to consider the nuclear power plant as an alternative to its traditional power plant operations, it needed to change its strategy. A whole new set of problems arose, associated with the applicable technology and production processes. In addition, different personnel having an advanced scientific outlook were required. These conditions of flux necessitated a stronger emphasis on differentiation among functions performed in the electric utility, creating conditions to which the traditionally established managerial hierarchy would be unable to respond adequately. As a result, electric utilities were forced to create separate organizational divisions for nuclear energy, having a different managerial approach from the original organization's.

Such so-called contingency studies indicate that a general recommendation to move toward a democratic leadership approach would violate the need to consider the nature of the organization's environment. Robert Dubin has pointed out that, "Structuring of the supervisor's work responds to technical and organizational imperatives," and "Consideration for the worker as individual decreases as the technology becomes more complicated and the production schedule demands higher rates of continuous output."[19] Consequently, supervisory behavior in the high-production work environment may well be characterized by "initiating structure" and "production orientation" (or autocratic leadership style) and still be highly effective in achieving productivity from the work group. In contrast, "consideration" and "employee orientation" (or democratic leadership style) would be appropriate in a work situation emphasizing quality or professionalism, when the manager needs to be sensitive to his subordinates' needs to call forth their best efforts. However, this analysis cannot be regarded as conclusive.

Selznick has suggested that the executive becomes a "statesman" as he makes the transition from administrative management to institutional leadership. "To institutionalize" in Selznick's terms refers to an organization's becoming infused with value beyond the technical requirements of the task—and always remaining responsive to adaptive change and evolution due to social needs and pressures.[20] Managers must expect to meet change in the environment, or may deliberately proceed to create such change to ensure survival for their organization. For this purpose, the characteristics of the organic or continually changing and adaptive organization should be regarded as indispensable and part of the strategy to meet the needs of change.

Writers on management and organization continue to suggest that some variation of leadership that encourages a participative approach to management will yield greater effectiveness and efficiency in organizational performance, despite the fact that the brute evidence does not yet support this claim. However, we must recognize the general preference in our society for allowing men to participate in matters that will affect their lives and livelihoods, thus explaining the general tendency toward advocating democratic participation in organizations.

Contingency Theory of Leadership

Business and industry have attempted to find the best qualified men for the available jobs. For this reason, they have sponsored selection and training of these men to develop in them a leadership style that might be required in their jobs. Fred Fiedler

[19]Robert Dubin, "Supervision and Productivity: Empirical Findings and Theoretical Considerations," in Dubin et al., eds., *Leadership and Productivity*, p. 24.

[20]Phillip Selznick, *Leadership in Administration* (Evanston, Ill.: Row, Peterson, 1957), pp. 4, 12, 17.

and his associates, long active in the area of leadership studies, have argued that the emphasis on leadership development has been wrong all along.[21] There is little hope or justification for attempting to change a manger's style of leadership, they hold. Such a change would necessitate changing his personality and would be extraordinarily difficult. Rather, Fiedler suggests that it is more important to establish the particular quality of an individual's leadership style and then structure or find a situation that will enable him to function at his best.

To reach his objective, Fiedler has developed the LPC (Least Preferred Co-worker) scale. A high score on the LPC indicates that the individual, when asked to judge the qualities of a co-worker he least preferred to work with in comparison with all individuals he has worked with in the past, does not see much difference between the least preferred co-worker and the others. The "high LPC individual" appears to be generally person-oriented in his attitudes. This compares with a "low LPC individual," who rates the least preferred co-worker significantly different from the other individuals with whom he has worked. The low LPC individual is characterized by Fiedler as being task motivated rather than human-relations oriented.

According to Fiedler's work, the LPC score is related indirectly to the effectiveness of group performance—that is, in relation to the favorableness or unfavorableness of the situation from the point of view of the leader and his power and influence over the group. Favorableness of the situation is a reflection of (1) the leader-member relations (good interpersonal relations would give the leader more power and influence), (2) the task structure (firmly established task structure with procedures and rules would make the leader's command over the group easy), and (3) the leader's position power in relation to the group (high position power provides the leader with more force and influence).

Figure 10.4, based on a number of studies in different settings, shows that a controlling or task-oriented leadership style (low LPC score) is appropriate for the "most favorable" (Octant I) and "least favorable" (Octant VIII) situations, as reflected in the performance of the group. The group that is characterized as "intermediate in favorableness' (Octants IV and V) performs best under a permissive leader (high LPC). Rating the dimensions of favorableness, the leader's interpersonal relations rank first in importance; next, the task structure; and the leader's position power, last. For the most favorable situation, the low LPC leader is apparently effortlessly in command and may see an opportunity to gain wider recognition and importance from the group's productive performance. Such a leader is therefore ignoring the internal social needs of the group, since these already function in his favor. The

[21]Fred E. Fiedler, "Engineer the Job to Fit the Manager," *Harvard Business Review* 43 (September–October 1965), pp. 115–22. See also Fiedler's *A Theory of Leadership Effectiveness* (New York: McGraw-Hill, 1967).

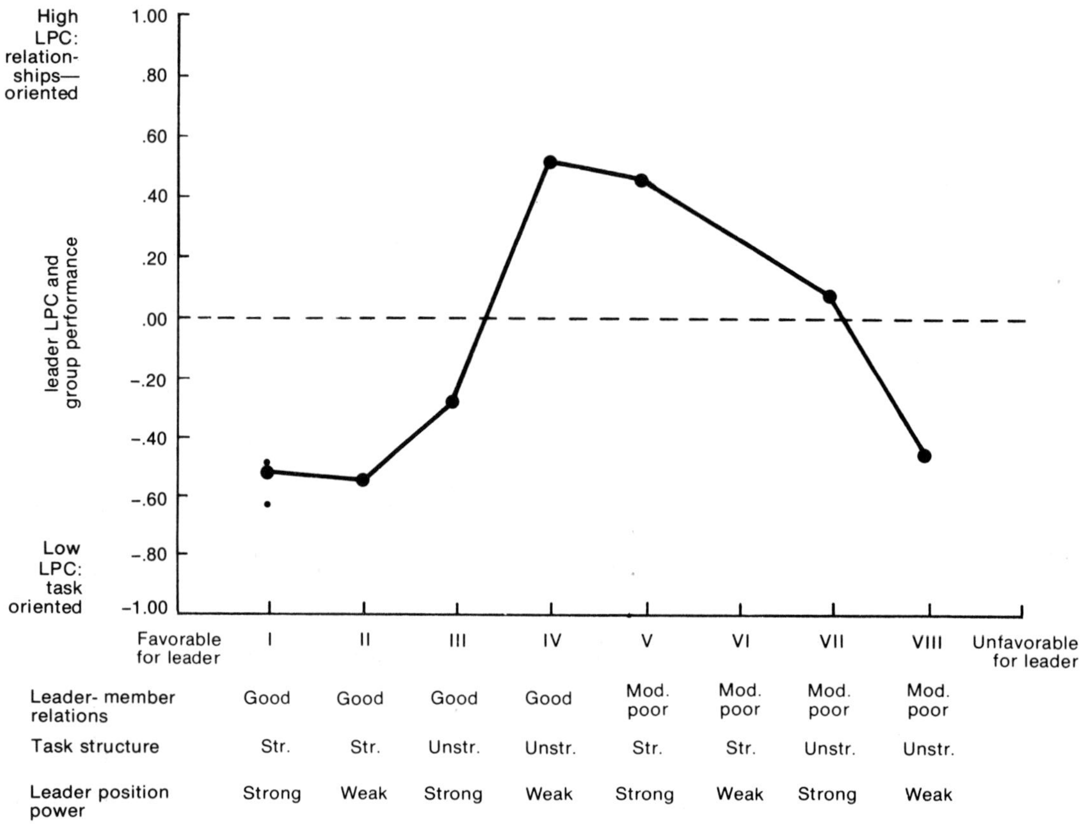

Figure 10.4 Correlations between Supervisor LPC and Team Performance in Various Work Situations (Obtained by Hunt, 1966)

Source: Fred E. Fiedler, *A Theory of Leadership Effectiveness* (New York: McGraw-Hill, 1967), p. 176.

"least favorable" group requires as a leader a task master to bring about their performance.

In the "intermediately favorable" situation, the high LPC leader can elicit performance from the group by soliciting cooperation from its members. He must rely on this approach. Leader-member relations in the organization are not (or cannot be) as good as they might be, and therefore require that he exerts effort in that direction. Or, by resorting to the groups's willingness to cooperate, he may succeed in defining the structure further to assure performance of the group. Thus, instead of emphasizing changing the personality of selected candidates for managerial jobs to equip them with the needed leadership style, through his contingency model Fiedler suggests that a given leadership style be matched up with the situation. A necessary change can be accomplished in the situation through altering either leader-member relations, the task structure, or the position power granted by the organization.

SUMMARY

This chapter has indicated that the qualities of leadership have been a continual source of interest, as writers and researchers have tried to explain what makes a "leader" in the context of the "great man" approach, individual traits, leadership style, and situational demands.

Investigations of Likert and his associates, at the University of Michigan, attempted to show that high morale and high satisfaction caused high productivity. Although they failed to prove this, they concluded that there was some evidence to support the proposition that employee-oriented leadership style, or supervision, led to high productivity. From a slightly different perspective, the Ohio State leadership studies provided additional confirmation in that same direction. They concluded that supervision that was characterized by consideration for both the employees and the task was likely to have high productivity.

In contrast, Fiedler and his associates suggested that an individual's leadership style should be regarded as a reflection of his personality not easily changeable, and therefore as given. It should match the situational requirements, in particular, those judged favorable or unfavorable from a leader's point of view. This contingency approach to leadership is tied to studies by Woodward; Burns and Stalker; and Lawrence and Lorsch; whose conclusions point to an approach to management and organization tailored to the particular enterprise and its internal and external environment.

Common to these investigations is the notion that the leader acts in a social or interpersonal setting. It is his responsibility to embody and encourage the realization of organizational objectives and goals. To achieve these goals, he seems to coordinate the effort of employees who are willing to cooperate in return for his support of their personal needs. As a formal leader, he simultaneously represents the maintenance needs of his workers and the goal and task achievements of the organization.

Alongside the manager's involvement with the organization's technical processes and task structure, leadership has to do specifically with people, regardless of their position or level. The qualities of leadership are critically involved in the interpersonal and integrating relationships that are established throughout the organization. Success in ensuring the growth and development of the membership will reflect the quality of leadership and the continuing viability of the organization. As the executive transcends mere administrator management to assuming true institutional leadership—that is, when the organization's values, objectives, and goals have become internalized and activated by its members—managers attain the level of "statesmen," whose values provide direction for society.

DISCUSSION QUESTIONS

10.1. What is the difference between power and influence?

10.2. Compare and contrast the "great man" approach to understanding leadership with the "traits" approach.

10.3. Give an example of a leadership duty that would support the "situational" approach.

10.4. What is a leader expected to do?

10.5. Which leadership style do you believe is the best? Why?

10.6. What conclusions did the University of Michigan leadership studies reach about leadership?

10.7. How do the conclusions reached in the Michigan studies compare with those reached in the Ohio State University studies?

10.8. Using the managerial grid illustrated in Figure 10.3, how would you rate a supervisor whom you regard as striking an ideal balance between technical efficiency and organizational climate?

10.9. How does improved technology affect leadership style?

10.10. Can an individual's leadership style be changed? If so, how? Do you think such a change should be attempted?

APPENDIX
Admiral Horatio Nelson

Admiral Nelson knew that sooner or later, England had to face a battle with the Napoleonic naval forces. However, ill health had forced Nelson to duty on shore. When word came that a battle was looming as the French tried to enter the English Channel, the strength of public opinion virtually forced Nelson out of retirement to take command of the battle, which everyone knew he wanted to fight.

The French naval forces under Admiral Villenueve had taken refuge in the Spanish harbor Cadiz. Napoleon was indecisive. He sent messages back and forth regarding what should be done, one order nullifying a previous order. At this time, there was an order on the way to Cadiz to relieve Villeneuve of command because of his cowardliness in battle. Suspicious that Napoleon would again contradict himself, and apparently with foreknowledge of his personal impending disaster, Villeneuve set out to sea without waiting for further orders from Napoleon. He wanted to break through the British line with his thirty-four ships and the seven Spanish ships that were supposed to join him. The crews were inexperienced, and the Spanish had been giving him a hard time—two Spanish ships were worth one French ship in battle, according to orders he had received.

The spirit of the British naval units had been gloomy. A first encounter with the French navy had not been completed. British Vice-Admiral Calder was a cold and haughty figure who did not contribute much spirit to the captains of his ships. On his arrival at the scene, Nelson found no flagship; Calder had taken it with him on returning to England, where he had demanded a court martial to set the facts straight on the battle he had fought.

Nelson then proceeded to invite each ship's captain individually on board his own *Victory*, giving full reception due a captain in command of his ship, and sparing nothing to make the meal a memorable event. During these meetings, Nelson set out in detail his plan for the battle, including the role to be played by his visitor's ship. The next day, following the captain's return to his own ship, he received a cordial note from Nelson that included a memorandum detailing the maneuvers the captain was to follow during the battle. Nelson's basic guideline was: "No captain can do very much wrong if he places his ship alongside that of the enemy."

The battle at Trafalgar has gone down in history as one of the great sea battles. The British were victorious with their twenty-seven ships, though the victory cost the life of its leader, Admiral Horatio Nelson.

1. Who are the leaders in this case?
2. How do you explain the action taken by Villeneuve?
3. What appears to be the contrast between Calder and Nelson according to the leadership models we have talked about?
4. What seem to be the dimension(s) on which Admiral Nelson's leadership rested?

SELECTED READINGS

Dubin, Robert et al., eds. *Leadership and Productivity*. San Francisco: Chandler, 1965.

Hersey, Paul, and Blanchard, Kenneth H. *Management of Organizational Behavior*, 2nd ed. Englewood Cliffs, N.J.: Prentice-Hall, 1972.

Selznick, Phillip. *Leadership in Administration*. Evanston, Ill.: Row, Peterson, 1957.

Chapter 11 Conflict and Change in Organizations

Conflict takes many forms in organizations. There is the inevitable clash between formal authority and power and those individuals and groups affected. There are disputes over how revenues should be divided, how the work should be done, and how long and hard people should work. There are jurisdictional disagreements among individuals, departments, and between unions and management. There are subtler forms of conflict involving rivalries, jealousies, personality clashes, role definitions, and struggles for power and favor. There is also conflict *within* individuals—between competing needs and demands—to which individuals respond in different ways.

Conflict sometimes has a destructive effect on the individuals and groups involved. At other times, however, conflict can increase the capacity of those affected to deal with problems, and therefore it can be used as a motivating force toward innovation and change. Conflict is encountered in two general forms. *Personal conflict* refers to an individual's inner workings and personality problems. Many difficulties in this area are beyond the scope of management and more in the province of a professional counselor, but there are some aspects of personal conflict that managers should understand and some they can possibly help remedy. *Social conflict* refers to interpersonal, intragroup, and intergroup differences. Although personal and social conflict cannot always be separated, as we shall see, these classifications nevertheless provide a convenient way for us to discuss conflict in this chapter.

PERSONAL CONFLICT

Argyris has pointed out that there is a basic incompatibility between the authority and structure of formal organizations and the human personality.[1] Human behavior cannot be separated from the culture that surrounds it.

A top management executive who said that "the trouble with the workers today is that 5 percent of them work, 10 percent of them think they work, and 85 percent of them would rather die than work," was asked to repeat the same sentence and add at the end of it the phrase "in my plant." The executive complied with the writer's request and obtained a new outlook on the problem of understanding human motivation.[2]

[1]Chris Argyris, *Personality and Organization* (New York: Harper & Row, 1957).
[2]Ibid., p. 48.

If people malfunction as workers or as occupants of executive suites, we cannot argue that "this is the nature of man" and let it go at that. Rather, we need to conclude that this is the way certain workers and executives behave in a particular environment.

Maturity-Immaturity Theory

According to Maslow, Argyris, McGregor, Rogers, and other writers of the so-called growth schools, there is a basic tendency in the development of the human personality toward self-fulfillment, or self-actualization. This implies that as an individual matures, he wants to be given more responsibility, broader horizons, and the opportunity to develop his personal potential. This process is interrupted whenever a person's environment fails to encourage and nurture these desires.

Formal organizations are rational structures that, based on their assumption of emotions, feelings, and irrationality as human weaknesses, try to replace individual control with institutional control. Thus the principle of task specialization (discussed in Chapter 2) is seen as a device that simplifies tasks for the sake of efficiency. As a consequence, however, it uses only a fraction of a person's capacity and ability. The principle of chain of command centralizes authority but makes the individual more dependent on his superiors. The principle of normal span of control, which assigns a maximum of six or seven subordinates to report to the chief executive, reduces the number of individuals reporting to the head of the organization or to the manager of any subunit. Although this simplifies the job of control for the manager, it also creates more intensive surveillance of the subordinate, and therefore permits him less freedom to control himself.

Under such conditions, subordinates are bound to find themselves in conflict with the formal organization, and sometimes with each other. They advance up the narrowing hierarchy where jobs get fewer, and "fewer" implies competing with others for the decreasing number of openings. Task specialization tends to focus the subordinate's attention on his own narrow function and divert him from thinking about the organization as a whole. This effect increases the need for coordination and leads to a circular process of increasing the dependence on the leader.

Subordinates adapt to these conditions in the organization in several ways. In the extreme, they may find the situation intolerable and leave the organization. Or they may strive to advance to positions of higher authority, there to adopt the controlling style they are trying to escape. They may respond to organizational pressures and threats by defensive reactions such as aggression against their supervisors and co-workers, fixated behavior or apathy, compromise and gamesmanship, or psychological withdrawal and daydreaming. All of these defense mechanisms reduce a person's potential for creative, constructive activity on the job. Finally, employees may organize unions or unsanctioned informal groups whose norms of behavior are opposed to many of the organization's goals. As a sort of self-fulfilling prophecy, all of these reactions to the constraints of the formal organization merely serve to reinforce and strengthen them.

The conflict between the formal organization and the individual will continue to exist wherever managers remain ignorant of its causes or wherever the organizational structure and the leadership style are allowed to become inconsistent with the legitimate needs of the psychologically healthy individual. Everyone recognizes the necessity for order and control in organizations. Those of us who enter management, however, must learn to recognize in addition that order and control can be achieved only at the expense of individual freedom.

Role Conflict

Another facet of personal conflict has to do with the multiple roles people play in organizations. Consider the case of Sam Wilson, a salesman for a large soap company located in the central part of the United States. Mr. Wilson enjoyed being with his wife and children. They liked going on picnics and trips together and dreamed of the day when they would own a small cabin at a nearby lake. They planned to spend weekends there when they could and to use the cabin for vacations. Mr. Wilson's success as a salesman soon made it possible to purchase the cabin. Being a good salesman, however, required him to spend a great deal of time on the road, often including weekends. When he could take time off during the week to compensate for working on weekends, the children were in school. He found himself facing a dilemma: Which role to choose? He might continue to work hard and become a sales manager, thus remaining in sales but travelling less. He might let his job suffer and spend more time with his family. He might neglect his family. Or he might try to be both a good salesman and a good family man at the same time—a nearly impossible task. In his attempt to solve this problem, he probably experienced a great deal of anxiety.

Behavioral scientists sometimes describe an organization as a system of position roles. Each member of the organization belongs to a role set, which is an association of individuals who share interdependent tasks and thus perform formally defined roles, which are further influenced both by the expectations of others in the role set and by one's own personality and expectations. For example, in a common form of classroom organization, students are expected to learn from the instructor by listening to him, following his directions for study, taking exams, and maintaining appropriate standards of conduct. The instructor is expected to bring students high-quality learning materials, give lectures, write and conduct tests, and set a scholarly example. Another in this role set would be the dean of the school, who sets standards, hires and supervises faculty, maintains a service staff, readers and graders, and so on. The system of roles to which an individual belongs extends outside the organization as well, and influences his functioning within it. As an example, a man's roles as husband, father, son, and church member are all intertwined with each other and with his set of organizational roles.

As a consequence, there exist opportunities for role conflict as the various roles interact with one another. In the role theory discussed by Katz and Kahn, Sam Wilson's dilemma is an example of *person-role* conflict.[3] Wilson finds his personal needs and the role requirements to be a good salesman opposed by another set of personal needs and role requirements to be a good husband and father. Other types of role conflict occur when an individual receives inconsistent demands from another person; for example, he is asked to serve on several time-consuming committees at the same time that he is urged to get out more production in his work unit. Another kind of role strain takes place when the individual finds that he is expected to meet the opposing demands of two or more separate members of the organization. Such a case would be that of a worker who finds himself pressured by his boss to improve the quality of his work while his work group wants more production in order to receive a higher bonus share.

These and other varieties of role conflict tend to increase an individual's anxiety and frustration. Sometimes they motivate him to do more and better work. Other times they can lead to frustration and reduced efficiency. The effects of frustration and defensive reactions were discussed in relation to motivation, in Chapter 9.

Counseling

When personal conflict leads to frustration and loss of efficiency, counseling may prove to be a helpful antidote. Although few organizations can afford the luxury of having professional counselors on the staff, given some training, managers may be able to perform this function. Nondirective counseling, or "listening with understanding," is little more than being a good listener—something every manager should be.[4] Sometimes the simple process of being able to vent one's feelings—that is, to express them to a concerned and understanding listener, is enough to relieve frustration and make it possible for the frustrated individual to advance to a problem-solving frame of mind, better able to cope with a personal difficulty that is affecting his work adversely.

Frank Jones, an office supervisor, became so irritated with an employee who had a habit of humming to himself on the job that he posted a notice on the bulletin board banning "whistling, humming, singing, and other forms of wasting time." The employees responded by getting together and agreeing to step up their music making. Surreptitious humming and whistling in the office increased. Jones's irritation also increased. As a result of discussing the situation with his wife, Jones came to realize that his anger at the humming employee was the result of frustration over pressure

[3]Daniel Katz and Robert L. Kahn, *The Social Psychology of Organizations* (New York: John Wiley & Sons, 1966), pp. 171–93.

[4]For a discussion of this, see Henry P. Knowles and Borje O. Saxberg, *Personality and Leadership Behavior* (Reading, Mass.: Addison-Wesley, 1971). See also Chapter 8 in this book.

from his own boss to increase the efficiency of his office. He had made the humming employee a scapegoat for his inability to solve the efficiency problem. He removed the notice from the bulletin board and decided to seek the advice of appropriate staff people. In this case, an understanding wife helped Frank Jones through an episode of frustration that might have created serious conflicts with his employees.

Maier lists the following requisites of nondirective counseling:

An acceptant attitude. *This means conveying warmth and empathy toward the person who is in a frustrated state of mind. Frustration often brings feelings of shame and rejection to the person affected. Perhaps more than anything else, he or she needs to feel that he is accepted as a "true member of society."*

Belief in a person's ability to solve his problem. *This attitude on the part of the listener conveys respect toward the person who is struggling with a problem and helps him take responsibility for finding a solution. If the troubled person can help himself, the solution is likely to be more lasting and perhaps more suitable than if he is told what to do.*

Listening skills. *This is a process which allows a troubled person to "vent his feelings," getting rid of hostile and regressive impulses in an acceptable and useful way. Once these feelings are "sanitized," people are likely to become more rational and better able to cope with the problems which led to frustration.*

Reflecting feelings. *In addition to the* orientation *of an acceptant attitude and belief in the person's ability to solve his problem, counseling is facilitated if the manager-counselor has something concrete to do. This is provided in a technique called "reflecting feelings." Summary statements, particularly of expressed feelings, show that the counselor is actively listening and that he is trying to understand the problem as the other person experiences it. This kind of response and acceptance encourages the frustrated person to explore his problem further.*

Use of questions. *Once the frustrated person has returned to a problem-solving frame of mind, as evidenced by a calmer, less passionate view of the problem, the manager-counselor can help him in rational ways. Exploratory questions may be raised and various avenues of solution considered, although care must still be used to avoid criticism. "The counselor must never be seen as one who takes sides or as one who becomes critical and impatient."*[5]

The nondirective approach is one effective way for managers to deal with frustrated subordinates and co-workers. There are other more direct and more diagnostic ways that might be used

[5]Norman R.F. Maier, *Principles of Human Relations* (New York: John Wiley & Sons, 1952), pp. 418–22. Nondirective counseling is based on the client-centered therapy of Carl Rogers. See Carl R. Rogers, *Client-Centered Therapy* (Boston: Houghton Mifflin, 1951).

in appropriate circumstances. The great strength of the nondirective approach, however, lies in its simplicity, its effectiveness, and the fact that it deliberately avoids the manager-counselor's diagnosing and interpreting emotional problems, which would call for special psychological training. No one has ever been harmed by being listened to sympathetically and understandingly. On the contrary, this approach has helped many people to cope with problems that were interfering with their effectiveness on the job.

CONFLICT WITHIN GROUPS

Conflicts between people in work groups, committees, task forces, and other organizational forms of face-to-face groups are inevitable. As we have mentioned, these conflicts may be destructive as well as constructive. According to Mills, conflict arises in groups because of the scarcity of *freedom, position,* and *resources.*[6] People who value independence tend to resist the need for interdependence and, to some extent, conformity within a group. People who seek power therefore struggle with others for position or status within the group. Rewards and recognition are often perceived as insufficient and improperly distributed, and members are inclined to compete with each other for these prizes.

In our culture, winning is more acceptable than losing, and competition is more prevalent than cooperation, all of which tends to intensify intragroup conflict. Group meetings are often conducted in a win-lose climate—that is, individual or subgroup interaction is conducted for the purpose of determining a winner and a loser rather than for achieving mutual problem solving. The win-lose conflict in groups may have some of the following negative effects:[7]

1. Divert time and energy from the main issues
2. Delay decisions
3. Create deadlocks
4. Drive unaggressive committee members to the sidelines
5. Interfere with listening
6. Obstruct exploration of more alternatives
7. Decrease or destroy sensitivity
8. Cause members to drop out or resign from committees
9. Arouse anger that disrupts a meeting
10. Interfere with empathy
11. Leave losers resentful
12. Incline underdogs to sabotage
13. Provoke personal abuse
14. Cause defensiveness

Conflict in the group need not lead to negative results, however. The presence of a dissenting member or subgroup often results in more penetration of the group's problem and more crea-

[6]Theodore M. Mills, *The Sociology of Small Groups* (Englewood Cliffs, N.J.: Prentice-Hall, 1967), pp. 14–15.

[7]J. William Pfeiffer and John E. Jones, eds., *The 1973 Annual Handbook for Group Facilitators* (Iowa City, Iowa: University Associates, 1973), p. 106.

tive solutions. This is because disagreement forces the members to think harder in an attempt to cope with what may be valid objections to general group opinion. But the group must know how to deal with differences that may arise. As pointed out in our discussion of decision making in Chapter 5, avoiding conflict by depending on a leader to decide, using voting procedures, or succumbing to intimidation by a leader who feels threatened by conflict often deprives the group of the creative force of dissent.

True interdependence among members leads automatically to conflict resolution in the group. Interdependence recognizes that differences will exist and that they can be helpful. Hence, members learn to accept ideas from dissenters (which does not imply agreeing with them), they learn to listen and to value openness, and they learn to share a mutual problem-solving attitude to ensure the exploration of all facets of a problem facing the group.

INTERGROUP CONFLICT

Conflict between groups is a sometimes necessary, sometimes destructive, event that occurs at all levels and across all functions in organizations. Intergroup conflict may help generate creative tensions leading to more effective contributions to the organization's goals, such as competition between sales districts for the highest sales. Intergroup conflict is destructive when it alienates groups that should be working together, when it results in win-lose competition, and when it leads to compromises that represent less-than-optimum outcomes.

Intergroup conflict occurs in two general forms. **Horizontal strain** involves competition between functions: for example, sales versus production, research and development versus engineering, purchasing versus legal, line versus staff, and so on. **Vertical strain** involves competition between hierarchical levels: for example, union versus management, foremen versus middle managment, shop workers versus foremen.[8] A struggle between a group of employees and management is an example of vertical strain or conflict. A clash between a sales department and production over inventory policy would be an example of horizontal strain.

Certain activities and attitudes are typical in groups involved in a win-lose conflict.[9] Each side closes ranks and prepares itself for battle. Members show increased loyalty and support for their own groups. Minor differences between group members tend to be smoothed over, and deviants are dealt with harshly. The level of morale in the groups increases and infuses everyone with competitive spirit. The power structure becomes better defined, as the "real" leaders come to the surface and members rally around the "best" thinkers and talkers.

In addition, each group tends to distort both its own views and those of the competing group. What is perceived as "good" in

[8]Katz and Kahn, *The Social Psychology of Organizations*, pp. 447–48.

[9]Robert R. Blake, Herbert Shepard, and Jane Mouton, *Managing Intergroup Conflict in Industry* (Houston, Texas: Gulf Publishing, 1964), pp. 18–33.

one's own position is emphasized, what is "bad" is ignored; the position of the other group is assessed as uniformly "bad," with little "good" to be acknowledged or accepted. Thus, the judgment and objectivity of both groups are impaired. When such groups meet to "discuss" their differences, constructive, rational behavior is severely inhibited. Each side phrases its questions and answers in a way that strengthens its own position and disparages the other's. Hostility between the two groups increases; mutual understandings are buried in negative stereotypes.

It is easy to see that under the conditions described above, mutual solutions to problems cannot be achieved. As a result, the side having the greater power wins; the other side loses. Or the conflict may go unresolved, and undesirable conditions or circumstances continue. Or the conflict may be settled by a higher authority.

None of these outcomes is a happy one. Disputes settled on the basis of power, such as through a strike or a lockout in a labor-management dispute, are often deeply resented by the loser. Such settlements may be resisted and the winner defeated in underground ways that are difficult to detect and to counter. When this happens, neither side wins; both are losers. If the conflict is left unresolved, as when both sides withdraw from the scene, intergroup cooperation and effectiveness may be seriously impaired to the detriment of the entire organization. Disputes that are settled by higher authority also may cause resentment and what is called "lose-lose" consequences. Such settlements are invariably made on the basis of incomplete information—without data that the conflict itself obscures—and therefore are poor substitutes for mutually reasoned solutions. Again, both sides have lost. A specific approach to resolving intergroup conflict is outlined in the next chapter on organization development.

As we have suggested, conflict affecting organizations can occur in individuals, between individuals, and between groups. Conflict is not always destructive, it may be a motivator. When it is destructive, however, managers need to understand and do something about it. A rational process for dealing with the conflict (such as that shown in Figure 11.1) should be programmed. Such a process should include a planned action response on the part of the manager or the organization, rather than relying on a simple reaction or a change that occurs without specific action by management. In the remainder of this chapter and in the following chapter, we will discuss some aspects of the change process in organizations.

CHANGE

Management is presumed to be guided by a vision of the future. The manager reflects in his decision-making activities the values of the organization as they have developed through time, from the original founder-owner to the present top-management personnel. In navigating a path between the values of the organization and its objectives and goals, management has expectations

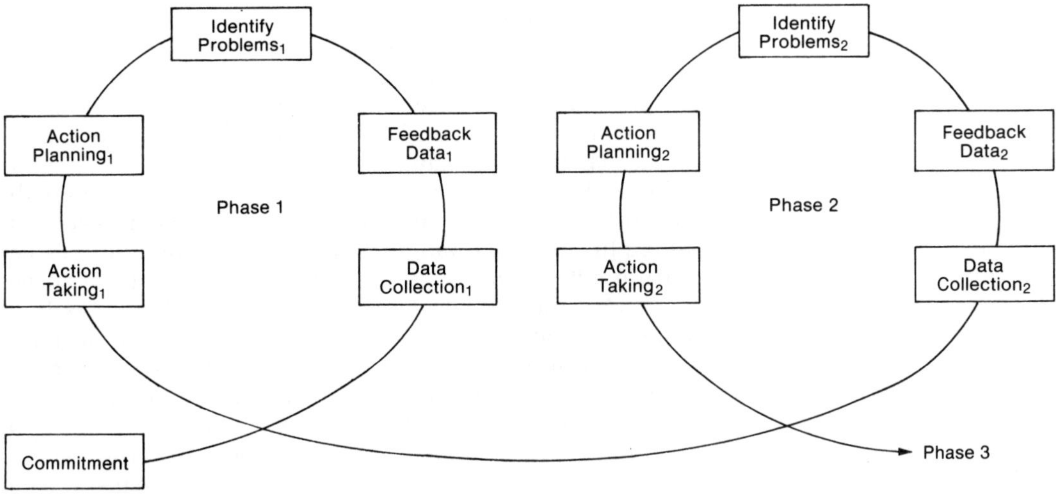

Figure 11.1 A Rational Process for Resolving Organizational Conflicts

concerning the organization's effectiveness and efficiency and frequently initiates changes within the organization. On other occasions, changes in the external environment—market demand, technology, or the political, social, or economic environment—require making appropriate changes in the activities of the organization. The organization faces these demands for change through the men and women who make up its membership, since organizational change ultimately depends on the willingness of employees and others to change their attitudes, behavior, their degree of knowledge and skill, or a combination of these.

Personality Change

In many contexts, the manager appears to be asking an employee to change his personality to conform to new requirements. This was the challenge that faced Freud as he used psychoanalysis in treating his patients, who had varying degrees of neurosis. The relationship between the psychoanalyst and his patient is one in which the individual seeking help is asked over a period of time to reach into his memory, and through free associations to communicate with the psychoanalyst about past incidents and events. The patient is "peeling off" layer after layer of experience stretching over a lifetime, with the support, understanding, and guidance of the psychoanalyst.

As the patient unfolds his life story, he may reach a particular incident or event that may be a root cause of his problem in living a "normal" life. If such an event in his past took place in preverbal infancy or before, he may never be able to recall it. Nevertheless, the psychoanalytic relationship itself, because it involves a high degree of trust and confidence and an opportunity to admit to a sympathetic and understanding listener hidden anxieties and fears, may provide the therapeutic experience that is needed for the patient to decide that he has solved his problem.

Though there are various theories of psychotherapy in addition to traditional Freudian psychoanalysis, the processes of therapy are often quite similar. In any case, changing an individual's personality is rarely possible in the short run. After all, it presumes changing the meaning that an individual has assigned to his life experiences. The Neo-Freudian point of view, however, assumes that an individual is amenable to personality change until he dies, though any change must become increasingly difficult as the individual adds layer upon layer of experience as he moves along life's path. Changing the core of an individual's personality within a "reasonable" time may offer little hope of success.

A study based on the Korean War illustrates fundamental change in individuals over time. The United Nations forces encompassed a number of nationalities: U.S., British, and Turkish, among others. The prisoner-of-war records indicate that British and Turkish troops captured in action by the North Koreans made a number of attempts at escaping from the North Korean prisoner-of-war camps. In addition, these troops appeared to be able to preserve their morale, and few succumbed to the North Korean invitation to turn against their comrades in arms.

The experience of U.S. troops seemed quite different. To begin with, there were few recorded attempts at escape from the prisoner-of-war camps. In addition, U.S. captured troops appeared to run a high rate of illness, and records show many deaths due to illnesses that normally would not have been fatal.

Apparently, the North Korean camp command had embarked on a deliberate program of achieving change in the prisoners of war, trying to turn them against their own comrades and thus make them useful for propaganda purposes. To accomplish this, individual prisoners were isolated from their fellows and thus removed from whatever influence might have been exerted by their officers. They were then subjected to an intensive program of orientation and indoctrination, which cast doubt on the integrity of their fellow prisoners, on the purposes behind U.S. involvement in the war with North Korea, and on the values of American society compared to those of North Korea.

One wonders why the U.S. soldier exposed to this change process did not resist successfully, as did most of the British and Turkish troops. The explanation seems to be that U.S. soldiers resisted through a process of apathy and withdrawal rather than through defiance and aggression, which may explain the high death rate they experienced. They seemed to lack a sufficiently firm value base with which to oppose the North Korean brainwashing process. As the American prisoner of war was deprived of the support of fellow prisoners and officers in charge, he experienced personal anomie (disorientation, or a lack of normative standards of conduct)—consequently he possessed no norms or values with which to fight the new ideas he was being asked to absorb under conditions of duress and crisis. The upshot of these events was the incorporation of a serviceman's creed into the training of U.S. armed forces in the hope that this would provide future soldiers a degree of strength on which to act and behave if taken prisoner.

For a manager to change the personalities of his employees, complexities are clearly involved. The preceding account of North Korea's program of planned change during the Korean War illustrates two of the basic problems involved. First, the process, if it is to succeed, would be intensive, costly, and drawn out. Second, individuals will strongly resist change if they see no need for it. Nevertheless, we continue to hear about managers who want to increase employee motivation to achieve higher productivity, greater commitment to the organization's objectives and goals, or a reduction in absenteeism and tardiness. In the wake of our discussion, we may reasonably raise questions about how successful a manager will be if he expects to change a worker's attitudes by changing his personality.

On the other hand, the manager may be able to specify what results (behavior) he wants. Perhaps the manager need not concern himself with changing employees' personalities as long as he can define and properly reward the employee behavior that he wants. (See the discussion of behavior modification later in this chapter.)

Change in Groups

Kurt Lewin, whose contributions to the analysis of change were introduced in Chapter 2, pioneered in the investigation of group behavior. He first became involved in this pursuit when he used the tools of behavioral science to study the problem of changing food habits in the United States during World War II. The government enlisted his help to get people to eat other parts of slaughtered animals than those normally regarded as edible. Lewin faced the challenge of convincing homemakers that they should include such items as heart, tongue, and liver in the family menu.

In addition to conducting an exhaustive investigation into the channels of distribution and the problems of introducing the new items to the housewife at the meat counter, Lewin conducted experiments designed to discover the most effective way to get people to change their meal planning. He compared the results of having homemakers attend lectures and watch demonstrations of cooking with the new ingredients against other ways of providing information about the desirability of menu changes. In addition, he asked housewives to attend discussion groups that used a consensus process in deciding to try out the new recipes. Group discussion with consensus decision making turned out to be the most effective way to bring about an actual change in behavior.

Group Psychotherapy

In other work with groups, a British military hospital discovered almost by accident the applicability of group psychotherapy to soldiers experiencing symptoms of shell shock during World War II. Traditional psychoanalysis had always assumed that therapy required private interaction between a psychoanalyst and a patient. However, due to expense and limited availability of psychoanalysts and psychiatrists, individual therapy was not possible. The important discovery was that a group of patients under the surveillance and guidance of a trained person were able to obtain the benefits of therapy simply by exploring *with each other* their war experiences.

This led to the introduction of group psychoanalysis and group psychotherapy as an approach in dealing with mental illness, when physical impairment of the brain was not involved. Mental hospitals generally have accepted the lessons learned from the original therapy groups (or "T-groups") in the British hospitals. Discovering the effectiveness of group therapy helped redefine all forms of neurosis as loss of control of oneself resulting from a lack of significant communication in the form of feedback from others. We can see the practice of this method today in such lay therapy groups as Alcoholics Anonymous.

The application of group psychoanalysis to industrial work groups was explored extensively by Elliott Jacques, who became a consultant to the Glacier Metal Company from the Tavistock group of behavioral scientists in Great Britain. Jacques defined his role as "facilitator of change." To carry out this role, he legitimized his entry into the company by visiting and gaining personal acceptance by top management; by the work council representing middle management, the staff, and the workers; and by the trade union. He advocated the use of group discussions to explore problems in the work environment. These sociotherapeutic groups were composed of individuals who were involved in the situation and who were thereby able to learn each others' viewpoints as well as explore relevant alternatives to solve the problem. Jacques saw his role as one of drawing the group's attention to questions that needed to be raised, but left it to the group to decide on action to be taken. He labelled this approach "social analysis," because it covered the whole organization, vertically and horizontally.

The Tavistock group of behavioral scientists, who were broadly involved in psychiatric and managerial consulting, introduced a further dimension to the sociotherapeutic approach with an investigation of changes in coal mining in Great Britain.[10] Traditionally, coal mining had involved an integrated team of pairs of miners who performed all jobs necessary for getting the coal out. Thus, on a three-shift operation there would be a total of three pairs, or six men working the same coal face. The men chose their own team members. In addition, pay was computed on the basis of total output by the team of six men.

The introduction of new machinery advanced the method of long-wall coal mining. Now the total face of the coal seam covered an area of up to 600 feet. To reach this, teams of men were formed to remove the underside of the coal seam so that coal could be blown down from the face a certain number of feet. Other teams were formed to set up new supports along the coal face, while yet another team moved the conveyor belt up to the new coal face. The work was now divided so that each shift did only one of the above

[10]Eric Trist, *Organizational Choice* (London: Tavistock Publications, 1963).

operations in getting the coal out. Pay was computed for each shift separately. As a result, management experienced unexpected problems in maintaining production. Absenteeism was high among the miners; long and involved discussions were continually going on about any shift's particular pay basis; there was little cooperation among the shifts or among members of a team.

Consultants from the Tavistock group suggested a new organizational design that would retain the advantages of the traditional six-man team. They suggested that each shift be viewed as a team responsible for all operations, including some degree of job rotation to provide the miners with multiple skills. This scheme retained the advantage of long-wall mining in addition to applying the benefits of modern technology. Also, and very importantly, it restored the social organization that had been disrupted when the new machinery had been installed, an organization highly valued by the men.

This sociotechnical approach viewed the change as one involving consideration for human relations, for technology, and for the organizational structure. From this discussion, we again assert that the manager must be aware at all times of the multilayered environment within which he makes decisions. Any change must be seen in relation to the physical-technical, job-task, and human environments—that is, as a total system.

A Behaviorist's View

A great deal of effort is being devoted to change processes sponsored by adherents of the various approaches to group dynamics. At the same time, a current version of the familiar stimulus-response approach to change is receiving some renewed attention. In a series of experiments, B. F. Skinner has shown that pigeons, provided a system of rewards, were able to learn increasingly complex pecking patterns. From these experiments, Skinner concluded that the key to changing performance in desired directions is by using a system of positive rewards. Using positive reinforcement, it is possible to ensure a repetition of successful performance.

Skinner's approach, called "behavior modification," has been used in mental hospitals, institutions for the mentally retarded, and with juvenile delinquents. His method is to establish a "token" economy, in which the patients or students are asked to do certain things, learn certain lessons, and perform certain activities, for which they are given appropriate rewards in the form of tokens symbolizing some desired object or event. For a prison inmate, this may involve points toward home furlough, for a student it may involve exchange that can be used at the nearby store, and so on.

A manager can easily see that he uses a considerable range of similar rewards in his own organization. Pay itself, the possibility of an extra day off for work well done, or tickets to a critical football game are examples of positive reinforcement. Skinner believes that such rewards should follow closely on successful performance, and that the effect is related only to *positive* rewards.

Negative or punitive rewards do not achieve the same results. The Skinnerian approach suggests that change in behavior is a function of reward rather than a function of change in personality and attitude.

Resistance to Change

Managers are agents of change, in addition to having supervisory responsibilities. The resistance of individuals and groups to changes imposed by what they consider an external authority is a common phenomenon. Recently, for example, the Delta Company decided to overhaul its employee classification system and adjust its pay scales to reflect more closely conditions prevailing in local industry. A team of outside experts was hired to do the job. After several months' study, the experts drew up a plan and presented it to the company's board of directors. The board decided to implement the new system immediately. All employees who would be affected were notified through their supervisors that they would be switched over to the new plan at the beginning of the next month. Through their representatives, the employees concerned immediately demanded that the new system be substantially altered to correct a number of inequities. When the company resisted, the employees staged a slowdown and threatened a strike if their demands were not met. The company (in this case) backed down, and there followed a long and costly process of reviewing and revising the original plan to one that was finally acceptable to both sides.

Why do people resist change? A favorite trick of a sociology professor was to introduce the subject of social change by asking students sitting in the left half of the classroom to scramble seats with those sitting in the right half. This simple maneuver can be a dramatic demonstration of the feelings of uneasiness that accompany changes imposed on established ways of doing things. Students would ask, What is he up to? Is this move preliminary to a pop quiz? Is he trying to break up some of the classroom cliques, which had served a variety of purposes for the students? What diabolical objective does he have in mind? When the professor finally asked how the students felt about changing seats, the students' responses provided an interesting basis for a long and fruitful discussion over what change might mean to people.

Let's use the Delta Company's case to explore some of the reasons why people resist change. Change seems to be resisted:

1. *If it is imposed from the outside.* In the Delta Company, a team of "outside experts" designed the new classification plan. It appears possible that the management group as a whole was also perceived as "outsiders" by the employee group. We tend to view outsiders with suspicion. We are not sure they understand us or our problems, or that they have our interests at heart. We may even suspect them of hidden motives. For these reasons we tend to reject their attempts to influence or change our group.

2. *If it threatens or seems to threaten traditional norms and values.* We do not have the details of Delta's proposed classification plan, but if it had the effect of modifying or seeming to modify established ideas of the way things ought to be done or of what people believed was fair or proper, it would be resisted. People cling to old ways because old ways of doing things are comfortable and seem to support security and order. People are apt to respond to new ways with, "This is the way we always did it. Why should we change something that works?" Their sense of fairness and rightness may be involved. "The old way is just. Your way is unjust." New ways may also seem to imply criticism of the old—hence, criticism of a person's judgment or ability.
3. *If it threatens or seems to threaten group relationships.* In the Delta case, if some job titles were changed or if the skills required were altered—and again, we must emphasize that such changes need only be *imagined*—those involved may feel a loss of rank in the group. Such changes affect self-esteem, power relationships, and influence, and may have adverse effects on job performance and other behavior. They may also cause dissension in work groups, reducing group cohesiveness and members' satisfaction. Even one's relations outside the group—for example, with a person's family and friends—may fall under the influence of the change.
4. *If it threatens or seems to threaten personal economic outlook.* It is possible that the Delta Company's new classification system resulted in changes in pay and in the mix of jobs, adding some jobs while eliminating others. These changes would create obvious problems for those affected.
5. *If it threatens or seems to threaten changes in skill requirements.* Some people will be afraid that they cannot acquire the new skills needed to meet classification requirements.

Throughout the above discussion, we have emphasized that the effects of various changes on individuals and groups need only be imagined to create resistance. In other words, people seem just as concerned about the uncertainties accompanying proposed change as they are about the actual change. McMurry ascribes this fear to the neurotic anxieties that nearly everyone acquires as a result of his life experience. Pointing to the fact that many attacks on a proposed change come from people who have no real reason to resist it, McMurry suggests that such reactions may be generated by vague fears and anxieties that lie just beneath the surface of the personality. Speaking of these fears in the individual, he says:

When any new challenge to his status or security occurs, it accentuates his existing anxieties and feelings of insecurity. These added fears almost inevitably upset his emotional balance. His latent fears, having been reinforced, once more threaten to become pain-

fully conscious. This must be avoided at any cost. Hence, he has powerful incentive to rid himself of the source of danger to his status and security.[11]

Managers, consequently, are not always dealing with objective reactions when people resist change. They may be dealing with supercharged emotions that (as we have suggested in the section on frustration and nondirective counseling) need to be expressed and worked out before those involved can get down to rational problem solving. Of course, these feelings are more intense if change is proposed in an already hostile and generally noncommunicative atmosphere.

Much of this emotional reaction may be eliminated or held within reasonable bounds if, when introducing change, managers understand the nature and causes of resistance and plan their approach to change accordingly. If at all possible, subordinates should be clued in as partners along every step of the way. For one thing, this approach helps to avoid the "outsider" problem. When people are informed in advance and are helped to understand what is taking place, many unfounded fears simply do not materialize. Think of what might have happened in the classroom if the sociology professor had explained in advance what he was up to! All of that speculation and apprehension would have vanished. So, of course, would his demonstration, but the point should be clear.

In a classic experiment, Coch and French demonstrated the effectiveness of employee participation in change decisions.[12] They introduced a change in work procedures in a clothing factory in two ways. In one group, the operators were told about the change by staff people and were given opportunities to ask questions. This was designated the "no-participation" method. In a second group, the operators met with the staff people, who explained the need for the change. The operators then discussed with the staff what improvements might be made. This was designated the "total-participation" method. (A third, intermediate method was used, but did not add anything to the final results.) The "total-participation" method proved to be significantly better than the "no-participation" method. In the latter, symptoms of resistance developed immediately, taking the form of hostility toward supervision, job turnover, and low production. In the "total-participation" group there was no hostility, no quits during the test period, and production exceeded the previous rate.[13]

[11]Robert N. McMurry, "The Problem of Resistance to Change in Industry," *Journal of Applied Psychology* 31 (December 1947), pp. 589–93.

[12]Lester Coch and J.R.P. French, "Overcoming Resistance to Change," *Human Relations* 1 (1948).

[13]Subsequent discussions have raised questions about the interpretation of the results and the actual events surrounding this experiment. A replication of the experiment in a Norwegian setting indicated different results. This was interpreted as a reflection of different expectations in relation to customary managerial behavior—authoritarian in Norway in contrast to more democratic management in the United States. See our discussion of contingency leadership theory in Chapter 10.

Lawrence, in commenting on the Coch and French experiment, cautioned that participation was not something that could just be turned on and off. Rather, he said, "Participation is a feeling on the part of people, not just the mechanical act of being called in to take part in discussions."[14] Lawrence also argued that the Coch and French experiment demonstrates that people resist *social* change rather than *technical* change; that is, the alterations in status and relationships that accompany changes in work methods and technology are what people react against.

Everyone, manager and worker alike, responds to the threat of change with sometimes active and sometimes passive, or covert, resistance. The reasons for this appear to be mainly psychological and social rather than technical. Man likes to continue to do things in ways that have seemed to him to be satisfying and successful in the past. He finds fulfillment of strong social and ego needs as an accepted member of a well-defined group. Change often seems to attack these arrangements. The individual may also have a submerged layer of neurotic insecurity that has no apparent basis in reality. Change often threatens to bring aspects of this suppressed experience to the surface, and the individual's natural reaction is to resist this for reasons not immediately clear to him. Managers can learn to cope with these problems of change, not through manipulation or "pseudoparticipation" but through building a climate of mutual respect and trust with their employees.

SUMMARY

Organizational conflict has been discussed in this chapter under two broad classifications, *personal* and *social*. Although these are often difficult to separate, in practice, personal conflicts are struggles within an individual between opposing values, attitudes, beliefs, motives, and roles; social conflict refers to conflicts between two or more persons or groups.

A basic incompatibility often exists between the needs for conformity and control in organizations and individual needs for autonomy and expression. Managers should be aware of this basic dilemma in order to anticipate and avoid human problems or to respond constructively and responsibly to their effects. One very effective way to stay in touch with employees is through active listening, or "listening with understanding." A common symptom of conflict, personal or social, is frustration resulting from the obstruction of needs, as discussed in Chapter 9. In some cases, counseling is helpful, and nondirective counseling is an especially useful and feasible approach for managers to practice.

Conflicts within work groups are often caused by struggles over control, status, and scarce resources. Conflicts between groups in organizations have similar origins. The constructive resolution of such conflicts can most often be achieved through a rational process of problem solving, coupled with a willingness to explore issues and alternatives and to *listen* to each other.

[14]Paul Lawrence, "How to Deal with Resistance to Change," *Harvard Business Review* (May–June 1954), pp. 49–57.

Conflict often implies a change in the status quo. Such changes ultimately depend on the capacity of people to alter their attitudes, behavior, knowledge, skill, or, most often, a mix of these. Changing personalities is an uncertain, costly, and difficult process, that requires special expertise. Hence, behavior change or modification, in which change is defined in performance terms and individuals learn that desired performance brings positive benefits, may be a more suitable approach for managers to use.

People often resist new modes of behavior, not because they cannot learn them but because they run counter to their values, result in social consequences that are hard to accept, or seem to threaten them with personal economic loss. One way to avoid these sometimes unanticipated and adverse reactions is to get people involved in *planning* the changes that will affect them.

DISCUSSION QUESTIONS

11.1. Discuss the different forms of conflict in organizations.
11.2. What are the causes of intragroup conflict? Of intergroup conflict?
11.3. Argyris suggests that formal organizations and the human personality are basically incompatible. Discuss. How can this incompatibility be reduced?
11.4. What is meant by nondirective counseling? How is it helpful to managers in dealing with conflict?
11.5. Discuss win-lose competition in groups. What are some of the negative results of such competition?
11.6. What is involved in personality change? How does this relate to the problem of management?
11.7. Why do people resist change? What remedies have been suggested?

EXERCISE

11.1. Think of your school as an organization. What forms of conflict can you identify among its members (administration, faculty, staff, and students)?
   a. Select an area of conflict, identify its probable causes, and draw up a plan for changes that you feel are needed to reduce the conflict.
   b. How would you proceed to implement your plan?

APPENDIX
Memo from the General Manager

Cal Johnson, general manager of the assembly department of TMC's refrigeration division, was perplexed. For several months, a number of the men working on mechanical assembly had been grumbling about their jobs. Mostly they mentioned things like the monotony of the assembly work, long hours at the conveyor, low pay, and noisy working conditions.

The comments of two of the older and more experienced employees were typical. Sam Campanella told Cal Johnson that he was ready to ask for a transfer if something wasn't done about his job. "A guy can go crazy just screwing on doors and fitting gaskets all day with no variety!" Harry Kranz threatened to quit outright at the end of the month if things didn't get better. "Why can't we be given more different things to do? Any Lincoln Park monkey could do what I do! Why can't you give us bigger jobs like putting together a whole compressor assembly or even a whole refrigerator? We'd be a lot happier and turn out more and better units than we're doing now!"

As Cal Johnson reflected on these reactions from his workers, he opened a memorandum he had just received from the vice-president:

March 12, 1976

To: Production Managers and General Foremen

From: John Garrison

Subj: Work Specialization Program

Our industrial engineers tell me that we can both cut costs and speed up production through further specialization of assembly-line jobs. I have scheduled a meeting on this subject Thursday, March 29, for all concerned. Jim Carpenter of Industrial Engineering will be there to present his views.

John Garrison

1. If you were Johnson, would you discuss the situation with your production manager and/or with Garrison before the meeting? If so, what would you plan to say? Would you explore the problem with your assemblers?
2. What would you plan to say at the 29 March meeting? Would you be willing to compromise with Industrial Engineering?
3. Suggest some practical approaches to this potentially serious conflict between the needs of the company and those of the assemblers.

SELECTED READINGS

Argyris, Chris. *Personality and Organization.* New York: Harper & Row, 1957.

Bennis, Warren, Benne, Kenneth D. and Chin, Robert. *The Planning of Change*, 2nd ed. New York: Holt, Rinehart and Winston, 1970.

Blake, Robert R., Shepard, Herbert A., and Mouton, Jane S. *Managing Intergroup Conflict in Industry*. Houston, Texas: Gulf Publishing, 1964.

Coser, L. *The Functions of Social Conflict.* New York: The Free Press, 1964.

Sherif, M., and Sherif, C. W. *Social Psychology*. New York: Harper & Row, 1969.

LEADERSHIP &
INVENTORY
CASH FLOW
SALES PRODUCTION
CONFLICT

Chapter 12 Organization Development and Renewal

Suppose a manager believes he is having "people problems" in his plant. Things just aren't going well. Efficiency is slipping and there have been a lot of arguments, grievances, and generally poor effort and low morale on the shop floors. The manager has been reading some books on leadership and has even attended a leadership workshop at a nearby university. These have been useful but have not really provided an answer. The only thing he knows for sure is that he has a problem and needs help. Someone suggests that he contact an OD consultant. What, he wonders, is OD? In this chapter we will describe what **OD**, or **organization development**, is and how it works.

Managers and theorists continue to seek improved understanding and utilization of the human aspects of organization. Scientific management, industrial psychology, and the human relations movement have made important contributions. More recently, the systems approach combined with advances in the behavioral sciences have brought new knowledge and a new perspective to the study and practice of management. Organization development, or OD, is the latest process in this long line of cumulative knowledge and practice. A procedure for improving organizational performance, OD draws on these forerunners for theory and intervention techniques; but it has some significant differences from these other approaches and has added a theory and style of its own.

EARLY DEVELOPMENT

Kurt Lewin, whom we have discussed previously, played a key role in the evolution of organization development as it is known today. As early as World War II, Lewin experimented with a collaborative change process (involving himself as consultant and a client group) based on a three-step process of planning, taking action, and measuring results. This was the forerunner of action research, an important element of OD, which will be disucssed later. Lewin then participated in the beginnings of laboratory training, or T-groups, and, after his death in 1947, his close associates helped to develop survey-research methods at the University of Michigan. These procedures became important parts of OD as developments in this field continued at the National Training Laboratories and in growing numbers of universities and private consulting firms across the country.

The failure of off-site laboratory training to live up to its early promise was one of the important forces stimulating the development of OD. Laboratory training is learning from a person's "here and now" experience as a member of an ongoing training group. Such groups usually meet without a specific agenda. Their purpose is for the members to learn about themselves from their spontaneous "here and now" responses to an ambiguous hypothetical situation. Problems of leadership, structure, status, communication, and self-serving behavior typically arise in such a group. The members have an opportunity to learn something about themselves and to practice such skills as listening, observing others, and functioning as effective group members.

As formerly practiced (and occasionally still practiced for special purposes), laboratory training was conducted in "stranger groups," or groups composed of individuals from different organizations, situations, and backgrounds. A major difficulty developed, however, in *transferring* knowledge gained from these "stranger labs" to the actual situation "back home." This required a transfer between two different cultures, the relatively safe and protected environment of the T-group (or training group) and the give-and-take of the organizational environment with its traditional values. This led the early pioneers in this type of learning to begin to apply it to "family groups"—that is, groups located *within* an organization. From this shift in the locale of the training site and the realization that culture was an important factor in influencing group members (along with some other developments in the behavioral sciences) emerged the concept of organization development.

A CASE HISTORY

The Cambridge Clinic found itself having difficulty with its internal working relationships. The medical director, concerned with the effect these problems could have on patient care, contacted an organizational consultant at a local university and asked him for help. A preliminary discussion among the director, the clinic administrator, and the consultant seemed to point to problems in leadership, conflict resolution, and decision processes. The consultant suggested that data be gathered so that a working diagnosis could be made. The clinic officials agreed, and tentative working arrangements were concluded.

The consultant held a series of interviews involving all members of the clinic staff, the medical director, and the administrator. Then the consultant "thematized," or summarized, the interview data to identify specific problem areas. At the beginning of a workshop about a week later, the consultant fed back to the clinic staff the data he had collected. The staff arranged the problems in the following priorities.

1. Role conflicts between certain members of the medical staff were creating tensions that interfered with the necessity for cooperation in handling patients.
2. The leadership style of the medical director resulted in his put-

ting off decisions on important operating matters. This led to confusion and sometimes to inaction on the part of the medical and administrative staffs.

3. Communication between the administrative, medical, and outreach (social worker) staffs on mutual problems tended to be avoided. Open conflicts over policies and procedures were thus held in check, but suppressed feelings clearly had a negative influence on interpersonal and intergroup behavior.

Through the use of **role analysis** (discussed later in the chapter) and other techniques suggested by the consultant, the clinic staff and the medical director were able to explore the role conflict and leadership problems and to devise effective ways of coping with them. Exercises designed to improve communication skills and a workshop session on dealing with conflict led to progress in developing more openness and trust throughout the clinic.

An important result of this first workshop was the creation of an action plan that set forth specific steps to be applied to clinic problems by clinic personnel during the ensuing period. The consultant agreed to monitor these efforts and to assist in any way he could. Additional discussions and team development sessions were held with the director and the medical and administrative staffs.

A second workshop attended by the entire clinic staff took place about two months after the first. At the second workshop, the clinic staff continued to work together on the problems of dealing with conflict and interpersonal communication. During the last half-day of the meeting, the staff developed a revised action plan covering improvement activities to be undertaken in the following weeks and months to improve the working relationships of the clinic.

A notable additional benefit of this OD program was that the clinic staff learned new ways of monitoring the clinic's performance as an organization and of coping with some of its other problems. Six months later, when the consultant did a follow-up check on the organization, the staff confirmed that interpersonal problems were now under better control and that some of the techniques learned at the two workshops associated with the OD programs were still being used.

WHAT IS OD?

Organization development is a *contractual relationship*[1] between a *change agent* and a *sponsoring organization* entered into for the purpose of *using applied behavioral science* in a *systems context* to *improve organizational performance* and the *capacity of the organization to improve itself*. Let us examine the separate terms of this definition in some detail. (The reader may wish to

[1]As will be explained, the change agent may be a staff or line member of the organization who is schooled in OD theory and technique. In such a case, the "contractual relationship" is an in-house agreement that should probably be explicit with respect to all of the conditions involved except the fee.

refer back to the events in the case of the Cambridge Clinic described above.)

Contractual Relationship. Although neither the sponsoring organization nor the change agent can be sure at the outset of the exact nature of the problem or problems to be dealt with or how long the change agent's help will be needed, it is essential that some tentative agreement on these matters be reached. The sponsoring organization needs to know generally what the change agent's preliminary plan is, what its own commitments are in relation to personal commitments and responsibility for the program, and what the change agent's fee will be. The change agent must assure himself that the organization's, and particularly the top executives', commitment to change is strong enough to support the kind of self-analysis and personal involvement requisite to success of the program. Recognizing the uncertainties lying ahead on both sides, a termination agreement permitting either side to withdraw at any time is usually included.

Change Agent. A change agent in the sense used here is not a technical expert skilled in such functional areas as accounting, production, or finance. He is a behavioral scientist who knows how to get people in an organization involved in solving their own problems. His main strength is a comprehensive knowledge of human behavior, supported by a number of intervention techniques (to be discussed later). The change agent can be either external or internal to the organization. An internal change agent is usually a staff person who has expertise in the behavioral sciences and in the intervention technology of OD. Beckhard reports several cases in which line people have been trained in OD and have returned to their organizations to engage in successful change assignments.[2] In the natural evolution of change mechanisms in organizations, this would seem to approach the ideal arrangement. Qualified change agents can be found on some university faculties, or they may be private consultants associated with such organizations as the National Training Laboratories Institute for Applied Behavioral Science (Washington, D.C.) or University Associates (San Diego, California), and similar organizations.

Sponsoring Organization. The initiative for OD programs comes from an organization that has a problem. This means that top management or someone authorized by top management is aware that a problem exists and has decided to seek help in solving it. There is a direct analogy here to the practice of psychotherapy: The client or patient must actively seek help in finding a solution to his problems. This indicates a willingness on the part of the client organization to accept help and assures the organization that management is actively concerned.

[2]Richard Beckhard, *Organization Development: Strategies and Models* (Reading, Mass.: Addison-Wesley, 1969), p. 114.

Applied Behavioral Science. One of the outstanding characteristics of OD that distinguishes it from most other improvement programs is that it is based on a "helping relationship." The change agent is not a physician to the organization's ills; he does not examine the "patient," make a diagnosis, and write a prescription. Nor does he try to teach organizational members a new inventory of knowledge which they then transfer to the job situation. Using theory and methods drawn from such behavioral sciences as psychology, sociology, cultural anthropology, administrative theory, organizational behavior, economics, and political science, the change agent's main function is to help the organization define and solve its own problems. The basic method used is known as **action research.** This approach, which is described in detail later, consists of a preliminary diagnosis, collecting data, feedback of the data to the client, data exploration by the client group, action planning based on the data, and taking action.[3]

Systems Context. OD deals with a total system—the organization as a whole, including its relevant environment—or with a subsystem or systems—departments or work groups—in the context of the total system. Parts of systems, for example, individuals, cliques, structures, norms, values, and products are not considered in isolation; the principle of interdependency, that is, that change in one part of a system affects the other parts, is fully recognized. Thus, OD interventions focus on the total culture and cultural processes of organizations. The focus is also on groups, since the relevant behavior of individuals in organizations and groups is generally a product of group influences rather than personality.

Improved Organizational Performance. The objective of OD is to improve the organization's capacity to handle its internal and external functioning and relationships. This would include such things as improved interpersonal and group processes, more effective communication, enhanced ability to cope with organizational problems of all kinds, more effective decision processes, more appropriate leadership style, improved skill in dealing with destructive conflict, and higher levels of trust and cooperation among organizational members. These objectives stem from a value system based on an optimistic view of the nature of man—that man in a supportive environment is capable of achieving higher levels of development and accomplishment. Also essential to organization development and effectiveness is the scientific method—inquiry, a rigorous search for causes, experimental testing of hypotheses, and review of results. Finally, the democratic process is viewed as having a legitimate, and perhaps dominant, role in the highly effective organization.

[3]Wendell French and Cecil Bell, Jr., *Organization Development* (Englewood Cliffs, N.J.: Prentice-Hall, 1973), p. 18.

Organizational Self-Renewal. The ultimate aim of the outside OD practitioner is to "work himself out of a job" by leaving the client organization with a set of tools, behaviors, attitudes, and an action plan with which to *monitor* its own state of health and to take corrective steps toward its own renewal and development. This is consistent with the systems concept of feedback as a regulatory and corrective mechanism.

ACTION RESEARCH

French and Bell define organization development at one point as "organization improvement through action research."[4] If one idea can be said to summarize OD's underlying philosophy, it would be action research as it was conceptualized by Kurt Lewin and later elaborated and expanded on by other behavioral scientists. Concerned with social change and, more particularly, with effective, *permanent* social change, Lewin believed that the motivation to change was strongly related to action: If people are active in decisions affecting them, they are more likely to adopt new ways. "Rational social management," he said, "proceeds in a spiral of steps, each of which is composed of a circle of planning, action, and fact-finding about the result of action."[5] As we have discussed before, Lewin's description of the process of change involves three steps:[6]

Unfreezing: Faced with a dilemma or disconfirmation, the individual or group becomes aware of a need to change.
Changing: The situation is diagnosed and new models of behavior are explored and tested.
Refreezing: Application of new behavior is evaluated, and if reinforcing, adopted.

Figure 12.1 summarizes the steps and processes involved in planned change through action research. Action research is depicted as a cyclical process of change. The cycle begins with a series of planning actions initiated by the client and the change agent working together. The principal elements of this stage include a preliminary diagnosis, data gathering, feedback of results, and joint action planning. In the language of systems theory, this is the input phase, in which the client system becomes aware of problems as yet unidentified, realizes it may need outside help to effect changes, and shares with the consultant the process of problem diagnosis. In the Cambridge Clinic case, this would cover the period from the time difficulties were perceived by the medical director and his associates to the beginning phases of the first workshop. It would include also seeking outside aid, the interviews, the assembly, feedback, and ranking of the interview data according to priorities,

[4]Ibid., p. 18.

[5]Kurt Lewin, "Group Decision and Social Change," in E.E. Maccoby, T.N. Newcomb, and E.L. Hartley, eds., *Readings in Social Psychology* (New York: Holt, Rinehart and Winston, 1958), p. 201.

[6]Ibid., pp. 210–11.

Figure 12.1 Systems Model of Action-Research Process

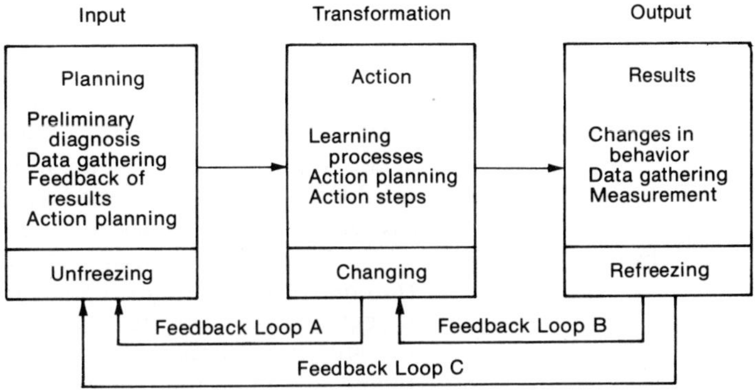

tentative problem diagnosis, and the joint decision to employ role analysis and other interventions as part of the action phase.

The second stage of action research is the action, or transformation, phase. This stage includes actions relating to learning processes (perhaps in the form of role analysis) and to planning and executing behavioral changes in the client organization. As shown in Figure 12.1, feedback at this stage would move via Feedback Loop A and would have the effect of altering previous planning to bring the learning activities of the client system into better alignment with change objectives. Included in this stage is action-planning activity carried out jointly by the consultant and members of the client system. Following the workshop or learning sessions, these action steps are carried out *on the job* as part of the transformation stage. In the Cambridge Clinic case, this stage would cover the learning activities at the first workshop and action steps tried out during the period between the first and second workshops.

The third stage of action research is the output, or results, phase. This stage includes actual changes in behavior (if any) resulting from corrective action steps taken following the second stage. Data are again gathered from the client system so that progress can be determined and necessary adjustments in learning activities can be made. Minor adjustments of this nature can be made in learning activities via Feedback Loop B (see Figure 12.1). Major adjustments and reevaluations would return the OD project to the first, or planning, stage for basic changes in the program. In the Cambridge Clinic example, only minor adjustments were required prior to the second workshop. In checking back with the clinic six months after the second workshop, the data showed a need for further discussion and planning between the client organization and the change agent, an example of the operation of Feedback Loop C. These discussions are now taking place.

The action-research model shown in Figure 12.1 closely follows Lewin's repetitive cycle of planning, action, and measuring results. It also illustrates other aspects of Lewin's general model of change. As indicated in the diagram, the planning stage is a period of *unfreezing*, or problem awareness. The action stage is a period of *changing*, that is, trying out new forms of behavior in an effort to understand and cope with the system's problems. (There is inevitable overlap between the stages, since the boundaries are not clear-cut and cannot be in a continuous process.) The results stage is a period of *refreezing*, in which new behaviors are tried out on the job and, if successful and reinforcing, become a part of the system's repertoire of problem-solving behavior.

Action research is problem centered, client centered, and action oriented. It involves the client system in a diagnostic, active-learning, problem-finding, and problem-solving process. Data are not simply returned in the form of a written report but instead are fed back in open joint sessions, and the client and the change agent collaborate in identifying and ranking specific problems, in devising methods for finding their real causes, and in developing plans for coping with them realistically and practically. Scientific method in the form of data gathering, forming hypotheses, testing hypotheses, and measuring results, although not pursued as rigorously as in the laboratory, is nevertheless an integral part of the process. Action research also sets in motion a long-range, cyclical, self-correcting mechanism for maintaining and enhancing the effectiveness of the client's system by leaving the system with practical and useful tools for self-analysis and self-renewal.

OD INTERVENTIONS

This section describes some of the principal learning processes or interventions in the "action" stage (see Figure 12.1) of organization development. **Interventions** are structured activities used individually or in combination by the members of a client system to improve their social or task performance. They may be introduced by a change agent as part of an improvement program, or they may be used by the client following a program to check on the state of the organization's health, or to effect necessary changes in its own behavior. "Structured activities" mean such diverse procedures as experiential exercises, questionnaires, attitude surveys, interviews, relevant group discussions, and even lunchtime meetings between the change agent and a member of the client organization. Every action that influences an organization's improvement program in a change agent–client system relationship can be said to be an intervention.

There are many possible intervention strategies from which to choose. Several assumptions about the nature and functioning of organizations are made in the choice of a particular strategy. Beckhard lists six such assumptions.[7]

1. The basic building blocks of an organization are groups (teams). Therefore, the basic units of change are groups, not individuals.
2. An always relevant change goal is the reduction of inappropriate competition between parts of the organization and the development of a more collaborative condition.

[7]Beckhard, *Organization Development*, pp. 26–27.

3. Decision making in a healthy organization is located where the information sources are, rather than in a particular role or level of hierarchy.
4. Organizations, subunits of organizations, and individuals continuously manage their affairs against goals. Controls are interim measurements, not the basis of managerial strategy.
5. One goal of a healthy organization is to develop generally open communication, mutual trust, and confidence between and across levels.
6. People support what they help create. People affected by a change must be allowed active participation and a sense of ownership in the planning and conduct of the change.

Interventions range from those designed to improve the effectiveness of individuals through those designed to deal with teams and groups, intergroup relations, and the total organization. There are interventions that focus on task issues (what people do), and those that focus on process issues (how people go about doing it). Finally, interventions may be roughly classified according to which change mechanism they tend to emphasize: for example, feedback, awareness of changing cultural norms, interaction and communication, conflict, and education through either new knowledge or skill practice.[8]

One of the most difficult tasks confronting the change agent is to help create in the client system a safe climate for learning and change. In a favorable climate, human learning builds on itself and continues indefinitely during man's lifetime. Out of new behavior, new dilemmas and problems emerge as the spiral continues upward to new levels. In an unfavorable climate, in contrast, learning is far less certain, and in an atmosphere of psychological threat, it often stops altogether. Unfreezing old ways can be inhibited in organizations because the climate makes employees feel that it is inappropriate to reveal true feelings, even though such revelations could be constructive. In an inhibited atmosphere, therefore, necessary feedback is not available. Also, trying out new ways may be viewed as risky because it violates established norms. Such an organization may also be constrained because of the law of systems: If one part changes, other parts will become involved. Hence, it is easier to maintain the status quo. Hierarchical authority, specialization, span of control, and other characteristics of formal systems also discourage experimentation. The change agent must address himself to all of these hazards and obstacles. Some of the things which will help *him* are:

1. A real need in the client system to change
2. Genuine support from management
3. Setting a personal example: listening, supporting behavior
4. A sound background in the behavioral sciences
5. A working knowledge of systems theory

[8]French and Bell, *Organization Development*, ch. 8.

6. A belief in man as a rational, self-educating being fully capable of learning better ways to do things

The interventions to be described in the following pages are simply examples of many that are used. Those chosen are survey feedback, role analysis technique, and an intergroup building exercise.

Survey Feedback

Survey feedback is an intervention that is often used when involving the entire client system in a massive improvement program is considered desirable. In one application described by Floyd Mann of Michigan's Survey Research Center, a company-wide survey of employee and managerial attitudes and opinions was used.[9] The survey covered (1) attitudes and perceptions of nonsupervisory employees toward their work, promotion opportunities, supervision, and fellow workers, (2) supervisors feelings about various aspects of their jobs, and (3) management's philosophies, roles in policy formation, and problems of organizational integration.

Data from the questionnaire were then fed back to the president and his immediate subordinates, and then down through all levels to supervisors and employees. Each unit manager or supervisor discussed the data with his subordinate group. The data presented to each group were those pertaining to that group and to subunits that fell within their responsibility. During the discussions, the groups sifted results, established meanings and implications, and then formulated constructive administrative actions designed to deal with identified problems. Helping at these discussions (as resource people only) were members of the consulting team from the university's research center.

On the basis of this and other experiences in the use of survey feedback, Mann concluded that feedback is a powerful force for creating and supporting change. Such feedback deals with forces in the individual in addition to forces in the organizational situation, both of which influence behavior and change. Through survey feedback's interlocking series of conferences, it is capable of harnessing the power of the formal organization in the interest of change. Survey feedback takes place within, and is concerned with, important aspects of the total organizational culture and therefore deals with data that are meaningful and relevant to social behavior. Finally, two features of survey feedback help overcome resistance to change: participation by all those concerned and the use of objective information as opposed to gossip and hearsay.

Role Analysis Technique (RAT)

French and Bell describe a role analysis technique designed by Dayal and Thomas which is useful in team building.[10] Organizations contain a number of "teams" that are often the target of

[9]Floyd Mann, "Studying and Creating Change: A Means to Understanding Social Organization," in Alton C. Bartlett and Thomas A. Kayser, eds., *Changing Organizational Behavior* (Englewood Cliffs, N.J.: Prentice-Hall, 1973), pp. 100–17.

[10]French and Bell, *Organization Development*, pp. 117–19.

improvement programs—for example, the boss and his immediate subordinates, the board of directors, the production division managers, project teams, and various task groups and committees. Confusion over roles, role overlap, and role conflict often interfere with the effectiveness of team effort. Role analysis technique, or RAT, is a process that enlists the aid of all team members in defining one another's roles.

Team member A in an open session with the other team members describes his role as he sees it. He first indicates what his responsibilities are, how he attempts to carry them out or, in the case of a newly formed team or group, how he *expects* to carry them out, and how he sees his role fitting into the general role structure of the group. Next, he describes his perceptions of his role relationships with each of the other role incumbents in the group and states how he intends to handle these relationships. Finally, each of the other members of the team discusses his concept of A's role, and the discussion continues until differences have been explored and clarified or until they have been noted as problems to be worked out after all member roles have been discussed.

The same process is followed for every member of the team. In this way, each member takes part with the others in defining roles, in isolating role conflicts, and in working out any difficulties. Again, through active, collaborative, deliberate exploration and discussion by the people directly involved, team members learn what others expect of them and what they expect of each other. Although this technique is useful in established groups, it is particularly valuable in new groups, where confusion over roles may be masked by the members' preoccupation with their newly assigned tasks.

Intergroup Building Exercise

In the previous chapter, we discussed the potential destructiveness of intergroup conflict in organizations, such as between different functional units and between different levels. As we pointed out, these struggles may be difficult to resolve because group loyalties have powerful catalytic effects that block objective thinking and tend to isolate the members of one group from the members of the other.

Managers usually attempt to settle intergroup problems through negotiation, compromise, isolation, or coercion. Negotiation rarely works well because it is usually handled by representatives. Such people are either bound by membership ties and obligations to their group's original position or, if they attempt to act independently or innovatively, they are rejected by their group. Compromise, isolation, and coercion are ways of avoiding the real issues, some of which may remain hidden and continue to fester.

One successful approach that has been used by OD practitioners is based on an intergroup building exercise.[11] This ap-

[11]This approach was first suggested by Robert Blake, "Psychology and the Crisis of Statesmanship," *The American Psychologist* 14 (1959), pp. 87–94. See also Robert Blake et al., *Managing Intergroup Conflict in Industry.*

proach fits the third-party facilitator concept of OD, as well as the action research concepts and the basic focus of OD on the group rather than on individuals. The reasoning behind this approach is that, if stereotyping of each group by the other can be reduced, problem-solving behavior will have a better chance of success.

In this exercise, each group is asked first to agree to try an approach that may help reduce the hostile climate and perhaps make it possible to deal more effectively with their differences. Most groups that recognize their interdependence are willing to agree to this for the sake of a common stake in organizational goals. Once this commitment is made, each group is then asked to meet in private and prepare a list of its impressions of the other group. Some of the questions they are directed to ask themselves are: In what ways does the other group irritate you? How does the other group interfere with your doing your job? What are some things you would like them to stop doing? What are some things about the other group you like?

The groups are now brought together and share their lists of likes and dislikes. The role of the facilitator or OD consultant becomes critical at this point. He must help the groups, first by encouraging them to help each other understand points at issue, second by helping them explore blocking feelings and sentiments, and finally by helping them focus on substantive issues—that is, issues founded in fact. In this way, some difficulties will be eliminated while others will be sharpened and brought into better perspective.

Eliminating some perceptual distortions helps reduce tensions and antagonisms and makes it easier for the groups to think clearly about the issues. At this point, they are again asked to meet in private and review the refined list of issues that emerged from the intergroup session. In this meeting, the list is usually reduced further and some ordering of priorities takes place. The groups are then brought back together to build a final list. The items of this final list are ranked in importance by both groups working together, and an action plan is drawn up for dealing with the problems.

This approach to intergroup problems is an attempt to deal with the culture of conflict as it actually exists in an organization. The characteristic features of intergroup conflict are stereotyped behavior, destructive relationships, and an absence of means to cope with these problems. The intergroup building exercise attempts to face each of these difficulties squarely and to involve group members directly in working out their problems.

SOME CRITICISMS OF OD

Critics of OD point out that it suffers from a value bias: democratization of the organization. They also believe that despite its protestations to the contrary, OD is mainly concerned with the social-psychological dimensions of organization and pays little attention to the technical side of task performance—methods, machinery, finance, and the legal and public environment—as parts of the total system.

In response to the criticism of value bias, one of the pioneers and authorities of OD, Chris Argyris, asserts that the old values of the formal organization—task specialization, hierarchical authority, the Protestant ethic—are still valid. "The difficulty is that, in the past, they had been used to guide the executives' behavior in all situations."[12] Clearly there is an implied need here for a contingency view of the appropriate ways of dealing with organizational problems. We see nothing in OD's philosophy or approach that precludes such a view.

The other criticism, that OD seems concerned with human relationships as if they existed in a technological vacuum, seems perhaps to be more valid. French and Bell acknowledge this bias and look to future alliances with "practitioners in such fields as management science, personnel and industrial psychology, operations research, and industrial engineering in order to provide a broader range of options for organizational interventions."[13]

SUMMARY

In this chapter, we have presented an overview of some elements of organization development. OD was defined as a formal arrangement (with many open-ended features) between a behavioral scientist and a client system. Such an arrangement is entered into for the purpose of improving the organization and acquiring self-renewal knowledge and skills from the interaction between the practitioner and organizational members. We have emphasized that OD is culture centered rather than structure or function centered and is more concerned with groups than with individual behavior. It is also systems oriented—aware of the interdependence of the parts of the organization and wary of attempts to deal with isolated problems. Its underlying philosophy is humanistic and its principal technique can be expressed as an elaboration of Lewin's action research approach to social change.

DISCUSSION QUESTIONS

12.1. Define OD.

12.2. Discuss the uses of an action-research program in accomplishing change in organizations.

12.3. How would you describe the functions of a change agent?

12.4. What are some of the criticisms of OD? How can they be overcome?

12.5. Discuss Lewin's model of change. How would it apply to classroom learning situations?

12.6. Discuss the role analysis technique (RAT) as an OD intervention.

12.7. What is the value base for OD, that is, how do its practitioners view such matters as leadership, the nature of man, the nature of learning, and so on?

12.8. Sketch and describe a systems model of the action-research process.

[12]Chris Argyris, *Interpersonal Competence and Organizational Effectiveness* (Homewood, Ill.: Dorsey-Irwin, 1962), p. 285.

[13]French and Bell, *Organization Development,* p. 195.

APPENDIX
Slocum Products Company

Thomas Slocum, president of Slocum Products Company, a small manufacturer of household plastic tableware, returned from a one-week sensitivity training workshop sponsored by a local university. He was enthusiastic about his experience and told one of his associates that in his opinion, "Every manager should attend a T-group."

His associate, Jim Cassady, plant superintendent, found himself puzzled about what Slocum said. He asked himself why Mr. Slocum believed that Slocum Products managers needed to learn something about their own behavior. He also wondered whether other company executives would benefit from such an experience. If so, should they all go to the same training group or would it be better to go to separate sessions? What would be the objective of such training and what would it cost?

Jim remembered a recent conversation he had had with his general foreman, who had attended a church-sponsored T-group. Mike had remarked that he believed, "It was pretty much a waste of time for most people," and that "Only the real talkative ones got anything out of it." Cassady had recently read several articles about the organization development approach to organizational problems. He wondered if this approach might not be better.

The following day, Cassady received a telephone call from Slocum inviting him to a meeting in the latter's office for the purpose of discussing informally the question of management training and T-groups. All of the company executives were to be there. Cassady wondered how to approach this meeting and what to say.

1. What material should Jim Cassady prepare for this meeting?
2. What recommendations would you make if you were Jim Cassady?

SELECTED READINGS

Beckhard, Richard. *Organization Development: Strategies and Models*. Reading, Mass.: Addison-Wesley, 1969.

Bennis, Warren. *Organization Development: Its Nature, Origins, and Prospects*. Reading, Mass.: Addison-Wesley, 1969.

Blake, Robert R., Shepard, Herbert A., and Mouton, Jane S. *Managing Intergroup Conflict in Industry*. Houston, Texas: Gulf Publishing, 1964.

French, Wendell, and Bell, Cecil H., Jr. *Organization Development*. Englewood Cliffs, N.J.: Prentice-Hall, 1973.

Schein, Edgar. *Process Consultation: Its Role in Organization Development*. Reading, Mass.: Addison-Wesley, 1969.

Chapter 13
Control of Operations

The underlying theory of control applies to all systems, whether they are biological, social, political, economic, or technical. Only the application, and not the nature, of control changes. If we understand what control is and what role it plays in the operation of systems, if we know what characteristics to design into control systems, and if we can judge the proper degree of control to exercise, we can become more effective managers.

In 1916, Henri Fayol formulated one of the first definitions of control as it pertains to management: "Control consists of verifying whether everything occurs in conformity with the plan adopted, the instructions issued, and principles established. It ['s] object [is] to point out weaknesses and errors in order to rectify [them] and prevent recurrence."[1] Robert J. Mockler presented a more comprehensive definition of managerial control:

Management control can be defined as a systematic effort by business management to compare performance to predetermined standards, plans, or objectives in order to determine whether performance is in line with these standards and presumably in order to take any remedial action required to see that human and other corporate resources are being used in the most effective and efficient way possible in achieving corporate objectives.[2]

We shall define control as "that function of the system that adjusts operations as needed to achieve the plan, or to maintain variations from system objectives within allowable limits." The control subsystem functions in close harmony with the operating system (see Figure 13.1). The degree to which they interact depends on the nature of the operating system and its objectives. For example, the *sensitivity* of a system refers to the degree of variation from the norm that occurs before a correction is made. **Stability** concerns a system's ability to maintain a pattern of output without wide fluctuations. **Rapidity of response** pertains to the speed with which a system can correct variations and return to expected output.

A political election can illustrate the concept of control and

[1]Henri Fayol, *General and Industrial Management*, trans. Constance Storrs (New York: Pitman Publishing, 1949), p. 107.

[2]Robert J. Mockler, ed., *Readings in Management Control* (New York: Appleton-Century-Crofts, 1970), p. 14.

Figure 13.1 The Elements of a Control Subsystem

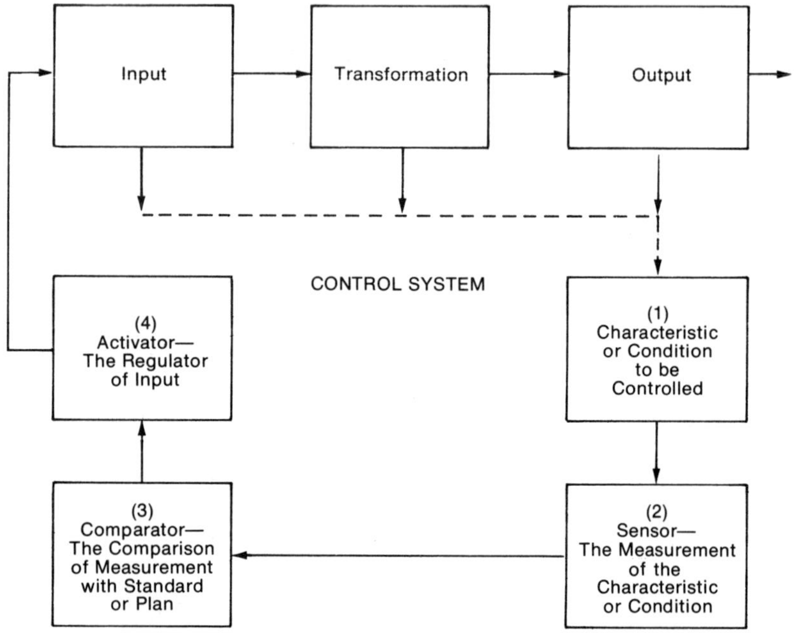

the importance of feedback. Each party organizes a campaign to get its candidate selected and outlines a plan to inform the public about both the candidate's credentials and the party's platform. As the election nears, opinion polls furnish feedback about the effectiveness of the campaign and about each candidate's chances to win. Depending on the nature of this feedback, certain adjustments in strategy and/or tactics can be made in an attempt to achieve the desired result.

THE ELEMENTS OF CONTROL

The four basic elements in a control system—(1) the characteristic or condition to be controlled, (2) the sensor, (3) the **comparator**, and (4) the **activator**—occur in the same sequence and maintain a consistent relationship to each other in every system. As illustrated in Figure 13.1, the first element is the *characteristic* or condition of the operating system which is to be measured. We select a specific characteristic because a correlation exists between it and how the system is performing. The characteristic may be the output of the system during any stage of processing or it may be a condition that has resulted from the output of the system. For example, it may be the heat energy produced by the furnace or the temperature in the room which has changed because of the heat generated by the furnace. In an elementary school system, the hours a teacher works or the gain in knowledge demonstrated by the students on a national examination are examples of characteristics that may be selected for measurement, or control.

The second element of control, the *sensor*, is a means for

measuring the characteristic or condition. The control subsystem must be designed to include a sensory device or method of measurement. In a home heating system this device would be the thermostat, and in a quality-control system this measurement might be performed by a visual inspection of the product.

The third element of control, the comparator, determines the need for correction by comparing what is occurring with what has been planned. Some deviation from plan is usual and expected, but when variations are beyond those considered acceptable, corrective action is required. It is often possible to identify trends in performance and to take action before an unacceptable variation from the norm occurs. This sort of preventative action indicates that good control is being achieved.

The fourth element of control, the activator, is the corrective action taken to return the system to expected output. The actual person, device, or method used to direct corrective inputs into the operating system may take a variety of forms. It may be a hydraulic controller positioned by a solenoid or electric motor in response to an electronic error signal, an employee directed to rework the parts that failed to pass quality inspection, or a school principal who decides to buy additional books to provide for an increased number of students. As long as a plan is performed within allowable limits, corrective action is not necessary; this seldom occurs in practice, however.

Figure 13.2 illustrates four examples of control including the specific elements related to each instance. In the first example, when a person wishes to drive to another part of the city to visit a friend, he or she checks the address and then uses the street signs and house numbers as characteristics. The eyes (sensors) follow the names of the various streets and compare progress against the programmed route or map. If a wrong turn is made, a correction is necessary and the driver returns the automobile to the proper route by an adjustment in direction.

Relationship between the Elements of Control and Information

Information is the medium of control, because the flow of sensory data and later the flow of corrective information allow a characteristic or condition of the system to be controlled. To illustrate how information flow facilitates control, let us review the elements of control in the context of information.

Controlled Characteristic or Condition

As noted earlier, the primary requirement of a control system is that it maintain the level and kind of output necessary to achieve the system's objectives. It is usually impractical to control every feature and condition associated with the system's output. Therefore, the choice of the controlled item (and appropriate information about it) is extremely important. There should be a direct correlation between the controlled item and the system's operation. In other words, control of the selected characteristic should have a direct relationship to the goal or objective of the system.

Figure 13.2 Examples of the Elements of Control

Activity or Organization	Measurable Characteristic	Sensor	Comparator	Activator
Driving to visit a friend	Geographic location	Perception of location	Driver with the aid of landmarks	Driver
Coaching a baseball team	Number of games won or lost	Secretary of the league	Coach, alumni, sports writers, and public	Coach
Managing a United Good Neighbors team	Dollars raised	District reports by UGN representatives	Chairman compares reports against specific goals	UGN Chairman
Managing a manufacturing plant	Quantity, quality and cost of output	Foreman and supervisors	Production control, quality control, and budget groups	Manufacturing manager

Sensor

After the characteristic is sensed, or measured, information pertinent to control is fed back. Exactly what information needs to be transmitted and also the language that will best facilitate the communication process and reduce the possibility of distortion in transmission must be carefully considered. Information that is to be compared with the standard, or plan, should be expressed in the same terms or language as in the original plan to facilitate decision making. Using machine methods (computers) may require extensive translation of the information. Since optimal languages for computation and for human review are not always the same, the relative ease of translation may be a significant factor in selecting the units of measurement or the language unit in the sensing element.

In many instances, the measurement may be sampled rather than providing a complete and continuous feedback of information about the operation. A sampling procedure suggests measuring some segment or portion of the operation that will represent the total.

Comparison with Standard

In a social system, the norms of acceptable behavior become the standard against which so-called deviant behavior may be judged. Regulations and laws provide a more formal collection of information for society. Social norms change, but very slowly. In contrast, the standards outlined by a formal law can be changed from one day to the next through revision, discontinuation, or replacement by another.

Information about deviant behavior becomes the basis for controlling social activity. Output information is compared with the standard or norm and significant deviations are noted. In an industrial example, frequency distribution (a tabulation of the number of times a given characteristic occurs within the sample of products being checked) may be used to show the average quality, the spread, and the comparison of output with a standard.

If there is a significant and uncorrectable difference between output and plan, the system is "out of control." This means that the objectives of the system are not feasible in relation to the capabilities of the present design. Either the objectives must be reevaluated or the system redesigned to add new capacity or capability. For example, the traffic in drugs has been increasing in some cities at an alarming rate. The citizens must decide whether to revise the police system so as to regain control, or whether to modify the law to reflect a different norm of acceptable behavior.

Activator

The activator unit responds to the information received from the comparator and initiates corrective action. If the system is a machine-to-machine system, the corrective inputs (decision rules) are designed into the network. When the control relates to a man-to-machine or man-to-man system, however, the individual(s) in charge must evaluate (1) the accuracy of the feedback information, (2) the significance of the variation, and (3) what corrective inputs will restore the system to a reasonable degree of stability. Once the decision has been made to direct new inputs into the system, the actual process may be relatively easy. A small amount of energy can change the operation of jet airplanes, automatic steel mills, and hydroelectric power plants. The pilot presses a button, and the landing gear of the airplane goes up or down; the operator of a steel mill pushes a lever, and a ribbon of white-hot steel races through the plant; a worker at a control board directs the flow of electrical energy throughout a regional network of stations and substations. It takes but a small amount of control energy to release or stop large quantities of input.

The comparator may be located far from the operating system, although at least some of the elements must be in close proximity to operations. For example, the measurement (the sensory element) is usually at the point of operations.[3] The measurement information can be transmitted to a distant point for comparison with the standard (comparator), and when deviations occur, the correcting input can be released from the distant point. However, the input (activator) will be located at the operating system. This ability to control from afar means that aircraft can be flown by

[3]Remote detection and measuring devices (such as radar) are fundamental to some control systems, however.

remote control, dangerous manufacturing processes can be operated from a safe distance, and national organizations can be directed from centralized headquarters.

KINDS OF CONTROL

Control may be grouped according to three general classifications: (1) the nature of the information flow designed into the system (that is, open- or closed-loop control), (2) the kind of components included in the design (that is man or machine control systems), and (3) the relationship of control to the decision process (that is, organizational or operational control).

Open- and Closed-Loop Control

The difference between **open-loop control** and **closed-loop control** is determined by whether all of the control elements are an integral part of the system being regulated, and whether allowable variations from standard have been predetermined.[4] In an open-loop system, not all of the elements will be designed into the system, and/or allowable variations will not be predetermined.

A street-lighting system controlled by a timing device is an example of an open-loop system. At a certain time each evening, a mechanical device closes the circuit and energy flows through the electric lines to light the lamps. Note, however, that the timing mechanism is an independent unit and is not measuring the objective function of the lighting system. If the lights should be needed on a dark, stormy day the timing device would not recognize this need and therefore would not activate energy inputs. Corrective properties may sometimes be built into the controller (for example, to modify the time the lights are turned on as the days grow shorter or longer), but this would not close the loop. In another instance, the sensing, comparison, or adjustment may be made through action taken by an individual who is not part of the system. For example, the lights may be turned on by someone who happens to pass by and recognizes the need for additional light.

If control is exercised as a result of the operation rather than because of outside or predetermined arrangements, it is a closed-loop system. The home thermostat is the classic example of a control device in a closed-loop system. When the room temperature drops below the desired point, the control mechanism closes the circuit to start the furnace and the temperature rises. The furnace-activating circuit is turned off as the temperature reaches the preselected level. The significant difference between this type of system and an open-loop system is that the control device is an element of the system it serves and measures the performance of the system. In other words, all four control elements are integral to the specific system.

An essential part of a closed-loop system is feedback; that is, the output of the system is measured continually through the item

[4]The reader should not confuse open- and closed-loop control with open and closed systems. It was noted in Chapter 4 that the definition of open or closed systems pertains to their ability to adjust their basic design to meet changes in the environment; whereas open-loop or closed-loop control refers to the way the system is able to adjust inputs as it operates.

controlled, and the input is modified to reduce any difference or error toward zero. Many of the patterns of information flow in organizations are found to have the nature of closed loops, which use feedback. The reason for such a condition is apparent when one recognizes that any system, if it is to achieve a predetermined goal, must have available to it at all times an indication of its degree of attainment. In general, every goal-seeking system employs feedback.

Man and Machine Control

The elements of control are easy to identify in machine systems. For example, the characteristic to be controlled might be some variable like speed or termperature, and the sensing device could be a speedometer or a thermometer. An expectation of precision exists because the characteristic is quantifiable and the standard and the normal variation to be expected can be described in exact terms. In automatic machine systems, inputs of information are used in a process of continual adjustment to achieve output specifications. When even a small variation from the standard occurs, the correction process begins. The automatic system is highly structured, designed to accept certain kinds of input and produce specific output, and programmed to regulate the transformation of inputs within a narrow range of variation.

For an illustration of mechanical control, as the load on a steam engine increases and the engine starts to slow down, the regulator reacts by opening a valve that releases additional inputs of steam energy. This new input returns the engine to the desired number of revolutions per minute. This type of mechanical control is crude in comparison to the more sophisticated electronic control systems in everyday use. Consider the complex missile-guidance systems that measure the actual course according to predetermined mathematical calculations and make almost instantaneous corrections to direct the missile to its target.

Machine systems can be complex because of the sophisticated technology, whereas control of people is complex because the elements of control are difficult to determine. In human control systems, the relationship between objectives and associated characteristics is often vague; the measurement of the characteristic may be extremely subjective; the expected standard is difficult to define; and the amount of new inputs required is impossible to quantify. To illustrate, let us refer once more to a formalized social system in which deviant behavior is controlled through a process of observed violation of the existing law (sensing), court hearings and trials (comparison with standard), incarceration when the accused is found guilty (correction), and release from custody after rehabilitation of the prisoner has occurred.

The speed limit established for freeway driving is one standard of performance that is quantifiable, but even in this instance, the degree of permissible variation and the amount of the actual variation are often a subject of disagreement between the patrolman and the suspected violator. The complexity of our society is reflected in many of our laws and regulations, which establish the general standards for economic, political, and social operations. A citizen may not know or understand the law and consequently would not know whether or not he was guilty of a violation.

Most organized systems are some combination of man and machine; some elements of control may be performed by machine whereas others are accomplished by man. In addition, some standards may be precisely structured whereas others may be little more than general guidelines with wide variations expected in output. Man must act as the controller when measurement is subjective and judgment is required. Machines such as computers are incapable of making exceptions from the specified control criteria regardless of how much a particular case might warrant special consideration. A pilot acts in conjunction with computers and automatic pilots to fly large jets. In the event of unexpected weather changes, or possible collision with another plane, he must intercede and assume direct control.

Organizational and Operational Control

The concept of organizational control is implicit in the bureaucratic theory of Max Weber, discussed in Chapter 2. Associated with this theory are such concepts as "span of control," "closeness of supervision," and "hierarchical authority." Weber's view tends to include all levels or types of organizational control as being the same. More recently, writers have tended to differentiate the control process between that which emphasizes the nature of the organizational or systems design and that which deals with daily operations. To illustrate the difference, we "evaluate" the performance of a system to see how effective and efficient the design proved to be or to discover why it failed. In contrast, we operate and "control" the system with respect to the daily inputs of material, information, and energy. In both instances, the elements of feedback are present, but **organizational control** tends to review and evaluate the nature and arrangement of components in the system, whereas **operational control** tends to adjust the daily inputs.

The direction for organizational control comes from the goals and strategic plans of the organization. General plans are translated into specific performance measures such as share of the market, earnings, return on investment, and budgets. The process of organizational control is to review and evaluate the performance of the system against these established norms. Rewards for meeting or exceeding standards may range from special recognition to salary increases or promotions. On the other hand, a failure to meet expectations may signal the need to reorganize or redesign.

In organizational control, the approach used in the program of review and evaluation depends on the reason for the evaluation—that is, is it because the system is not effective (accomplishing its objectives)? Is the system failing to achieve an expected standard of efficiency? Is the evaluation being conducted because of a breakdown or failure in operations? Is it merely a periodic audit-and-review process?

When a system has failed or is in great difficulty, special diagnostic techniques may be required to isolate the trouble areas and to identify the causes of the difficulty. It is appropriate to investigate areas that have been troublesome before or areas where some measure of performance can be quickly identified. For example, if an organization's output backlog builds rapidly, it is logical to check first to see if the problem is due to such readily obtainable measures as increased demand or to a drop in available man hours. When a more detailed analysis is necessary, a systematic procedure should be followed.

In contrast to organizational control, operational control serves to regulate the day-to-day output relative to schedules, specifications, and costs. Is the output of product or service the proper quality and is it available as scheduled? Are inventories of raw materials, goods-in-process, and finished products being purchased and produced in the desired quantities? Are the costs associated with the transformation process in line with cost estimates? Is the information needed in the transformation process available in the right form and at the right time? Is the energy resource being utilized efficiently?

Perhaps the most difficult task of management concerns monitoring the behavior of individuals, comparing performance to some standard, and providing rewards or punishment as indicated. Sometimes this control over people relates entirely to their output. For example, a manager might not be concerned with the behavior of a salesman as long as sales were as high as expected. In other instances, close supervision of the salesman might be appropriate if achieving customer satisfaction were one of the sales organization's main objectives.

The larger the unit, the more likely that the control characteristic will be related to some output goal. It also follows that if it is difficult or impossible to identify the actual output of individuals, it is better to measure the performance of the entire group. This means that individuals' levels of motivation and the measurement of their performance become subjective judgments made by the supervisor. Controlling output also suggests the difficulty of controlling individuals' performance and relating this to the total system's objectives.

PROBLEMS OF CONTROL

The perfect plan could be outlined if every possible variation of input could be anticipated and if the system would operate as predicted. This kind of planning is neither realistic, economical, nor feasible for most business systems. If it were feasible, planning requirements would be so complex that the system would be out of date before it could be operated. Therefore, we design control into systems. This requires more thought in the systems design but allows more flexibility of operations and makes it possible to operate a system using unpredictable components and undetermined input. Still, the design and effective operation of control are not without problems.

The objective of the system is to perform some specified function. The purpose of organizational control is to see that the specified function is achieved; the objective of operational control is to ensure that variations in daily output are maintained within prescribed limits. It is one thing to design a system that contains all of the elements of control, and quite another to make it operate true to the best objectives of design. Operating "in control" or "with plan" does not guarantee optimum performance. For example, the plan may not make the best use of the inputs of materials, energy, or information—in other words, the system may not be designed to operate efficiently. Some of the more typical problems relating to control include the difficulty of measurement, the problem of timing information flow, and the setting of proper standards.

Measurement of Output

When objectives are not limited to quantitative output, the measurement of system effectiveness is difficult to make and subsequently perplexing to evaluate. Many of the characteristics pertaining to output do not lend themselves to quantitative measurement. This is true particularly when inputs of human energy cannot be related directly to output. The same situation applies to machines and other equipment associated with human involvement, when output is not in specific units. In evaluating man-machine or human-oriented systems, psychological and sociological factors obviously do not easily translate into quantifiable terms. For example, how does mental fatigue affect the quality or quantity of output? And, if it does, is mental fatigue a function of the lack of a challenging assignment or the fear of a potential injury?

Subjective inputs may be transferred into numerical data, but there is always the danger of an incorrect appraisal and transfer, and the danger that the analyst may assume undue confidence in such data after they have been quantified. Let us suppose, for example, that the decisions made by an executive are rated from 1 to 10, 10 being the perfect decision. After determining the ranking for each decision, adding these, and dividing by the total number of decisions made, the average ranking would indicate a particular executive's score in his decision-making role. On the basis of this score, judgments—which could be quite erroneous—might be made about his decision-making effectiveness. One executive with a ranking of 6.75 might be considered more effective than another who had a ranking of 6.25, and yet the two managers may have made decisions under different circumstances and conditions. External factors over which neither executive had any control may have influenced the difference in "effectiveness."

Quantifying human behavior, despite its extreme difficulty, subjectivity, and imprecision in relation to measuring physical characteristics is the most prevalent and important measurement made in large systems. The behavior of individuals ultimately dictates the success or failure of every man-made system.

Information Flow

Another problem of control relates to the improper timing of information introduced into the feedback channel. Improper timing can occur in both computerized and human control systems, either by mistakes in measurement or in judgment. The more rapid the system's response to an error signal, the more likely it is that the system could overadjust; yet the need for prompt action is important because any delay in providing corrective input could also be crucial. A system generating feedback inconsistent with current need will tend to fluctuate and will not adjust in the desired manner.

The most serious problem in information flow arises when the delay in feedback is exactly one-half cycle, for then the corrective action is superimposed on a variation from norm which, at that moment, is in the same direction as that of the correction. This causes the system to overcorrect, and then if the reverse adjustment is made out of cycle, to correct too much in the other direction, and so on until the system fluctuates ("**oscillates**") out of control. This phenomenon is illustrated in Figure 13.3. If, at Point *A*, the trend below standard is recognized and new inputs are added, but not until Point *B*, the system will overreact and go beyond the allowable limits. Again, if this is recognized at Point *C*, but inputs are not withdrawn until Point *D*, it will cause the system to drop below the lower limit of allowable variation.

One solution to this problem rests in anticipation, which involves measuring not only the change but also the rate of change. The correction is outlined as a factor of the type and rate of the error. The difficulty also might be overcome by reducing the time lag between the measurement of the output and the adjustment to input. If a trend can be indicated, a time lead can be introduced to compensate for the time lag, bringing about consistency between the need for correction and the type and magnitude of the indicated action. It is usually more effective for an organization to maintain continuous measurement of its performance and to make small adjustments in operations constantly (this assumes a highly sensitive control system). Information feedback, consequently, should be timely and correct to be effective. That is, the information should provide an accurate indication of the status of the system.

Setting Standards

Setting the proper standards or control limits is a problem in many systems. Parents are confronted with this dilemma in expressing what they expect of their children, and business managers face the same issue in establishing standards that will be acceptable to employees. Some theorists have proposed that workers be allowed to set their own standards, on the assumption that when people establish their own goals, they are more apt to accept and achieve them.

Standards should be as precise as possible and communi-

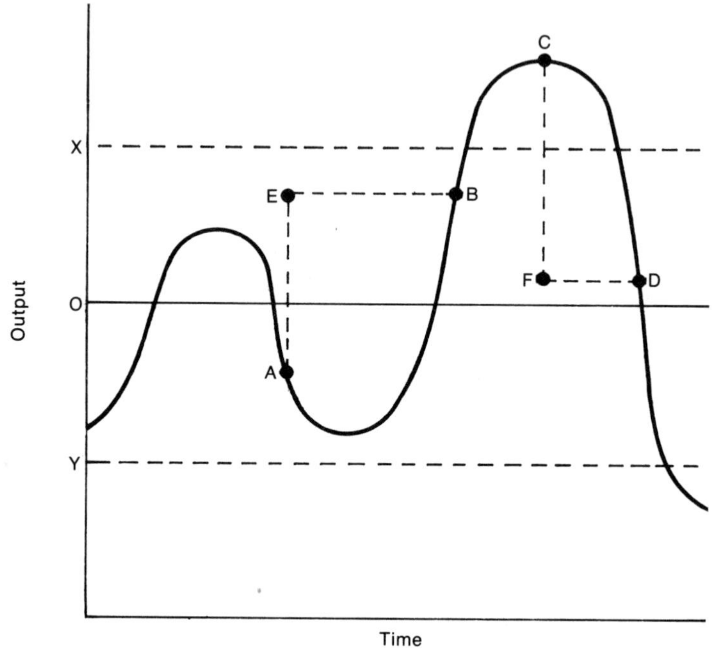

Figure 13.3 Oscillation and Feedback

Points

A=The time when output is measured
B=Corrective input is added
C=Error is noted
D=Input discontinued

Lines

EB and FD=Information time lag
O=Expected output or standard
X=Top limit of allowable variation
Y=Lower limit of allowable variation

cated to all persons concerned. Moreover, communication alone is not sufficient; understanding is necessary. In human systems, standards tend to be poorly defined and the allowable range of deviation from standard also indefinite. For example, how many hours each day should a professor be expected to be available for student consultation? Or, what kind of behavior should be expected by students in the classroom? Discretion and personal judgment play a large part in such systems, to determine whether corrective action should be taken.

Perhaps the most difficult problem in human systems is the unresponsiveness of individuals to indicated correction. As discussed in Chapter 11, this may take the form of opposition and subversion to control, or it may be related to the lack of defined responsibility or authority to take action. Leadership and positive motivation then become vital ingredients in achieving the proper response to input requirements.

Most control problems relate to design; thus the solution to these problems must start at that point. Automatic control systems, provided that human intervention is possible to handle exceptions, offer the greatest promise. There is a danger, however, that we may measure characteristics that do not represent effective performance (as in the case of the speaker who requested that all of the people who could not hear what he was saying should raise their hands), or that improper information may be communicated.

SUMMARY

Control is that function of the system that provides adjustments in conformance to the plan or maintains variations from system objectives within allowable limits. The theory applies to human, man-machine, and machine systems. There are four elements in every control system: (1) the control characteristic, (2) the sensor, or sensing device, (3) the comparator, or comparison between the standard and actual performance, and (4) the activator, or unit adjusting input to change operations.

Information is the medium of control, because it is the flow of measurement data and later the flow of corrective information that allow a characteristic or condition of a system to be controlled. Keeping in mind that information flow is the connecting link of any control cycle, the control characteristic, the means of measurement, the standard, and the corrective unit should be designed in consideration of the total system.

Control may be classified in three general categories: (1) as open- or closed-loop control, (2) as man or machine control, and (3) as organizational or operational control. In an open-loop system, not all of the control elements are designed into the system, and/or allowable variations are not predetermined. In a closed-loop system, all of the four control elements are an integral part of the system being regulated. The elements in a machine control system are easy to identify, whereas in a human system, they are more obscure. Nevertheless, these same elements occur and follow the same sequence in both kinds of systems, and also in the most common man-machine combinations.

Another means of classifying a control process is to separate that control which emphasizes the nature of the organizational or systems design from that which deals with daily operations. In other words, we evaluate the performance of a system to see how effective and efficient the design has been (organizational control), and we measure the daily output of a system to see whether or not there are unacceptable variations from plan (operational control).

Problems occur in control systems when we fail to design the proper values or relationships among the four elements. Problems are most frequent in human systems because expectations and output are more difficult to associate with the objectives of the system.

DISCUSSION QUESTIONS

13.1. Define control. What are its objectives? Illustrate several kinds of control using organizational examples.

13.2. What is the difference between open- and closed-loop control?

13.3. Describe the difference between organizational and operational control. How do these terms relate to the effectiveness and the efficiency of the system?
13.4. What kinds of control problems might be the most typical in machine systems? In human systems?
13.5. Why is information timing so important in the effectiveness of control systems?
13.6. How can standards be established in human systems so that they will be meaningful to both workers and management?
13.7. Describe a control system for a department store having the objectives of (a) regulating inventory levels and (b) protecting the inventory from pilferage.

APPENDIX Flexible Hours

At the turn of the century, Frederick W. Taylor proposed that once workmen accomplished their assigned tasks, they should be paid for a full day's work and allowed to go home without waiting for the official end of the work period. This was a difficult philosophy for management to accept because incidents of tardiness, absenteeism, or unauthorized leave had been generally regarded as indications of inefficiency. Management has penalized such infractions of the rules through warnings, loss of pay, and even dismissal for the worst offenders. The time clock has been used as an instrument to link the worker to the job and to measure employees' attentiveness to work schedules.

Recently, flexible working hours are receiving more attention. Many firms in Europe have decided that it is better to let employees determine their own starting and ending times. Those who normally would be scheduled to work from 8 A.M. to 5 P.M. with an hour for lunch are permitted to choose another combination of starting and quitting times. For example, some may start at 7 A.M. and finish their work at 4 P.M. Under this plan, all workers have a responsibility to put in a designated number of hours each day and to be present during certain "core" hours of the day. To illustrate, all might be required to be on the job between 9 A.M. and 3 P.M., but the remainder of their work day may occur either before or after the designated core hours, or in some combination of before and after. In some instances, it is possible for employees to work more than the designated eight hours a day and to build up a credit account of flexible hours. One airline, for example, permits a maximum of ten hours to be built up, which can be used to secure time off from work.

An increasing number of private and public organizations in the United States are adopting this flexible time plan. Their employees are no longer considered tardy if they arrive after 8 A.M. In most instances, employees seem to follow the customary working hours closely, but they do seem to increase their productive time for they tend to start work as soon as they arrive and not to stop working early simply to be ready to leave when the bell rings. In fact, it is typical for people to work beyond the normal quitting time so they can finish a job they have started.

1. What is the basic difference in management's assumption about workers under the traditional and the flexible time plans?
2. What effect would this plan have on the production control system? Could management exercise any control?
3. List some situations that might make it very difficult (or impossible) to use the flexible time plan.
4. Why do you think that few organizations have been willing to follow Taylor's recommendation, that once the assigned work has been completed, employees should be allowed to go home?

SELECTED READINGS

Anthony, Robert N. *Planning and Control Systems—A Framework for Analysis*. Boston: Graduate School of Business Administration, Harvard University, 1965.

Beer, Stafford. *Decision and Control.* New York: John Wiley & Sons, 1966.

Mockler, Robert J., ed. *Readings in Management Control.* New York: Appleton-Century-Crofts, 1970.

AIR CANADA

Chapter 14
An Application of Systems Management

Management is more than organization, planning, design, communication, leadership, motivation, change, and control. It is a process of operating a system as a total unit with full understanding of the relationships among all of the parts. Such interrelationships become more apparent during periods of trouble. The following history of one vital division of the Boeing company serves as a case study to illustrate this point.

The information about this situation was presented by J. E. Steiner, the vice-President and general manager of the 707/727/737 Division, The Boeing Company, Commercial Airplane Group, before the Pan-American Management Club of New York, on January 11, 1972. There were three parts to his speech: (1) reductions in cost, (2) increasing sales, and (3) the future. The first part of his speech is reprinted in this chapter. Parts of the last section of his speech are also include in the last chapter of this book. Action described here addresses disappearances of the *market*. The company had taken earlier action, starting in 1966, to overcome the cost deterioration which rapid expansion had produced. In mid-1966, with orders at an all-time high, the company found itself in a spiraling dichotomy; as business got better, efficiency got worse. Basic operating systems were bogging down, part shortages and travelled work were mounting, changes were being made on top of changes. Drastic action was taken; sweeping revisions were made to the operating systems; tried-and-true approaches to management, which had been historically couched in "that's how we have always done it" were abandoned and new and better methods were implemented. Major management changes were made with improved operating conditions and efficiency resulting. Then, suddenly, in 1968, two-thirds of the market simply went away.

PROBLEMS AND CHALLENGES: A PATH TO THE FUTURE

This a story of trouble and action—of an industry's trouble—of a company's, and in particular a division's actions—to assure that there *would* be, in fact, a path to the future. The future, it turns out, probably will never be quite the same as the past—and I'll cover some of that too. I've titled

Reprinted by permission of J.E. Steiner and The Boeing Company, Seattle, Washington.

the three parts of the story: Reduction in Cost, Increase in Sales, and Future.

The *market* part of the story really starts in 1968, when the first signs of impending doom became evident. We, at Boeing-Commercial, had been flying along filled with the euphoria of an expanding industry—the confidence which comes from having a good product line, good sales volume, an experienced management team, and an excellent physical plant. Commercial orders were running between two and three billion dollars a year. Our major product and plant decisions were behind us. We had invested close to a billion dollars in facilities over a 4-year period and had assembly areas and machine capability unmatched in the world.

Suddenly, in 1968, orders became harder to get, money tightened, the U.S. market, which was two-thirds of the total, simply *went away*. Figure 14.1 shows what happened to commercial orders, both in number and in value. Numbers are plotted at year-end and, as in all such cases, we did not really want to believe what was happening until the usual calendar year-end totals were added up and examined in relation to targets which had been established. Thus it was late in 1968 before we really knew we were in trouble. Action was instigated almost immediately to reduce the number of employees and to curtail the plans for expanding our physical plant.

In the "market disappearance" case, as is noted, the "period of assessment" took almost a year. It was not until the fall of 1969 that we realized that action beyond any we had taken in our company's history would be necessary for survival. Such drastic layoffs as we then contemplated were difficult to make, and impossible to do instantly, due to the accumulated vacation and other benefits which the employees had amassed. Thus, although drastic action took place in October and November of 1969, the real drop in manpower did not begin to occur until Christmas 1969. It was a bad Christmas for many.

Figure 14.1 Boeing Commercial Orders

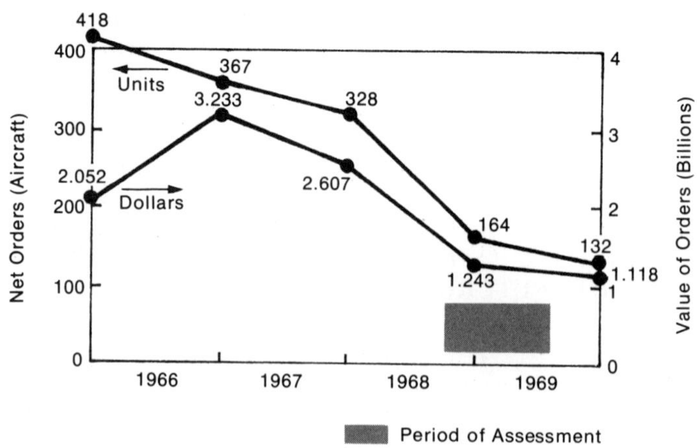

Figure 14.2 Boeing Commercial Employees (Commercial Airplane Group)

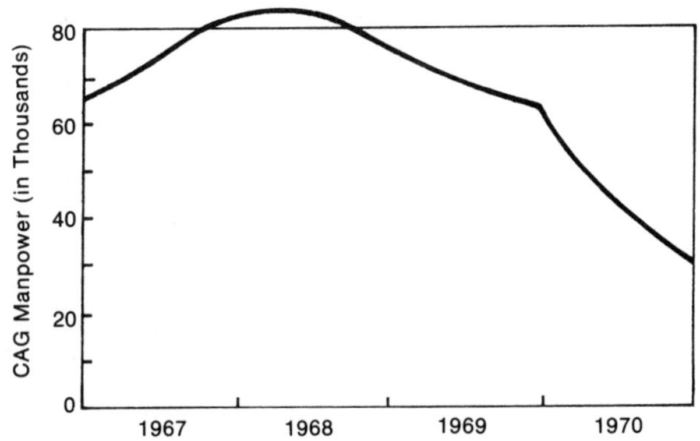

As shown in Figure 14.2, in 1970 we reduced the total number of commercial employees by over 50 percent, and that from a base which had already come down 25 percent in the previous period of about a year.

Reduction in Cost

Massive layoff plans are difficult in any company. At The Boeing Company, which had had relatively stable employment prior to the expansion of 1965 to 1968, and had had much of that expansion in terms of lower paid employees, the easiest type of reduction is the one shown at the left of Figure 14.3, that is, "layoff from the bottom of the deck." However, the average wage and salary structure resulting from that kind of action was untenable, so that the actual plan shown in Figure 14.3 was determined to be the one we would take. This meant hundreds, if not thousands, of salary reductions which had to be justified by organization realignments before they could be made. It meant reductions in all areas of the company including vice-presidents. In the case of top managers the least harmful manner, in many cases, was the retirement of personnel who had not yet reached retirement age, but who would probably reach it prior to reexpansion of the company and regaining of their former status. We had to take action which prepared the company for a low period and a future expansion, and yet was not "discriminatory" by age. In many cases, the

Figure 14.3 Manpower Reductions

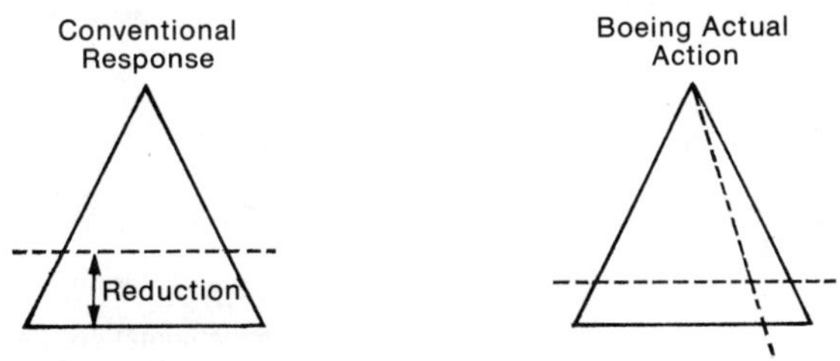

1. Reductions at all levels of company
2. Organizational consolidations and realignments
3. Salary downward adjustments consistent with reassignments

principal forcing function had to be the employee's knowledge of the situation and his loyalty to the ultimate welfare of The Boeing Company.

An interesting example of the sort of methods necessary is shown in Figure 14.4. During the period of expansion we had constructed two independent divisions to produce 737 and 707/727 aircraft. These divisions had their own plants, located six miles apart. Each was managed by a vice-president–general manager. Over a period of time a very worthwhile form of competition built up, so that each team regarded itself as "superior" to the other. In fact, the situation had progressed to such a point that, although combination of the divisions posed an obvious efficiency increase possibility, a "decision" to combine the two was *not* available. If such combination were to result in demoralization, the impact could be almost fatal to the company. Sizeable commitments for timely delivery existed for all three products, and the dates had to be met at a cost that would not increase our already extended financial position.

As noted, we appointed a six-man task force, kept all the vice-presidents out of the team's way, and did not appoint a chairman. We told them all we knew of the company circumstances and asked each to participate in a solution as if he were, in fact, the top manager of the company.

What resulted was both faster and more successful than anyone had expected. The task force, working seven days a week, came up with unanimous recommendations within a month. During the next month it came up with detailed plans, budgets, costs, etc.; notwithstanding the necessity to move the tooling over six miles, including major fixed-wing tooling, which normally is not "moveable" at all. All this was in spite of the fact that the six-man team knew that there would only be three survivors without demotion in position and salary. We then let the same six supervise the move itself, disclosing the "internal" origin of the plan in newsletters, speeches, etc. This was all the more interesting because the division whose site had to be abandoned had actually been slightly the more efficient of the two.

Figure 14.5 shows the consolidation of covered floor area. By 1969, as shown in Figure 14.5, the Commercial Airplane Group (CAG) occupied some 22.5 million square feet of covered

Figure 14.4 Example Combination of Formerly Competitive Organizations (737 and 707/727)

1. Appointed a 6-man task force (directors of engineering, manufacturing, and finance of both divisions).
2. No chairman. No interference from either general manager.
3. One month to initial report. Two months to conclusion and implementation.
4. Major cost saving attained in two months after implementation. Geographic and organizational moves completed six months after implementation.
5. Communication program stressed that plan was developed by six-man "internal" team.

Figure 14.5 CAG Facilities Consolidation

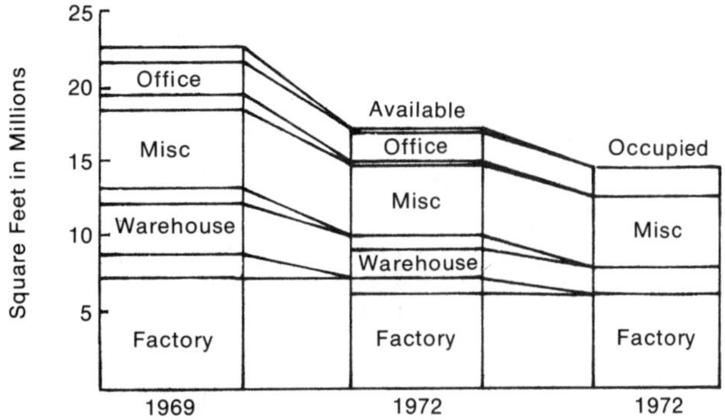

Misc: receiving, service areas, powerplant, laboratories, cafeterias maintenance, etc.

floor area of which 9.3 million was factory space. An aggressive program was undertaken to dispose of facilities that were excess to requirements. Lease holdings have been reduced 3.4 million square feet; and CAG has sold or reassigned, in a few cases to other company elements, 2.1 million square feet. Of the 17 million square feet currently available, 2.7 million are not occupied, but are being retained for future use. In many cases, such abrupt changes in organization and factory require commitment to plans in the face of forecasts by the operating groups that the plan will wreck their efficiency, ruin morale, and lose the production schedule. For example, in two cases we had to move out of new and very beautiful facilities back into old and more dismal ones. The manager involved had to maintain a degree of stubbornness and determination that has been unprecedented, at least in my experience. Another advantage of moving out of the new buildings and back into the old ones was the beneficial effect this had on the total tax bill.

An area of major cost recovery was the sale of surplus materials and equipment. Previously at Boeing, this had been handled on an organization-to-organization basis. In this case, circumstances were such that a central organization was established in 1970 to ensure that items offered for sale were, in fact, excess to the requirements of the total company, and also to add strength to the disposal plans. Accomplishments to date are shown in Figure 14.6, with approximately $21 million recovered so far. Of the total $31 million to the end of 1972, $16 million is capital equipment, i.e., machine tools, test equipment, office furniture, etc; $8 million is purchased parts, raw materials and supplies; and $7 million is scrap and salvage.

A disposal program of this size has required many methods: consigners, auctions, U.S. and foreign operations, contingent commissions, etc. In the process there were some surprises; for example, we sold used machine tools in good condition to such companies as General Motors and General Elec-

Figure 14.6 CAG Surplus Sales Material and Capital Equipment

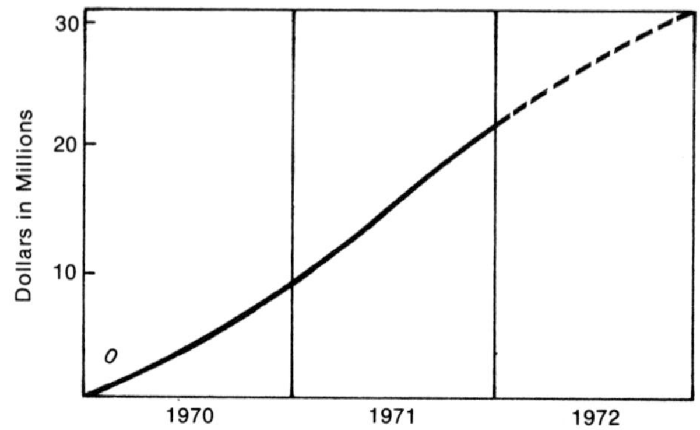

tric (which has added status to those who have suggested in the past that Boeing should look to the surplus market for some of its basic machine equipment). Auctions, although small, often resulted in surprisingly high returns. In some cases, agencies of the U.S. government bought the materiel and shipped it overseas as part of our international assistance programs. We sold new and used jet engines for surprisingly high values in relation to their purchase prices, aided of course by price increases that had occurred on similar engines in the interim. There were some comical instances, like the boiler we sold from one of our plants only to have the buyer's crew remove it and in its carelessness shut down our whole heating system. We also had buyers who knew so little about the machinery purchased, that they turned it into junk through misoperation the first time they used it. All in all, however, the surplus disposal program has been an outstanding success.

Another area of expense we concentrated on was the cost of computers. Historic and predicted cost is shown in Figure 14.7. In the process of the reduction, a company subsidiary, Boeing Computer Services, was formed in January 1971. The decreases shown in Figure 14.7 are principally accounted for by:

Figure 14.7 CAG Computer Utilization

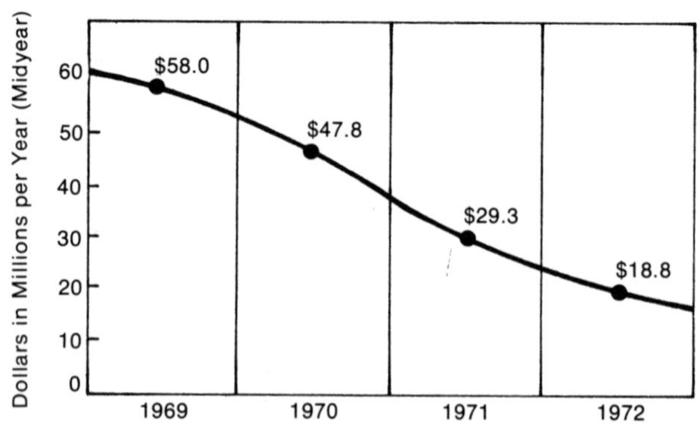

1. Screening of requirements and reducing where feasible.
2. More efficient seller/client relationships.
3. More efficient utilization of equipment.
4. Comprehensive evaluation of information systems using the computers (cancelled 14 major business systems, modified 69 business systems, and reduced research and development by 50 percent).
5. Reduced manpower through both arbitrary and nonarbitrary methods.

Figure 14.8 outlines an additional major area of cost savings, that of inventory control. For many years Boeing has had inventory reduction programs. However, these programs were made more urgent by the crisis of 1968 and 1969, and we took actions that we had not taken before. For example, we exhorted our suppliers to accept more of the inventory costs themselves, rather than expecting us to order early so that all of the costs of schedule adjustment capability were borne by us. A single such item (the traditional ordering pattern of jet engines) saved a peak of $28 million of inventory in one single action. In this, case, most of the $28 million was not a transfer, but an actual saving due to improving our planning function and that of our engine supplier.

The reduction in airplane reorder lead time was not, of course, an unmixed blessing. It made the airplanes easier to sell, and it reduced the inventory, but it also resulted in greater out-of-sequence work and greater costs to Boeing when airline customers were unable to meet their schedules for determination of buyer-furnished equipment and cabin configuration.

One of the stones around our neck at the end of 1970 was a large number of new, but unsold inventory aircraft. Most of them were on hand not because of speculative building but because we couldn't deliver them to customers who had purchased them but couldn't pay for them at the time of delivery. This inventory reached a peak number of 21 and was costing us at the end of 1970 $2 million *per month* in interest, insurance, taxes, and other costs. A whole separate task effort was made to unload these aircraft. To a greater or lesser extent, they were all customized, and when we sold them it was always necessary to revise the configuration to that desired by the

Figure 14.8 Inventory Control

1. Reduced airplane reorder lead time.
2. Shortened production flow time.
3. Reschedule installation point of high-value items.
4. Increased inventory turn-over rate.
5. Eliminated speculative building.
6. Conducted "White Tail" airplane sales campaign.
7. Conducted wall-to-wall inventory.

buyer. The operating organization was forced to step up to truly heroic airplane rebuilding feats, since such customers wanted "instant airplanes." For example, in one case we replaced body skin panels about 15 feet long, which ran from the top centerline of the body to a location 3 feet beneath the floor, to install an aft airstairs door and maintain commonality with the customer's other aircraft. We were under such pressure to dispose of those aircraft, that at one time I even moved some around behind a hangar where they couldn't be seen so easily.

Control of parts flowing to production lines at Boeing is done by computer. Several hundred thousand parts are involved and the computer selects the right ones at the right time when an airplane schedule is implemented. At appropriate times the computer must order parts from their sources in whatever amount and at whatever lead times have been fed into the computer. When production cycles are revised, the computer must make appropriate adjustments for either accelerating or decelerating the availability of necessary parts. The single most critical element to production efficiency is the availability of parts when required. Part shortages have a tendency to pyramid the amount of travelled work due to the sequence used in assembling aircraft. Thus, the number of parts that the computer *thinks* are in inventory becomes significant. Over a period of time, errors can occur. Corrections, when they are made, generally tend to run to too many rather than to too few parts so that the possibility exists that the actual inventory in some cases is significantly larger than necessary. However our experience has taught us that it is much more cost effective to have a small surplus of parts rather than part shortages. Renton is a 6.5 million square foot plant. We are currently conducting a "wall-to-wall" inventory which, from results obtained so far, will decrease our costs an additional amount by bringing the computer's base up to date and "finding" parts it doesn't know about.

In a move designed both to accelerate sales and decrease inventory, we reduced airplane reorder lead time for aircraft being built by the division as shown in Figure 14.9. This was a drastic reduction and emphasized "on-schedule" performance in every area in a manner previously unobtainable. However, in early 1971 when we did this, accomplishing the impossible was becoming habitual with our people. In the process we have obtained shorter reorder lead times from the engine manufacturer as well.

Late in 1970 we also started a low production rate efficiency study involving all of the division's elements and examining each of our systems from the standpoint of efficiency. Program management, engineering, manufacturing, quality, industrial relations, finance, computing, spares, and facilities were all involved in some 25 separate study and action efforts.

Figure 14.9 Airplane Reorder Lead Time (707/727/737 Division)

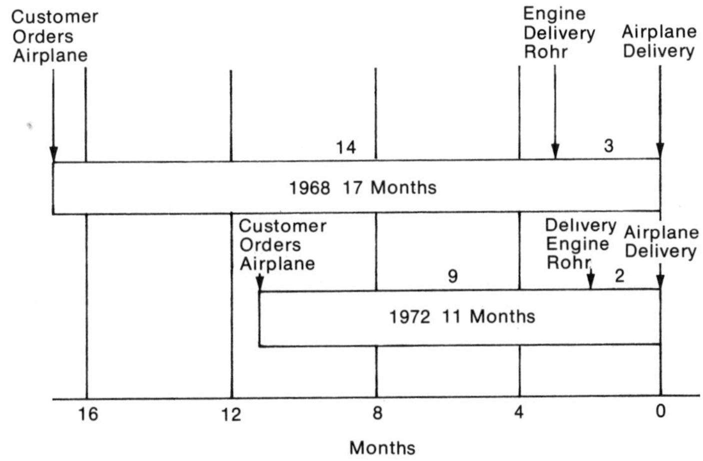

A large number of changes were pushed into the system by this method.

Figure 14.10 covers the general subject of industrial engineering, which was actually only a part of the low production rate investigations. An expansion of Figure 14.10 is shown in Figure 14.11 and describes the three fundamental phases of classical manufacturing efficiency improvement. A frustration involved in such work is the continuous "bumping" system established in our labor contract, which results in continuously shifting labor assignments as an organization contracts (even within our system, which doesn't allow bumping from outside job families). It was difficult to train workers for high efficiency in one job, when, in many cases, a week or two later they would be in another job. Nevertheless, our concentration on manufacturing efficiency improvement produced results. In an example case for section 46 of the 727, we eliminated some 153 operations in order to improve efficiency: hand trimming of shims, putting in rivets in one operation only to drill them out in the next one, and producing paperwork which had lost its meaning at low production rate. We found that our manufac-

Figure 14.10 Industrial Engineering

1. Manufacturing methods and systems technology oriented to maximum efficiency at high production levels.
2. In-depth study effort undertaken on low-rate production environment to maintain continued cost improvement, i.e.:
   - Time Studies
   - Method Studies
   - Work Simplification
   - Material Handling
   - Work Sampling
3. Other manufacturing roles involving programming, scheduling, manpower forecasting, and shop-load planning retain the traditional emphasis.

Figure 14.11 Manufacturing Efficiency Improvement

Phase I Get the worker to the work place.
Tool storage, parts supply, drawings, job paper, restrooms.
Phase II Get the worker to work at the work place.
Power supplies, motion economy, work simplification, posture, tool use, instruction methods, parts feed.
Phase III Improve worker method at the work place.
adjust•to recognize longer cycle (lower production rate).
Problem identification and correction, fixture consolidation, combined operations, greater automation.

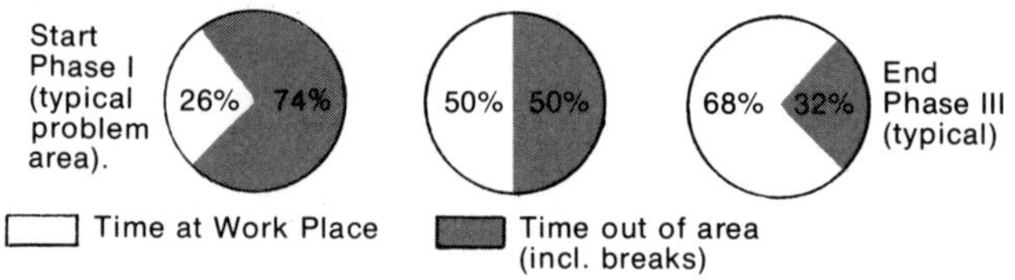

turing efficiency at high production rates in the past had succeeded in spite of a substantial load of correctable improvements. The lower part of Figure 14.11 indicates the magnitude of improvement which was possible in some cases. Obviously, not everything was this bad at the start, and some areas are even better than denoted at the end. The "work place" indicated is the very small local area in which the manufacturing worker actually functions (e.g., for the riveter, the local panel area and the rivet gun).

In another example, the size of the 747 and the floor area needed for its production aggravated travel to the point where, in one study, it was shown that for body join and installation, mechanics were required to walk a total of 72 miles per airplane. Tool location charts were constructed to expose the travel which fell into a regular pattern, and action was taken to reduce it.

With all of the studies going on for improved low production rate efficiency, combined organizations, and major moves of personnel and tooling, our need for "indirect" personnel was very large. The word *indirect* includes finance, facilities, planning, industrial engineering, industrial relations, program management, supporting services, and about 90 percent of all supervision: in short, everything that is not viewed as "direct." As shown in Figure 14.12, we were running about 60 percent indirect to direct ratio at the start of 1969. Reductions were started in 1969 which became more drastic as the crisis of 1970 approached. The percent reductions were obtained in the face of massive *direct* labor reductions going on at the same time. The result was the laying off of all indirect people who in *any* way could be dispensed with in order to retain the indirect people necessary to run the many studies necessary, make the moves that we had decided to make, and meet the indirect budgets assigned. During this period, we stopped most of our facility maintenance, quit mowing the lawn, quit weeding the

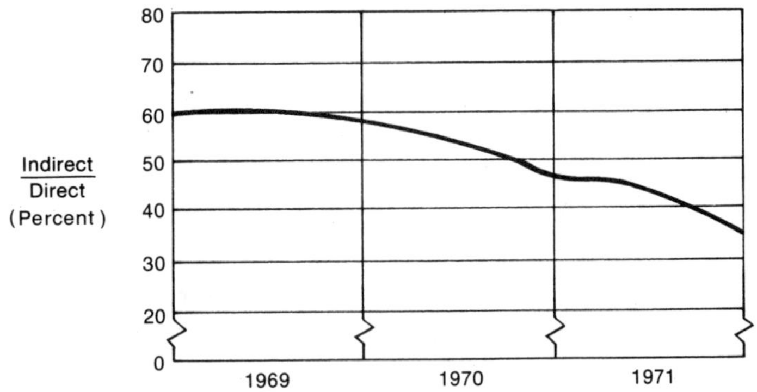

Figure 14.12 Commercial Airplane Group Direct to Indirect Ratio

flower beds, and supplied barely enough water to keep the shrubs alive (in some cases we didn't).

Under these circumstances it is very difficult to find the bottom, that is, the lowest actual level upon which an organization can in fact exist. Estimates by the operating manager, I found, were generally not accurate. Invariably his organization could exist with fewer people than he predicted. The last half of 1970 and all of 1971 were spent searching for the bottom. An example would be janitorial services, because at this point, an indirect head was an indirect head, and I would much prefer that he be used in planning manufacturing efficiency than in cleaning the work areas. We studied the washrooms to make unnecessary plumbing facilities unavailable, and therefore, not subject to cleaning. We were quite ruthless in attempting to find the bottom, and in this particular case, quit scrubbing the lavatory floors until a good case of fungus appeared, and we decided that, in this one function, we had apparently found the bottom. We added a janitor quickly and fixed it.

In another case, we decreased the maintenance on the factory doors until we finally had a line position change combined with a jammed door. We decided again we had found the bottom. Figure 14.12 is the result of the above philosophies being applied completely across the board with the single exception of safety, where we continued the effort believed by the organization involved to be adequate.

This sort of a situation could only be tolerated a few times in one's life, and leads to the sort of comment on Figure 14.13. It also requires a tough attitude which will even tolerate major employee complaints—like the one who wrote me quoting the 23rd Psalm and saying that the only reason that I wouldn't be afraid in the valley of the shadow of death was that I must be the biggest son of a bitch in the valley.

During this period we used all manner of methods to improve what was, of course, the deteriorating morale situation. The employee suggestion system and the Pride in Excellence award system were continued. To them was added a genuine

Figure 14.13 Notice to Employees

> The objective of all dedicated employees should be to thoroughly analyze all situations, to anticipate all problems prior to their occurrences, and to move swiftly to solve them when they do occur—However, when you are up to your ass in alligators it is difficult to remind yourself that your initial objective was to drain the swamp.

effort to communicate our problems and objectives to all supervisory personnel (who number about 1 in 10 employees). A monthly newsletter was sent to each from my office, and theatre meetings were (and are) held at regular intervals, involving every supervisor (which required several meetings), and indicating our total situation, our objectives, and our accomplishments. We solicit questions in writing before each meeting and, where possible, answer them during the meeting. We developed a telephone newsline system, primarily for the management but actually available to anyone who dials the number. It gives a one-minute run-down on a particular facet of our activity. The message is changed once per day. In all of this we were extremely candid, telling the employees more than we would have told them under normal conditions; but these conditions were far from normal, and we felt they had to understand what was necessary.

Figure 14.14 shows the budgeting cycle. We work to quarterly budgets that are agreed to in advance by each level of the management down to the grass roots. We allow adjustment for significant performance changes, e.g., a change in the schedule of airplane production. We do not allow adjustment for minor variations. I generally encourage the organizations to continuously underrun their targets. In this particular organization we have underrun 38 of the 52 weeks of 1971, or about 75 percent of the time, and have taken out indirect ratio from 52 percent at the start of 1971 to 37 percent.

While we have been heartless in replacing supervisors who could not meet schedules, budget, and quality, we have

Figure 14.14 Program Profit Planning (707/727/737 Division)

generally selected our management from within; and have used competent and experienced personnel having substantial service time and a demonstrated capability for the most difficult management assignments.

In the course of all of this, there were on a few occasions problems of quality. When they occurred, we involved the top quality and manufacturing managers directly in them, whether the situation was discovered internally or at one of our customer's facilities. The few problems that did occur were quickly corrected. There are no quality problems at the present time.

Figure 14.15 is typical of our cost performance. It is a normal curve on log-log paper of unit-time manhours plotted against airplane units for a typical program. What it shows is that we have continued our learning curve uninterrupted, even though we reduced the monthly production rate of the program by a substantial number, e.g., 10 to 2 per month. I believe this sort of efficiency can be maintained at production rates as low as less than one per month, retaining the computerized U.S. production methods and retaining excellent quality. A requirement, probably, is that the skill level of the employees be high, as it has been in our case. However, continuity at the same job has not been maintained as noted above and morale was maintained only with great difficulty. We have maintained costs that permit prices which are competitive in the world market. . . .

SUMMARY

The requirement was for a vast decrease in company size coupled with a substantial increase in efficiency. Manpower reductions at all levels, combination of organizations, elimination of leased facilities, and sale or reassignment of others, were accomplished. In addition, basic business systems were overhauled to reduce inventory, while at the same time decreasing computing services cost. A grass-roots return to classical industrial engineering yielded large dividends but re-

Figure 14.15 Manufacturing Cost Example

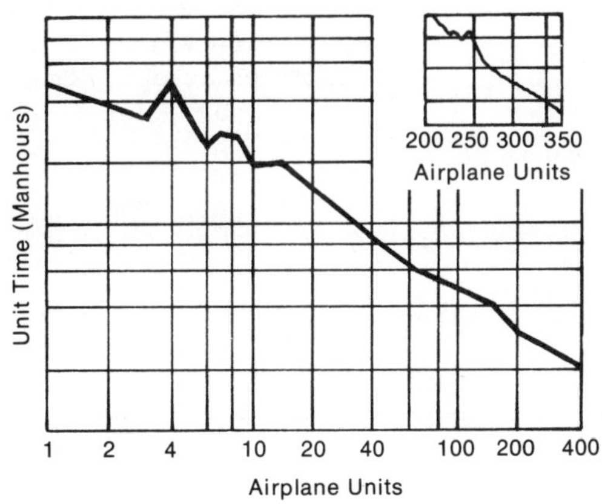

quired substantial "indirect" personnel. To meet this requirement and, in addition, decrease the ratio of indirect to direct personnel, most other indirect functions were reduced or virtually eliminated. The result was a tighter, more cost conscious, and more efficient commercial airplane organization.

Results were amazingly successful and proved that aircraft could be produced with high efficiency even at very low production rates. Conventional learning curves could be continued even with a 5- or 6-to-1 decrease in production rate, a conclusion not previously demonstrated in U.S. industry.

The massive efficiency improvement program preserved company integrity and prepared a foundation upon which product improvement, and therefore increased sales, could be built. A new and stronger division and company emerged.

DISCUSSION QUESTIONS

14.1. Do you believe that Boeing management reacted quickly to the crisis, or should they have started their recovery program earlier?
14.2. Why not lay off employees by seniority rather than following the layoff policy outlined in this chapter?
14.3. Discuss the task-force approach to reorganization and problem solving.
14.4. Why was it so important for Boeing to cut expenses?
14.5. Review and analyze Boeing's inventory control program.
14.6. How would "bumping" affect plant efficiency?
14.7. What is the effect of cutting indirect labor? What dangers are associated with such a policy?
14.8. Hypothesize about how this recovery program affected employee morale at The Boeing Company.
14.9. Using the systems approach, develop a model that illustrates the situation described in this chapter.

SELECTED READINGS

Argyris, Chris. *Organizational Change*. Reading Mass.: Addison-Wesley, 1970.

Carter, Martin, Weintraub, Walter, and Ray, Charles, eds. *Management: Challenge and Response*. New York: Holt, Rinehart and Winston, 1965.

Litterer, Joseph. *The Analysis of Organizations*, 2nd ed. New York: John Wiley & Sons, 1973.

UNITED FARM WORKERS
BOYCOTT GRAPES
N. J. INDUSTRIAL UNION COUNCIL
AFL-CIO
CONGRESS
INFLATION ↑
NOW MORE
WE ARE STEWING OVER
NOT EVEN A CHICKEN IN OUR POT

Part Three External Factors Influencing Managerial Decision Making

Decisions are affected by our frame of reference. The way we look at things, our priorities, and our goals are based on what we believe. Society in turn is a composite of the beliefs and ideologies of its individual members. It is extremely important to understand how societal values and attitudes influence the decisions that a manager makes and how groups of people exert pressure to secure decisions that will be compatible with their values and interests.

This section examines the values of society and the ideologies of special-interest groups and discusses the effects of these values on management's decisions.

There are many current social issues—for example, the environment, minority groups, and consumerism—that are relevant to today's managers. The manager who ignores the changing attitudes represented by these issues runs the risk of lawsuits or even punitive legislation. In our society, public opinion ultimately determines the ground rules within which managers and organizations must operate. Thus, to be effective, the modern manager must be alert to the internal problems of the organization as well as the external problems in the physical and the social environment. In this section, therefore, we will discuss how some of the major social issues of our time affect decision making in the organization.

Chapter 15
Values and Managers

Why do managers need to concern themselves with the values of society—or, more specifically, the values of their stockholders, workers, or consumers? Do societal values change? If so, how do these changes occur? This chapter considers these questions in relation to how social values shape management's environment and economic decisions. Management decision making inevitably involves social values—whether or not management is consciously aware of it.

A given set of information can be arranged in a number of different ways so that different conclusions are reached. The way one assembles the information, the type of decision system used, the decision criteria, and the goals to be achieved are all derived from the particular set of values each person possesses. The important question is why one value system is chosen over another. Often it is because of tradition—or political and economic interests.

Consumer buying patterns and habits, for instance, are ultimately determined by the individual's and society's values, which in turn are affected by economic, political, and social variables. The systems approach can help explain how values interact and influence each other and business decisions as well.

CHANGING VALUES

One factor that has interested some scholars is the *thrust* of a nondominant group to bring its values and its economic, social, and political position into greater prominence in the society.[1] In this way new values, which otherwise would never enter the dominant value system, are introduced into the mainstream of society. Groups effecting such changes have been called "thrusters." Events as well as groups can thrust new values into a society. In 1974, the fourfold rise of oil prices forced by the Arab cartel assigned a new value to oil and consequently to the use of energy in our industrial society. Other events that can induce new factors into a society and possibly change basic value structures (if the events persist) might include political elections, governmental coups, wars, inflation, depression, crop failures and other natural disasters.

[1]Neil W. Chamberlain, *The Place of Business in America's Future* (New York: Basic Books, 1973).

Values traditionally held by a group or nation seldom change without the impetus of some strong force. The penetration of a new value, therefore, must have some duration to survive and to win general acceptance in the framework of a larger value system. Persistence, then—or in physical terms, the force of the vector—is important in determining whether new values can change existing systems or become a part of them.

Social Forecasting

A matter of practical and immediate interest to most students of management is the development of social forecasting. Just as economic forecasting involves an attempt to predict the future state of the economy, social forecasting attempts to predict major social changes in the society. For example, corporations were totally unprepared to contend with the rise of consumer groups, civil rights legislation, and pressure from the environmentalists. Now, corporations such as General Electric attempt to anticipate which **special-interest groups** and what set of values are apt to emerge and, as a result, affect their business. Obviously, such forecasts are even more difficult to make accurately than are economic predictions. The objective, however, is to provide a number of alternatives, or "scenarios," with attached probabilities so that management can adopt the best strategy depending on what changes occur in their social and economic forecasts. Of prime consideration in each social forecast is the value system of the emerging interest group and the present profile of the society's dominant value system. General Electric is one firm that has been in the forefront of social forecasting.

Resolving Conflicting Values

Another critical question is how managers and societies make decisions that involve conflicting values. Individuals as well as managers face the problem of reconciling conflicting values. For example, a branch manager of a large national firm finds that the president of the company in New York has issued memos to him and statements to the press that the firm intends to follow all pollution regulations, and even to go beyond the present regulations and cease all industrial pollution. The manager also knows, however, that his future promotions and raises are determined primarily by the profitability of his branch of the firm. If he were to spend money to correct all pollution discharges, such capital costs would decrease or eliminate the profits of his branch for the next few years. Thus, despite the memos and press releases from the president, the branch manager makes only those changes necessary to keep the local pollution-control authorities off his back. Indeed, if he were confronted by the local press concerning the official company policy, he would not be sure what to say. The question in the local branch manager's mind is what the real values of his firm are. Traditionally, values have been directed toward profits. But now a new value has been expressed publicly by the firm's president. Should he believe it?

In another example, a branch banker has been instructed to make minority loans (which he considers risky and apt to affect his profit record) but ignores the directive until the bank management can convince him that another value besides profits will be considered in judging management success. It is conceivable that bank branches would be given a quota of so many dollars in loans to make to members of minority groups. If such goals were not met, then local managers would be asked to explain why. This approach would convince a local manager that the firm really has introduced a new (and conflicting) value into the system. Usually the local manager is quite right to need convincing, since top management may not have reconciled the conflict between profits and social goals before instituting the new policy and may continue to use traditional value guidelines (such as profits) to select and promote its officers. This suggests that when conflicting values exist, it is management's responsibility to determine the manner in which they can be resolved and applied in daily decisions. When no new operating technique is outlined, local managers have the right to believe that the past procedures, values, and guidelines still hold.

When we as individuals hold conflicting values, we are seldom rational in working out operational means to avoid inconsistencies in our behavior. Indeed, much of human behavior may be inconsistent. This is because most values are not totally formulated or not fully integrated into one's style of living. For instance, one may have to debate long and hard in deciding between two jobs—one that pays more money and another that is more interesting—if the criteria for a job are that it maximize both income and interest. In the case just mentioned, it may be impossible to maximize both conditions.

In making a decision, it is necessary to make a tradeoff. If you cannot maximize income and interest, would you be happier with more income and less interest, or more interest and less income? Individuals and businesses make such decisions every day. The more explicitly values and the tradeoff equation can be stated, the easier it is to make such decisions, and the more consistent one's decisions can be. Consistency assumes, of course, that values have not changed, and that by pursuing a given decision matrix, one's satisfaction is ultimately maximized within the given constraints.

Periods of crisis and unrest tend to stimulate interest in values. Often, crises are attributed to discarding traditional values and accepting new, untried, and radical values. The alteration or deterioration of customary life styles, behavior patterns, and basic institutions is considered by some to be a breakup of a society's value system. During such periods, considerable attention is devoted to analyzing the society's values and what are perceived as the new, emerging values. Charles Reich's book, *The Greening of America*, is a prime illustration of such analysis. The problem of analyzing values leads us directly into asking what we mean by "values" and how they are defined.

DEFINITION OF VALUES

Values serve an essential role for individuals and for society as a whole: *Values provide a mechanism for making choices*. For both society and individuals, it is necessary to have some framework within which to consider alternatives and select the best choice. In referring to either group or individual values, we are concerned more with the core set of values around which societies and individuals build their preferences than with simply listing preferences. **Core values** change less easily and create differences more difficult to resolve.

All along we have been assuming that actions are generally consistent with core values. Although this is true, periods do arise—particularly during transitional eras—when new sets of values challenge older ones for economic, social, political, or legal reasons. These periods generate considerable strain both in the society and in individuals as they attempt to reconcile actions stemming from conflicting values. In the past twenty years, civil rights and **affirmative-action legislation** have created conflicts between core values. For example, bussing school children in an attempt to integrate both black and white schools has been responsible for a great deal of conflict between the attitudes of **racism** and group superiority versus those of racial equality and non**ethnocentrism**. In 1975, the case of the enclave of South Boston Irish protesting the bussing of black students into their traditional white schools and their children being bussed to black schools is typical of this value conflict. It is not simply a question of black versus white (although that is certainly part of the issue); the South Boston Irish would object to any group's breaking their cohesiveness. This is a conflict of the core values of community racial ethnocentricism versus the legislated civil rights values of racial and ethnic equality. Much of the population of society at large holds both attitudes simultaneously. Needless to say, the confrontation between these opposing values—especially when compounded with a group's identity values—is bound to cause conflict for the society as well as for the individuals involved.

The next question that arises is whether social values are the same as personal values. Obviously they are not, since there are so many subgroups within the total society and so many subvalue systems within different age groups, even within the same family—and each subgroup has its own values. In the last century, one of the major debates in social science was between Karl Marx, who argued that the economic structure of society created its values, and Max Weber, who argued that personal values (such as religious ones) determine the economic and social structure of society. Richard Tawney, an English Socialist, argued that they were both right: that it was essentially the interaction between core values and economic institutions that creates societies' ultimate values and behavior. In recent years, Charles Reich has argued that psychological attitudes, or what he calls "consciousness," develop independently and can successfully modify institutions and society. On the other hand, historian Andrew Hacker believes that our own institutions eventually force us all to accept the values of our organizations. Whether or not Emerson was right in arguing that the organization was merely "the lengthened shadow of one man" is another important question for students of management to consider. The debate between values and institutions has raged for over a century and is still going on.

Social values can be categorized in at least four different ways: (1) as representing the spirit of the times, (2) as expressing the basic purposes of the society, (3) as defining norms of conduct, and (4) as sanctions for behavior.[2] Each of these roles will be discussed in the following section.

Representing the Spirit of the Times

According to the German philosopher Oswald Spengler, values express the essential "spirit of an age." They provide a particular way of looking at the world and at man himself. To Spengler, the core values of a society could be seen in the society's architecture. For instance, he suggested the religious orientation of European values in the fourteenth and fifteenth centuries was expressed by the Gothic architecture of the time. He argued that the spiritual and other-worldly nature of the Gothic pointed arches and spires, in contrast to the more refined and practical orientation of classical Greek architecture, were indicative of the essential differences in values between these two civilizations and eras. Spengler might have agreed with some modern architects, who have said that the great square boxes built in New York and most major American cities in the past several decades epitomize the values of a materialistic business society, whose major concern is to maximize the amount of floor space on a site as cheaply as possible. Thus all the adornment and design of medieval cathedrals or Greek temples have been replaced by strict, machinelike functionalism.

This analysis suggests that the priorities of materialism and technological achievement have placed the highest value on *efficiency* in modern American society. There is undoubtedly some truth to this, since business has been our central preoccupation, and business is primarily devoted to producing goods to maximize profits. Although in recent years, profit maximizing as the prime goal of the corporation has been questioned, nonetheless, few people would deny that efficiency has become an end in itself for much of our society. Thus, our boxlike architecture expresses our current preoccupation with the practical and the efficient, whereas the cathedrals of the Middle Ages expressed a central preoccupation with the religious and the spiritual. Archaeologists of the future would hardly be inclined to suggest that our greatest buildings and monuments were used for or influenced by religion. Instead, they would probably deduce that our society was firmly founded in commercialism. Our greatest engineering achievements —tunnels, bridges, rockets, missiles, and satellites—were either for commercial or military purposes. Archaeologists of the future

[2]Chamberlain, *The Place of Business in America's Future*, p. 14.

would have little trouble determining our major value orientation: a secular, commercial-military society in which the elites of business, government, and the military shared the highest status.

The famous sociologist Pitirim Sorokin has explained that the culture of a nation represents an integrated value system.[3] To change part of a value system results in changes in the rest of the society—laws, commerce, and mores. (This is a variation of the systems approach.) Some scholars argue that a particular outlook infuses all of a society's major institutions. If this so-called spirit of the times is secular and intellectual, then religious and emotional behavior are supplanted by respect for materialism and technology. This analysis helps one realize that a "modern industrial outlook" may differ radically from a "mystical, oriental" one or from a "religious-ascetic, nonscientific" one. Anyone doing business in other parts of the world soon realizes the basic differences among the value systems of foreign countries. Western businesses versus Eastern or developed nations versus underdeveloped all have different value orientations, to some extent. Today's "spirit of the age" is dominated by a secular industrial philosophy of modernity, although it often takes the form of a dual culture. Such dual cultures exist when "modernity" arrives at a beachhead in a metropolitan city in Africa or South America, while a few miles away in the jungle, the traditional society lives oblivious to the values and outlook of Western industrialism.

During the last half of the twentieth century, if the antimaterialistic values of such special-interest groups as the anticonsumptionists and the environmentalists were to become stronger and more permanent, they would represent a basic challenge to the general societal value of materialism. Perhaps the energy crises of the seventies will be more effective at changing the basic values and policies of our society.

Expressing the Purposes of Society

Social values can also be analyzed in relation to the basic purposes and goals of a society. For instance, putting a man on the moon was a specific goal of the United States, while the spirit of the age is typified in the value placed on science and technology, which permeated the society and made it possible to reach the moon. Again, if the values of **liberalism** and **individualism** are part of the spirit of our times, then pursuing antitrust as a policy to foster competitive markets is a logical social value.

Values as purposes and goals of society are, in a sense, more concrete than values that are considered part of the spirit of the times, since values that can be translated into specific governmental policies may appear more useful than more abstract values. Nonetheless, both are critical for society to operate. Similarly, these more specific values can be applied directly to the corporation (such as the value to maximize profits) and thus become specific goals of the firm.

[3]Pitirim Sorokin, *The Crises of Our Age* (New York: Dutton, 1945).

Defining Norms of Conduct

Beyond the more general levels of abstract social values, there are specific values that serve not only as goals but also as rules or norms of conduct or behavior. In the case of the firm, the goal of maximizing profits can be translated into a guide for day-to-day decision making. Thus, a general value—maximizing profits—can be brought down to an actual operational level. During transitional periods, it is possible that the assumed objective of the firm, profit maximization, is expressed quite differently at the operational level. Nonetheless, for values to be effective as both part of the spirit of the age and as social goals, they must be operationalized into conduct. If abstract values such as honesty and fairness in business dealings (which spring from either a religious ethic or a humanistic philosophy) are not operationalized, they will eventually be discarded.

Values as Sanctions of Behavior

Social values not only set the style of an era, identify the purpose of the nation and the organization, and develop operational rules of behavior to achieve these broader values, they also serve the purpose of coordinating and integrating activities within a society. Certain values may serve as sanctions for such behavior as applying coercion to those unwilling to follow core values of the society, for example, the protection of human life to maintain the security of the society. Social responsibility must make some use of sanctions (even if only in the name of public opinion). Philosophers have argued that a society must reach a general consensus on values before "right" conduct can be defined. Management must understand the accepted values of the society in addition to recognizing what values are in conflict or in transition to be able to set operational norms that are consistent with public opinion. In fact, the corporation can spare itself much of the cost and trouble of governmental legislation forcing the corporation to behave according to public desires by following public opinion dictates in the first place.

APPLICATION OF VALUES

Now that we are aware of some of the different dimensions of social values, let us look more carefully at the way values can apply to groups and individuals, in an effort to make the public and private policies of individuals and corporations more understandable. There are three basic sets of values of interest to us here: (1) *focal values*, or the values that constitute the "good life" for the consumer, manager, and worker; (2) *constitutional values*, or the values that determine, for instance, the relationship between worker and employer—which may range from absolute hierarchical authority at one extreme to total participative management at the other; and (3) *distributive values*, or how goods (wealth and income) are distributed within a society—Who gets what, and how? This last includes not only real income and wealth but also how status, prestige, and power are allocated.

Often, the values of a dominant group influence most of the values in a society, as the values of business did in this country until the great stock market crash of the 1930s. Traditionally, the major power groups came from either the military, the church, or the business community. Today other power groups have arisen, as a portion of political power has shifted from traditional elites to other special-interest groups.

These three types of values—focal, constitutional, and distributive—make up the basic value sets that determine groups' and individuals' behavior, establishing not only their identity but also making possible coherence and coordination with the rest of society. For instance, a focal value of materialism may necessitate a belief in constitutional individualism, which in turn results in distributive inequality. In contrast, a focal value of egalitarianism may be achieved only with a constitutional value of a strongly centralized government to achieve greater equality in income and wealth distribution. Or, a focal value of militarism results in constitutional values dictating hierarchical order and distributive values of inequality in rank and deference. Religious focal values can range from Catholic centralism to Lutheran autonomy—both with concomitant privileges of office and rank—to pure forms of communal living. The basic values must be commonly accepted within the group or the group will begin to unravel.

As we have suggested, the interest group that becomes dominant over time succeeds in impressing its values on society. As competing interest groups vie for power and/or leadership, the society's values either begin to splinter through pluralism, the competition is coerced into the dominant values, or the competition conforms as part of a "loyal opposition." Thus, the focal values of competing groups tend to be submerged, with differences occurring mainly in distributive values.

Governments tend to be supporters of the present interest or, at most, mediators of cautious change. Families, schools, and religious groups usually are the institutions that pass values on to succeeding generations. Thus, in countries such as China, where a drastic value reorientation is taking place, the state must try either to break up the traditional family unit or to force it to accept and teach the new values; religion and traditional family values are banned. In such a case, the schools are much easier to control than is the family and become change agents for the young. In fact, the most radical element in China, the Red Guard, is composed of teen-aged schoolchildren.

In Chapter 22, we will use this information as a basis for suggesting alternative scenarios that American capitalism may follow.

American Values of the Past

Marcel Proust, the great French novelist of the early twentieth century, wrote a seven-volume work entitled *Remembrance of Things Past*. In it he describes and dissects the values and behavior of his society, offering the reader the opportunity to contrast the values of the society just past with present values. Noting where changes have occurred highlights the new values; where old values have remained, we may now understand them better. In a sense, the past is the laboratory of the social sciences. To understand the dynamics of values in general and the contemporary values of business in particular, it is helpful to look briefly at the past as a touchstone of what has changed, what is changing, and what still endures.

American values were shaped largely by the early settlers in the Thirteen Colonies. The Declaration of Independence and the Constitution were the first American documents to describe the outlines of a society that valued independence and democracy and minimal governmental control. Underneath the political assumptions ran a value placed on private property and, moreover, the security of that property from the government, the military, and the church. Thus, built into the system were a search for private gain and the enthronement of private property—as both a goal and a motivation of the economic system.

The Puritan, or Protestant, doctrine that linked material gain and divine favor served as a practical justification for a work ethic that could provide salvation both in this world and in the next. This set the stage for increasing secularization—and a focus more on this life and material things than on the spiritual "next life." The rise of America's first sharp businessmen, the Yankee traders, was an outgrowth of Puritanism. New England ministers would eventually lecture their flocks that the earlier spiritual outlook had been replaced by an interest in worldly riches. But by then, the Puritan had become a Yankee trader.

In addition, American society, without the aristocracy of English or European feudalism to fall back on, had nothing on which to build a social hierarchy except the wealth of private property. Therefore, economic, social, and (historically) spiritual values all coincided to emphasize the pursuit of wealth and the value of private property. Indeed, private property was often valued above life itself. An apparent change in American values toward private property occurred during a riot in Washington, D.C., in the late 1960s, when President Johnson ordered police not to shoot looters, directly reversing the traditional sequence of placing a higher value on the security of private property than on the value on human life. (This gets us ahead of our story, however.)

By 1830, when the Frenchman Alexis de Tocqueville visited the United States and wrote his famous analysis of American society, the main value he found was a strong belief in equality. He explained how this single value was the underlying current of American laws and mores. At the same time that this value was increasing in strength, a dominant business class was emerging in American society.

The Civil War furthered industrialization in the North, although it destroyed the agricultural economy of the South for nearly a century. The main business and political currents of the country swept from the North Atlantic states, west across the northern middle states, to the West Coast—leaving the South in its own dire economic backwater. This created two societies and two economies: one based on business-industrial values and the other on agricultural values. Two sets of values and two economies, therefore, coexisted between the rest of the country and the South, until the prosperity following World War II began to incorporate the South into the rest of the industrial economy. As this happened, values in the South began to change and become more nearly like those of the North and the West. But for nearly a century, the United States was a dual economy with a dual value system.

By the twentieth century, the businessman was king. Calvin Coolidge's remark that, "The business of America was business" was quite apt. Only with the Great Depression of the 1930s did American business values finally face a grave public challenge. Business has never been the same since; the depression brought a collapse of confidence in business leadership and its values, which has never been fully restored. The era's disillusionment with business helped give birth to the **"welfare state"** (in which the government becomes responsible for an individual's welfare) and a new set of values in government. If World War II had not come along to unite the country and banish mass unemployment, it is interesting to consider what changes would have occurred in our society. Many believe that business values recovered their former status in society after the war. However, the rise of a counterculture value system in the 1960s, the massive business regulations passed by Congress and the increasing use of controls in the 1960s and 1970s, coupled with the dramatic fall in confidence in business recorded in public opinion polls in the 1970s, and compounded by the energy crisis point out vividly that although the Great Depression was a watershed between traditional capitalism and welfare capitalism, the present values of society are changing capitalism even further.

Thus, the value configuration in our society can trace the effect of egalitarianism on private property and unregulated business. From this historical perspective, it is apparent that the idea of equality in our society increasingly means equality of income and position and not just of opportunity or, as earlier, of equality before the law. As this interpretation of equality has changed, so has the emphasis on the security and place of private property in the society. More and more sanctions, taxes, and government regulations have restricted the use, ownership and security of private property to accord more with popular values, which stress equality of opportunity and, more recently, equality of condition. It follows that to achieve these value goals, governmental policies toward business have changed considerably over the past four decades.

American Values of the Present

The U.S. economy is a form of **welfare capitalism**. The most important interest may still be business, but business has lost much of its dominance. It no longer "runs the show" or "calls all the shots." Even though the economy is increasingly service oriented, the traditional industries are still the most powerful. But the values of society are no longer seen by the majority of the public as being identical with those of business. Thus other special-interests and power groups are assuming growing importance.

Focal Values

The traditional value of materialism as seen in private consumption and accumulation of private property is dominant in our society. The counterculture's nonmaterialistic values of the late sixties have not overwhelmed most Americans' basic materialistic values. Environmentalists, however, have successfully put forth a whole new vocabulary of values. In some instances, these new values have the force of law behind them and even contest the more dominant and traditional value of economic growth. Environmental impact studies must be made for new projects. And, on occasion, environmental values are actually taking precedence over economic growth. This is occurring even if the change tends to lower the standard of living, as in the dramatic rise in automobile prices, for instance. In the case of the automobile industry, many argue that short-sighted, unimaginative management is responsible for the fact that smaller cars, cheaper to buy and maintain, and still meeting pollution and safety requirements were not developed long ago. In this case, management seems to have been unaware of new value trends (and economic facts) that, if studied, would have dictated a fundamental change in direction and research for the auto industry. Currently, the decline in our traditional Protestant work ethic is part of the reason for the dramatic decline in U.S. productivity. If the trend continues, the result may be lower standards of living for most of us.

This changing relationship between the values of work and materialism is the consequence of many factors. First, youths whose backgrounds have not stressed the traditions of an industrial work ethic have been introduced into the industrial assembly line. Second, the spread of affluence in society has had the effect that many youths from middle-class homes find little meaning in working to accumulate more "objects" in their lives. As society becomes increasingly secular, only youths from backgrounds that have a tradition of the work ethic seem to work in a highly productive manner. Also, unions in many cases have established work barriers, and pay scales are seldom related to output. Nonetheless, it is apparent that the traditional values of thrift and hard work espoused by Benjamin Franklin are no longer taught with the same fervor in public schools. As hedonistic secular values compete with traditional work values, adverse economic effects occur as mentioned in regard to the level of productivity.

Constitutional Values

The growth in the size of our population (from a few colonies scattered along the Atlantic to a population of over 210 million stretching across a continent) and the parallel growth in size of our business organizations (such as General Motors, which has an annual sales volume equal to the GNP of major nations) have had major effects on the nation's constitutional values. The net result of this growth has been a move away from individualism and autonomy, which the individual and the firm took for granted in the past, toward a hierarchical form of organizational and governmental control. Sheer size—whether in population of a nation or in the number of employees in an organization—seems to dictate less freedom for the individual and more attempts to coordinate and control the population or the employees.

At the same time that corporations have grown larger and control has become hierarchical, the firm has seen the rise of more outside interest groups who challenge the authority of management. Environmentalists, civil rights groups, and consumer groups have all sponsored legislation (often successfully) that has removed an increasing number of decisions from management's hands. In one sense, this can be perceived as a rise in countervailing power. As the corporations have grown larger and more centralized, other power groups in society also have developed to challenge corporate decision making in many areas. The direct result of this interplay has been that the federal government has increased its role in society as an arbitrator between competing power groups. Government has become more and more a regulator and coordinator of the economy. As this happens, the corporation's autonomy declines and the value of individualism is replaced by reliance on a centralized government to make more decisions.

Thus, the value shift has been from expecting the individual to make decisions, then to corporate management, and now to expect the government to solve more of the individual's and society's problems. Whether or not this is good, it is a change. Just as the individual and his family were once expected to take care of the elderly, now the government has created Social Security for this purpose. Just as business was once responsible for decisions involving its own immediate environment, now governmental regulatory agencies protect the environment. In both cases, government became involved because of a breakdown or failure of the individual or the firm to handle situations in a manner satisfactory to the majority of the public. The result is a change in the public's values and expectations about who is responsible for solving basic problems. The trend has obviously been tilting steadily from smaller to larger units. There is little indication that the public has deserted this newer value system. The effects of this change on the future of business are discussed below, and also at length in the last section of this book.

Distributive Values

The traditional connotations of **equality**—as equality before the law, or equality of opportunity (particularly equality of education)—are being reinterpreted by minority and liberal groups to mean "equality of actual condition." This new interpretation implies a redistribution of income and wealth. The approval by many of a negative income tax and a floor to poverty level—higher than the average family income of most other nations—is evidence of the widespread acceptance of this newer meaning. (The authors do not attempt to argue the value of this.) Political interest groups at the polls will eventually determine whether such an interpretation of equality will become the dominant value. Certainly, equality in some form has been the major shaping value of our social and economic system.

Redistribution of money in the past three decades has mainly affected income rather than wealth. The laws that permit trusts to be established and make it possible for family fortunes to be inherited suggest that the already wealthy in the society have a distinct advantage over their poorer counterparts, and one that they have been able thus far to protect against the mass vote of the have-nots. Perhaps the major factor affecting the dispersion of wealth (though not income) is the level of the stock market. Since most of the great fortunes in this country are invested in stocks, a rising stock market makes wealth concentration greater, and a low or falling stock market makes it less so. Thus, one must be cautious about the base years used to prove whether or not the rich are getting richer and the poor are getting poorer.

In summary, business is still the single most powerful interest group influencing American values. Despite a general disillusionment with business, both business and the bulk of the public share a common materialistic value, judging others and themselves in terms of consumption and accumulation of private goods and property. Wealth still remains a major goal for the individual and for the firm. However, the loss of considerable respect and power by business cannot be ignored. Paralleling the decline of the work ethic in our society is a drop in economic productivity, and with it an increasing problem in our balance of payments and in maintaining our traditionally high standard of living.

Further, public values have shifted dramatically in the past half-century. Individualism and autonomy have been supplanted by the hierarchical organizations of business and government, which stress order and integration as major values. The net result is that government has been given the augmented role of solving individual, corporate, and societal problems. In the process, the value of equality is shifting in interpretation from equality of opportunity to equality of actual condition, as blocs of voters demand income and material goods not necessarily acquired in accordance with the traditional work ethic. This trend has major implications for productivity and even for the future of capitalism in the United States.

Predicting Future Values

How are social values in America apt to change during the next decade? Is there any basis for predicting? Is there a range of possibilities that might reasonably occur? Can we identify a range of possible "external vectors," or occurrences that may shape our society—such as wars, droughts, or plagues? Difficult as it is, business is forced to make predictions or else not stay in business. Buying an inventory, contracting to build a new plant, putting money into research are all decisions involving the future—either in the short run or the long run.

It is still a reasonable assumption that the value of materialism and the desire for private consumption and private wealth will continue to dominate American values for the rest of this century. (Barring a major change in our economic-political system.) This is not to say, however, that other values developing even now will not alter the present form of materialism. Environmental values will define the responsibility of corporations as they try to maximize profits and will limit private property even further. Private beaches and waterfront and other privately owned natural resources (possibly oil and mineral rights as well) may become state owned. Thus, the rights of private property will continue to shrink, even though private property is not likely to be totally disallowed soon. The standard of living of recent years will be difficult to match. The middle class will be harder pressed financially, and materialism and security could assume more importance rather than less.

Minority groups will continue to expand the interpretation of *equality* in the United States to mean "equality of condition." Through the passage of various welfare measures, they will be successful in redistributing money to the lower end of the income spectrum. The result will be that the range in the real standard of living will narrow. The standard of living will be lower particularly for those who rely on income and not on inherited wealth. All this suggests that it will become harder to maintain commonly accepted upper middle-class living standards in the face of higher deductions and taxes. At the same time, living standards at the lower end of the spectrum will improve, reducing the distance between the lower class and the middle class. Real income will be less important than the particular style of life that individuals pursue. The differences in style of life will accent the differences between basic values, such as hedonism and immediate gratification versus thrift and self-sufficiency. Conservative and radical life styles will not disappear—rather, within limited real income, such differences will be demonstrated by those families who stock their basements with two years' supply of food in contrast to thse who simply "live it up."

It is reasonable to expect the continued displacement of business values and business power as the dominant value and power structure of our society. Certainly, business will not be abolished; it will, however, become only one of a number of plural groups whose values will require compromises to remain viable. The government is apt to be less responsive in the future to the values and demands of business; even regulatory agencies, which have for generations been safely on the side of business, will become less flexible toward business.

SUMMARY

This chapter has pointed out that values serve a crucial role in decision making. They become the standard by which actions and goals are measured. Thus, given the same set of facts—but different values—the decision maker is apt to arrive at a totally different decision.

A systems approach to decision making means that values and all other major variables (economic, political, and social) are interrelated, so that it is necessary to look at the total set of data—values included—to understand how individuals act and how businessmen make decisions.

American society may be in a period of value change, particularly as the traditional work ethic, or value system, is replaced by a newer, hedonistic work ethic that may well be adversely affecting the national level of productivity. And as productivity declines, so does the standard of living.

Thus, values are closely related to other economic–political data in affecting business and production. Values therefore are important factors in understanding the decision-making process.

DISCUSSION QUESTIONS

15.1. Do society's values affect business? If so, how?

15.2. Why have the values of business been the dominant force in American life? Will they continue to be?

15.3. Do you find conflict between traditional and emerging American values? If so, which pairs of values do you find most conflicting?

15.4. Which values cause the greatest, or the fewest, differences between you and your parents?

15.5. What happens when values change?

15.6. Of the values discussed in the chapter, which do you believe are the least important and therefore subject to change?

15.7. How do you define values?

APPENDIX
The Little Leaguer

Recently, an incident was reported through the news media telling of the problems of a member of the Little League. His football team was sponsored by the fathers, who also had an agreement among themselves to participate personally in the clean-up chores. This Little Leaguer's father, a physician, sent a note with his son to the officer of the Little League explaining that he was too busy to become personally involved in the duties of a Little League father, but that he would be happy to pay for somebody else to substitute for him when it was his turn.

The group of fathers informed the physician that this was not acceptable. If he chose not to show up, his son could not continue as a member of the boys' group. The father refused to cooperate and the boy was excluded from participation in the football team and its activities, whereupon the father sued the Little League to have the court force reinstatement of his son according to his terms.

1. What values can you see exemplified in this case?
2. What assumptions did the father of the Little Leaguer make regarding the values of the other fathers?
3. Should the courts get involved? Why or why not?
4. Would it make any difference to you if the father-physician were a member of the Spanish-speaking minority group in the United States? Discuss.

SELECTED READINGS

Bell, Daniel, and Kristol, Irving, eds. *Capitalism Today.* New York: New American Library, Mentor Books, 1971.

Hickman, C. Addison, and Kuhn, Manford H. *Individuals, Groups and Economic Behavior.* New York: Dryden Press, 1956.

McGiffert, Michael, ed. *The Character of Americans: A Book of Readings.* Homewood, Ill.: Dorsey Press, 1964.

Monsen, R. Joseph, Jr. *Modern American Capitalism.* Boston: Houghton Mifflin, 1963.

Sutton, Francis X., et al. *The American Business Creed.* Cambridge, Mass.: Harvard University Press, 1956.

Warner, W. Lloyd, and Martin, Norman H., eds. *Industrial Man.* New York: Harper & Bros., 1959.

FRIENDS OF THE
FARM WORKERS
AFL-CIO
UNITED FARM WORKERS
BOYCOTT GALLO
N. J. INDUSTRIAL UNION COUNCIL
BOYCOTT GRAPES
N. J. INDUSTRIAL UNION COUNCIL

Chapter 16
Interest Groups and Management

Textbooks on management tend to ignore the crucial impact that major interest groups have on corporate decision making. The real-world influences that top managers must consider in developing policy and implementing it, which involve "political" influences both internal and external to the firm, are seldom discussed—either because they are considered an unfit subject for students to study (as was sex during the nineteenth century) or because they are thought to be outside the purview of a management text and are consequently left to courses in political science. Yet top managers spend most of their time on decisions that relate to these "political" problems. In large firms, middle management is generally responsible for specific issues involving production and sales, while top management must solve general problems concerning managers, takeovers, antitrust suits, political and charitable contributions, and especially placating major interest groups such as environmentalists and stockholders. In this chapter, we will identify who the major interest groups are and discuss the goals and points of view of these groups.

THE ROLE OF INTEREST GROUPS IN BUSINESS

Management faces many diverse questions and problems involving interest groups. For example, How can we be sure that the board of directors will vote favorably on proposed programs? How can we satisfy the critics among the directors? How can we keep dissident stockholders from filing lawsuits or starting proxy fights? How can we get lower and middle management to support their chief executive and prevent them from establishing unions? How should we deal with the leadership of unions already in existence? What strategy should we employ with affirmative action, particularly in relation to the recruitment and promotion of blacks, women, and other minorities? How to gain access to, and the support of, legislators and bureaucrats at the federal, state, and local levels? How to organize the firm's activities to cope with governmental and regulative problems? To whom and how much to contribute to political campaigns? How to avoid conflict with consumer and environmental groups? How to lobby for the kinds of environmental and consumer legislation that may be in the best interests of the stockholders and the consumers? How to cope with pollution and environmental problems without damaging firm profits unduly? How to meet with the press on all these issues and present the company's position? How to meet with security analysts and gain their support? These are only a few of the issues that management must face. Clearly, these are significant and difficult questions that should be dealt with in any study of managerial decision making.

How Does Management Influence Interest Groups?

How management influences interest groups is perhaps the most crucial problem for managers to consider. In each case, the technique used to influence those groups that tend to constrain managerial decision making may be different. For example, to deal with stockholders who are represented by a board of directors, management may be able to nominate those they wish selected for the board. Management has some further advantages in such a situation in having access to more information than either stockholders or directors can usually obtain. In this way, management can exact control and influence over an important power group. In those firms where "inside" boards are common—that is, directors are recruited from officers of the firm—management will doubtless control the company regardless of stockholders' interests. One might argue that since the officers are stockholders, they necessarily consider the stockholders' interests. Although this is true to a point, it is easy to see that management's and the shareholders' common interest may be divergent in some cases, at least in the short run. Fear of shareholder suits and proxy battles, however, tends to limit how far management can ignore shareholders' dissatisfaction.

Management's attempts to influence employees—middle management, unions, and minorities—are related basically to the employees' attitudes toward their relative position, the job environment, and the rate of pay within the firm as compared to similar firms. Remuneration tends to vary radically by industry, and so does labor unrest. Yet there is not necessarily a direct correlation between pay and labor dissension. Thus, it is apparent that management's relationship with employees and union leaders has an important psychological effect. Illegal influences undoubtedly exist in cases where organized crime controls unions and businesses, to the disadvantage of either the employees or the firm. This kind of influence is all too common.

The influence of business on government at federal, state, and local levels is usually a function of two things: access and wealth. Access to political leaders is necessary for the firm to present its interpretation of existing regulations and proposed legislation. Access is easily obtained to politicians if they are of similar ideology as management, if they are from a region where the firm may be a major employer, or if the firm has been a major contributor to their political campaigns. Management may employ **lobbyists**—that is, individuals representing organizations or interest groups, whose purpose is to influence governmental agencies and legislation—to aid and advise them. Favorable access varies from the chance merely to state one's case to the virtual assurance that one's case will be looked on (or voted on) favorably. Besides influencing legislation through overt political contributions, management must hope to contact those politicians who might be sympathetic to a particular position, perhaps because of the politician's own background, affiliations, or ideology.

Before proceeding further, it should be noted that *lobbyist* need not be a "bad" word. Because of the way the American political system works, it is necessary for industries and major corporations to be able to present their information directly to heads of committees and agencies that make legislation and enforce regulations. This process of communication should be the lobbyist's basic function. In reality, it is a critical function; indeed, those firms that fail to lobby are almost inevitably hurt—given a political system of pluralistic and competitive interests.

Wealth is the other key to influence in government. Wealth can be used to recruit skilled lobbyists, publicize policies, build organizations, and make political contributions. Recently, laws have been passed limiting contributions and requiring that they be made public. This action will limit the vote of dollars of the few (and as a result, the influence of business in government) in favor of the vote by ballot of the many.

Firms usually leave to the U.S. Chamber of Commerce or the National Association of Manufacturers (NAM) the attempts to influence public policy on topics that do not directly affect their firm or industry. Most industrial groups also maintain representatives in Washington, D.C. Legislation relating to specific issues of the firm or industry, however, is dealt with individually. Large firms generally have their own representative in Washington, and their own industry representative in most state capitals. But how does the small or middle-sized business influence public policy that may directly impinge on its interests? The most common solution is to hire a Washington law firm that specializes in representing business interests. The small corporation might also attempt to enlist the support of its local congressman. Since the small firm obviously cannot afford the most expensive legal help in the capital, it must rely mainly on what the congressman can do. If the firm is small, it may carry little clout with the congressman—thus, its main hope is to work with a group of some kind to increase its ability to bring pressure.

Groups are used by even the largest firms. Industry groups and special-interest groups engage in attempts to find common-interest grounds among different industries and perhaps also with unions to increase the total political impact on government.

Interest groups, as they combine and organize, inevitably wind up controlled by a few individuals. This is true whether it is a student-body group at a university, the NAM among businessmen, or the AFL-CIO among labor. The majority of most group memberships is seldom active. The actual control, therefore, passes to a self-perpetuating set of elected officers. The kind and direction of policy in organizations is similarly controlled by an inner circle. Therefore, for the firm to get the NAM or its industry group to take a position and pursue it means that the firm must first be able to rally the support of its own industry leadership. This generally requires contact and lobbying with the inner circle of leadership.

Floyd Hunter, in his interesting book *Top Leadership, USA*, found that the nation's elite in most occupational fields knew one another better than they knew the lower-strata members of their own organization or occupation. He found that an informal network of friendships, committees, clubs, customer relations, financial affairs, and social interactions was shared by national leaders. The same network exists among regional and local political leaders. Thus, we find leadership elites cut across occupational lines more easily than they cut across stratification in their own groups and field. Knowledge of how to gain access to group elites to present one's case is critical for success in influencing policy that may potentially affect one's own organization or firm.

Business as a Power Group

Business has been the leading power group from the beginning of our industrialized history. From the Civil War to the New Deal, neither agriculture nor labor was as influential in determining public policy. Over the past forty years, the power of business has ebbed considerably. Even though business would still be counted as the single most important group, it now must share control of public policy with other major power groups. In many ways, this loss of control over national policy is the major event in business history in recent decades. While business must struggle to influence presidents, congressmen, and government agencies, it now does so competing with labor, minorities, consumers, environmentalists, and other special-interest blocs. The following list best summarizes the main strengths and weaknesses of business.

Main Advantages in Influencing Public Policy

1. Ability to finance advertising (buying television or radio time); to send out letters, finance trips, make campaign contributions. But the new campaign contribution laws have seriously eroded the *legal* ability of business to make direct financial donations to political campaigns.
2. Access to elites who control groups and to public officials—all of whom may share clubs, be on the same committees, or have mutual friends.
3. Skillful Washington representatives—that is, business can hire talented lawyers and public relations firms to lobby for their interests when necessary.
4. Managers who generally possess organizational skills necessary for planning legislative campaigns. These top managers also have contacts and the ability to raise money for political campaigns.

5. Business hires labor—that is, it has power as an employer in local, regional and national firms.

Main Weaknesses in Influencing Public Policy

1. Lack of ideological and organizational unity in business organizations in contrast to labor's AFL-CIO.
2. Small numerical voting strength in comparison to labor. However, a greater percentage of businessmen vote, which helps in part to make up for their fewer numbers.
3. Minimal national political planning by national business groups to assist key local political elections. Labor is far better at planning and being active in critical elections.
4. Inadequate recognition that to gain congressmen's eventual vote, early financial aid for their elections is extremely helpful. Most firms tend to avoid participation in local elections and wait to influence elected candidates. By waiting, they may never gain the politicians' maximum support.
5. New laws restricting political contributions diminish the political power of business.
6. A strong populist ideology throughout the nation favors increasing governmental regulation of business, and may lead to an antibusiness political atmosphere.

All of the weaknesses suggest that business could achieve greater political influence by more and earlier planning and participation in the political process. Managers appear reluctant to pay much attention to the political process—whether from lack of interest or lack of sophistication—until an issue hits them directly. By that time, it is often too late to be effective. With recent legislative limitations on campaign contributions eliminating a major source of corporate political power, business will have to give greater thought to the political process earlier—as labor does—to attempt to maintain its political influence. As other interest groups gain greater advantages in determining public policy, the inevitable result will be legislation and regulations unfavorable to business. In the future, business will therefore need to work harder to gain the acceptance of its positions and arguments by the public and by Congress. Finally, the simple-minded and illegal practice of trying to "buy" politicians and votes can only backfire on businessmen who still maintain such naïve political methods of influencing public policy. It is critical for new managers to be aware of these issues.

MAJOR INTEREST GROUPS INFLUENCING MANAGERIAL DECISION MAKING

The following section considers the most important interest groups and their effect on managerial decisions. These include: (1) stockholders, or investors, in the corporation; (2) employees—the people working for the organization; (3) unions, or the representatives of the workers; (4) minority groups, including women; (5) government units at the federal, state, and local levels; (6) the military, and those organizations supplying goods and services for defense; (7) consumers and environmentalists; and (8) the press.

Stockholders

Although boards of directors are the major group of stockholders that executives worry about, dissident stockholder groups have become increasingly common. In recent years, the annual meetings of many well-known firms have become difficult times for managers, primarily because such meetings create a forum for direct confrontation between groups questioning managerial actions. Traditionally, if management could document reasonable company performance, they had little to worry about from stockholders. Failure to declare a dividend has always sounded the alarm to stockholders that something was wrong with the firm and perhaps new management was needed. Thus, managers have long been reluctant to provide stockholders with such an opportunity. Nonetheless, too frequently in the last few years many firms have had to omit dividends, with the result that they have had to hold off stockholders or at least pacify them in hopes of better performance, before stockholders have organized attempts to vote in a new board or until the old board decided to put in new chief executives. The more recent confrontations at annual meetings between environmentalists and other public-interest groups is a less familiar problem for managers.

For many years, being on a major firm's board of directors was considered simply a profitable honor. This is no longer true. With the increasing number of stockholders' suits against directors for not attending to their legal responsibilities of watching after the stockholders' interests, directors are forced to be more involved (or at least more watchful) over management. Former Supreme Court Justice Arthur Goldberg, for example, turned down a number of directorships when he felt he would be unable to scrutinize the activities and performance of the company adequately to protect himself fully against stockholders' suits. A new era seems to have dawned in the role of boards of directors. Even though management in many cases picks the directors, placing large numbers of employees on the board (making it, in effect, an inside board), as Standard Oil of New Jersey has done, still the trend is to select representatives from various groups in the society to be members of the board, as General Motors has done. In essence, then, boards of directors are becoming less passive and more representative of other interest groups than just stockholders. The implications of these changes for the future behavior and decision making of the corporation are profound.

The trend toward more active participation of the board of directors in management or top management supervision has both positive and negative aspects for efficient management. Some of the benefits are that this action can bring more expertise to bear on company problems. Also, managers may be more alert, knowing they will have to account more carefully for the firm's performance. A negative aspect of more active participation is that management could become overly cautious, which could slow major decision making by requiring greater board approval on a larger number of issues. When the board is more aware of what is going on in the firm and its members demand that they must clear more decisions and policy, this can obviously create great conflict between the board and management.

At the same time, the fact that the board's composition is changing, with more interest groups represented on it, means that the number of the firm's goals (beyond simply a profit orientation) will increase. The chief function of minority group members of the board so far appears to be to push for a greater role in the firm for minorities. The newer representatives on the board seldom represent stockholders but instead, special interest groups. Such a change in the composition of the board will undoubtedly introduce new performance concerns for management. In so doing, it also changes profit as the single focus of the corporation. Management will find it necessary to make tradeoffs between profit maximization and some of the newer goals introduced by board members who represent more than stockholder interests.

Again, the board members will want to protect themselves against a possibly increasing number of stockholders' suits, and as a consequence, insist that management take action to protect the board in addition to being sufficiently cautious to avoid lawsuits, as far as possible.

Employees

Modern corporations are structured in three levels of management—upper, middle, and lower—each level having its own perspective and goals. Thus, management can no longer be viewed as a single, homogeneous group having unified goals, demands, and expectations from the firm. Moreover, the bulk of blue collar employees and greater numbers of lower-level white collar workers are now members of unions. The interjection of union bargaining into company affairs specifically identifies union members as an independent interest group within the firm. In addition, workers from minority groups and women employees form separate special-interest groups (even if they are members of a union). Recent affirmative-action legislation (see Chapter 18) and lawsuits involving large damage settlements from corporations identify minority employees as a group having quite separate demands from those of union members, particularly since they often feel that unions do not adequately represent their special interests.

Top Management

The attitude of top management is that they *are* the firm. In a large sense this is true, since they make policy and basic operating and planning decisions for the firm. They are the spokesmen and the arbitrators between competing interests. In this capacity, top management is the best positioned of any group within the firm to satisfy their own interests: such as the often huge salaries, bonuses, expense accounts, and stock option plans that they have instigated for themselves (with the tacit approval of the board of directors). The chief executive officer's average length of time in office—between four and five years—seems surprisingly short. However, he has often been in the "upper" inner circle of management for many years before that. Nonetheless, the short tenure in office of the chief executive of a large corporation suggests another reason for his usually receiving very high remuneration. It may suggest also that a chief executive is able to achieve only a few rather specific goals in this length of time. Indeed, for the corporation to be "the lengthened shadow of one man" means that this person must be both extremely energetic and preferably around for a long time. Thus, the "lengthened shadow" theory of the firm may apply to a specific subset of top management rather than to the employees as a whole. Very often, top management's inner circle comes and goes with the chief executive, since most chief executives want their own "loyal" men around them. This may mean periodic housecleaning of the management's inner circle. As an interest group, however, this is the most powerful in the organization—for they control the firm, and it, therefore, largely reflects their own aims. No other group, with the possible exception of the board of directors, has such access to the levers of power in the organization.

Middle Management

Until recently, middle management was assumed to be simply an extension of top management. In some ways they are. However, as middle management has found that they, too, could be replaced by machines (computers) and that economic cuts and recessions and mergers could penetrate the security of their positions, the spectre that even this group might resort to unionization has become a growing threat (to top management, at least). Such unionization may be a long way off; however, it suggests that middle management often feels itself a separate interest group from top mangement. As such, middle management can be expected to pursue its own best interests within the framework of the corporation—interests that will often be quite different from the aims and goals of top management. To maintain greater control over their positions, therefore, middle managers will attempt to screen information upward and orders downward in an effort to keep as much control and security as possible over their positions and authority. Undoubtedly, middle management interests will come to be recognized as a distinct subset within the plural framework of the firm.

Lower Management

Often made up of blue collar assembly-line supervisors or managers, this group finds its interests and security protected in two ways—by the union and by "going by the book." In the first case, seniority, or length of service, becomes the particular criterion that unions generally use to protect and reward their members. "Going by the book" is this group's main form of protection against trouble. For this reason, it becomes all too easy to create a conservative bureaucracy, set on following the rules, among lower management. This consequence appears almost unavoidable, given the present state of large organizations. Nonetheless, the interests of lower management appear distinct, falling between those of the general work force and those of middle management—a level that, often due to lack of education, lower management may not be able to penetrate.

Unions

American unions are quite different from unions in most other countries. Whereas unions in England or Japan may be led by socialist-oriented labor leaders, labor leaders in this country are much less ideological. In contrast to England, American labor is less class conscious, nor is it Marxist, and thus it does not advocate workers' revolution to bring down the upper classes. Samuel Gompers, the original giant of American labor, summed up labor's philosophy in one word—*more*. Over the years, American labor has earnestly argued and struck, not to bring capitalism down, but to get a larger share of the fruits of the system. This pragmatic bent has been a distinct feature of American labor and also a distinct advantage to business and society, whether or not it is generally recognized.

In one regard, contemporary labor leaders have not followed Gomper's advice to avoid becoming a "tail" to any political party. In the past forty years, the American labor movement has generally allied itself with the Democratic rather than the Republican party. This would have been even more beneficial for the Democrats, except for one fact—only about half of all union members vote. Moreover, even among those who vote, a substantial minority responds to the cross-pressures of family tradition, religion, the candidate's personality, campaign advertising in the media, and the opinions of friends and groups of which they are members. Thus, many union members cast ballots opposite to the desires of union leaders. Still, the majority of union voters follow the Democratic candidates. This political phenomenon traces back to the New Deal and President Franklin Roosevelt, when labor achieved the peak of power under the Wagner Act, passed in 1936. The Republican's passage of the Taft-Hartley Act of 1947 and the Landrum-Griffin Act of 1959 diminished some of labor's power, barring such things as the closed shop, which had assisted labor in its rapid growth.

Unionism no longer generates the same loyalty and militance it did during the depression of the 1930s. The relative decline in traditional sources of union membership suggests that sources for increasing membership lie not with unskilled workers but in newer fields that are more difficult to unionize and more independent in voting for a prescribed political party or candidate. The further growth of unions lies in the areas of white collar, education, and government employees. The pragmatic ideology of labor in the United States boils down most simply to demands for higher wages, better working conditions, fringe benefits, and government welfare and medical plans obtained through collective bargaining and political lobbying.

As far as management is concerned, however, labor has too much power, for it is often able to place detrimental constraints on management decision making. Unlike those in Germany, however, American labor leaders have never actively campaigned to be elected to the board of directors or to have a vote in company decision making. In Germany, such involvement is now required by law. American management has occasionally been accused of giving away the business to labor for the sake of creating a good labor relations record to aid in management's own promotion. This criticism applies particularly to branches and plants of large firms where labor negotiations go on directly, and where top management evaluates managers according to company-worker relations while overlooking the cost of labor settlements. As long as management could pass on such wage increases in higher prices to the consumer, profits would be unaffected. For certain periods following World War II, this has often been the case.

There is a basic difference between the ideology of labor and that of most businessmen. Labor basically favors an extended governmental role over the direction and control of the economy and in welfare programs. Business' opposition to this policy comes from the strong realization that such an extension of government's role will only decrease its own decision-making powers and increase its taxes. This basic ideological difference over the proper role of government tends to enforce the conflict between the two groups, particularly as seen during attempts at collective bargaining.

Minorities

The blacks' burst to power in the 1950s and the recent surge of women's groups have introduced several new power groups into the manager's balance of power equation. Not until the Supreme Court decision of 1954, in *Brown* v. *the Board of Education*, which nullified the doctrine of "separate but equal," did blacks achieve their legal due, which began the restructuring of their position in society. In 1972, Title 7 of the Equal Opportunity Employment Act was expanded to provide new guidelines for the role of women within the corporation. Discrimination was illegal and very costly. For example, the multimillion-dollar settlement Bell Telephone had to make to women and other minority employees whom they had not promoted gave ample evidence of at least the legal and economic necessity of recognizing the rise of minorities as a major new interest group within the corporation. For business, this event was fortuitous since it enabled them to employ the talents of a labor pool nearly double that from which they had formerly drawn for many jobs.

Top managers now face the difficult problem of maintaining morale in their nonminority male workers while at the same time attempting to comply with governmental interpretations regarding the promotion of previously overlooked groups to managerial positions. In a sense, then, the rise of a new power group means a loss for already established groups. Management must balance interests and power between the competing groups in the corporation. The rapidity of the rise of new power groups (such as women) has made it difficult for management to change long-term policies quickly enough. The settlement of some of the nation's largest banks with the government to promote half women and half men to officer rank necessitates radical changes in traditional personnel and promotion policy. Although the labor pool of women in banks already exists, their training, background, and experience may be more limited than that of their male counterparts. Rapid promotion to satisfy governmental agencies or courts will not be achieved without major changes in the corporations. Obviously, social forecasting might have been able to anticipate such governmental action. Even if this could not have been foreseen, an evaluation of business policy in light of basic value trends in this country would have suggested the need for alterations in traditional ways of hiring and promoting. The constitutional values of equality and equal treatment in their newer interpretations suggest that top managers ought to spend considerably more time evaluating business policy in relation to basic societal values to see where their own business behavior may be vulnerable. The threat and the actuality of lawsuits make leisurely managerial decisions no longer possible. Changes in business in response to emerging power groups are now far faster than in the past. Corporations that continue to move at the previous slower rates of change face considerable risk.

Government

Federal Government

National policy in the federal government is made by three basic groups: congressional committees, administrators of federal agencies, and representatives of private-interest groups. These three groups comprise the legislative and executive branches of government and private interests. Clearly, then, business deals not just with "government" as a single power group but with the congressmen as a voting power group who can change and create laws, and also with individual administrators of various federal agencies that directly affect business. The federal agencies are far more important than some businessmen often assume. First, the agencies provide Congress with the information necessary for all the legislative hearings, and they may also aid in drafting proposed legislation. In addition, once legislation is passed in general outline, the federal agencies interpret it and draw in the details for administering it. Thus, sympathetic agencies may be far more important to business than sympathetic congressmen. Most private-interest groups are dependent on the administration of legislation by governmental agencies—farmers are allied with the agriculture department, unions with the labor department, most industry groups with their regulative agencies (for example, airlines with the Civil Aeronautics Board). Economists often argue that the regulative agencies are more sponsors of the groups they were created to regulate than regulators. That is, the government agency often becomes a spokesman for the clientele groups who support it and often of those industries it was designed to regulate. Many agencies depend on their clientele groups to lobby in Congress to support their budget requests. Thus, mutual support between some agencies and their supporting groups becomes the rule.

This is less the case, however, with other, more general agencies—for example, business may have more difficulty dealing with the antitrust division of the Justice Department. The common tactic is for private-interest groups to attempt to get those individuals sympathetic to their position and interests assigned to the respective regulative agency. In so doing, the thrust of much regulative congressional legislation is blunted. The close relationship between the military and the major defense contractors with which the Pentagon does business is another example of what might be called "careful cultivation of interests."

This is only one side of the picture, however. Max Weber explained long ago why bureaucracies were necessary both for government and for business when he said that, "The fully developed bureaucratic mechanism compares with other organizations exactly as does the machine with the nonmechanical modes of production."[1] In large, complicated, technologically dependent societies, a vast governmental bureaucracy inevitably develops. In small, less complex societies, where interpersonal relations are direct, bureaucracies need not develop. Since American society resembles the former description, one might understand why between two and three million civilian employees are required to run the federal government. Indeed, the federal bureaucracy is so vast that it has become the largest and most complex business in the nation. The federal government is both the largest single employer and the biggest spender in the economy. This complex machine is run by administrators. The federal bureaucracy, therefore, is one of the most powerful of all interest groups. Thus, the fundamental relationship between business and the federal government is crucial to the success or failure of the manager trying to get favorable interpretations of rulings for his firm, or trying to do business in any form with the government. The governmental bureaucracy is unique in that it regulates, buys from, and on occasion sells to, the same firm.

The very size of the federal bureaucracy is, of course, part of the reason for its power as an interest group. Its growth in size has been chiefly due to programs created in response to the following three concerns: a series of wars (World War I, World War II, the Korean War, and the war in Southeast Asia), efforts to control depression and inflation, and social concerns ranging from social security to civil rights. An increasing need to plan for the growth of cities and the use of energy and natural resources is adding a new

[1]Max Weber, "Technical Advantages of Bureaucratic Organizations," in H.H Gerth and C.W. Mills, eds, *From Max Weber: Essays in Sociology* New York: Oxford University Press, 1946), p. 214.

fourth to the list of governmental programs. As a consequence of these administrative needs, a federal civil and military bureaucracy that would have been undreamed of at the turn of the century has emerged. Indeed, the bureaucracy has become a necessary structure for government to cope with new and changing public expectations.

In summary, then, business may gain influence with the federal government by winning ideological support from the president, gaining supporters in Congress, or attempting to "capture," or create friendly relations with, relevant agencies. All of this can be done either legally or illegally. Firms can be perfectly legal and effective in getting their cases before Congress and the relevant agencies. However, following legal paths requires more knowledge of the governmental process than was required in merely writing a check for a campaign contribution.

Further, the federal bureaucracy is composed of powerful interest groups (federal agencies) capable of affecting business. The agencies' powers stem from: (1) their ability to initiate and shape legislation (through public policy), (2) their role in informing and advising Congress and testifying before congressional committees (indeed, the agencies often have a monopoly on certain kinds of information), and (3) their role in interpreting and applying legislation. All this indicates clearly that the federal bureaucracy is one of the key power points in our society and of critical importance to business.

Although the Congress is generally considered the source of legislation in our society, a corporation that focuses solely on Congress misses much of the essential power base. It is essential for the corporation to recognize the respective roles of the president and his agencies, as well as that of the Congress. Public policy and the ground rules governing business are made by the executive branch, Congress, and private-interest groups.

State Government

Except in the most populous states, it is considerably easier for business or any group to present one's case to appropriate individuals in state governments, or to obtain legislative votes or the chief executive's ear. The smaller size of the state bureaucracy makes personal contact easier, campaign contributions less, and with fewer professional staff advisors, legislators and even agencies are more in need of information that can be supplied by outside interest groups. Large firms or a strong industry group, therefore, can have considerable influence at the state level.

Although many of the same comments about the federal bureaucracy apply to the state bureaucracy, the main difference is that the state bureaucracy handles many fewer areas. State regulations involve far fewer regulatory agencies; consequently, the state bureaucracy is less powerful in its effect on business. This observation varies, however, depending on the particular business. For example, state-chartered banks find their focus of interest in the state banking department rather than in Washington, as do nationally chartered banks. Obviously, the state banks face regulations emanating from their state capital rather than from the national capital. In such cases, state agencies are critical for business. In total numbers, however, there are far fewer state regulative agencies than federal ones. In the two most heavily populated states, California and New York, lobbying by business and by all other major interest groups is little different from that carried on at the federal level. The state government in both these states is vast, complex, and has an effect on great sums of money in many different industries. One can make the generalization that the larger the economic scope of an agency, the more extensive the private-interest group's lobbying efforts, and the more powerful the government agency is as an interest group itself.

It has been common practice for lawyers who have been appointed to state utility commissions charged with setting rates for the public to go to work for utility companies after their term of office has expired. The same is true for state banking commissioners. In this case, the commissioners have generally been bankers to start with and return to banking when their term is over. As a result, many state regulatory agencies tend to be sympathetic to the group they regulate. The problem for business at all governmental levels is that agencies such as antitrust or environmental affairs may be less business oriented and thus less sympathetic to business interests.

An additional consideration in dealing with state legislators and agencies is that the state pay scale does not attract or keep the most qualified politicians or bureaucrats. With few exceptions, the federal level of government pays substantially better than does the state or local level. This problem, coupled with the lower level of professional staff support at the state and local levels, makes the operation of many state and local governments relatively primitive. This is both an advantage and a liability to business in dealing with state and local government.

Local Government

The importance of local government to business varies greatly, depending on the business of the firm. Many firms find that weak city or county governments have relatively little effect on their operations. Other firms, however, may find that if they are involved in a pollution or zoning problem requiring a city variance, and in which the results could be costly for the firm, the local governmental decisions may suddenly be a key factor in their operations. In such a case, the most effective means of gaining a hearing from the mayor or city council is to have supported them in their race for office. Campaign contributions at that stage—before the firm has a specific political need—are far more appropriate than offers of political contributions when an issue affecting the firm is at stake, at which time contributions could be viewed as bribes. It is essential to "avoid the very appearance of evil." Only by having the foresight to support a potentially successful candidate early can a firm have much assurance that its case may be heard, particularly if it appears to be an unpopular case. If management has not had the foresight to be so involved, then the best strategy when an issue is at stake is to hire a lawyer with political access, who can effectively argue the firm's case before the city bureaucracy and elected officials. In addition, it is necessary to understand who among the elected officials might be most sympathetic to one's case, in the hope that a sympathetic official might act as a defender of the firm's position. How foreign this may seem to a manager whose only interest and knowledge involves production! Unfortunately, such managers may often be ill suited for top managerial positions.

The Military

The military is a key power group for industries that supply goods and services for defense. Some firms have necessarily made themselves heavily dependent on the military, but most industries (outside of aerospace, electronics, or chemicals) are less directly affected. Many firms who do not think of themsleves as selling to a military source are affected, nonetheless—by proximity to military bases and the general economic level created by military spending and maintaining troops. The consequences of spending between eight and nine percent of the nation's GNP on military-related items spreads throughout the total economy; any time billions of dollars are spent, power is created. Thus, the military is a major power group within the economy, and more particularly, in relation to business. How the military spends billions of dollars for equipment and maintenance of its manpower is critical for many firms. Obviously, most firms that deal directly and regularly with the military develop contacts and relationships that give them some avenues of access.

Not to be overlooked is the power held by senators and congressmen who are members of, or who head, defense appropriation committees. Powerful congressmen have been able to secure important economic advantages for their states by bringing in military bases and seeing that a certain number of military contracts were awarded to firms in their home state. Thus, the relationship between business and the military can be aided substantially through the sympathetic auspices of important legislators. For example, the head of a large national consulting firm accurately predicted several years ago that the government would have to come to the aid of Lockheed. When reminded that the government had not done so for Boeing and the SST, he replied, "But Washington does not have nearly as many congressional seats or electoral votes as California." He correctly forecast that major defense firms located in states having more political power would be treated differently than are firms in states having less power.

Indeed, this generalization about political power would apply to all firms dealing with the government or making special requests, whether or not one is sympathetic to this type of governmental decision making. All other factors being equal (which they seldom are), the large firm would generally be safer locating in a state with major political clout. If one analyzes the dispersion of military bases and defense spending, it becomes apparent that defense spending is heavily weighted toward a selected few states that have had powerful congressmen or many electoral votes to help tip the scales for governmental decisions in their favor. Military spending does not appear to be made on a random basis.

Consumers and Environmentalists

During the 1960s, consumerism and environmentalism became popular movements in the United States. **Consumerism** began primarily with the publicity generated by Ralph Nader in his confrontation with General Motors over automotive safety. Since that time, consumer and environmental groups—and lawsuits—have proliferated. Most managers who had always viewed the consumer as a passive voter, who from advertising would vote dollars into the market for their product, suddenly found a much more critical, often antagonistic, group facing them. Questions regarding the safety, reliability, side effects, environmental effects, and prices of products were suddenly being raised—questions to which management had seldom had to respond to before. These groups seemed more critical of business than even such governmental agencies as the Food and Drug Administration (FDA), the Interstate Commerce Commission (ICC), the National Air Pollution Control Administration (a HEW department), the Federal Water Pollution Control Administration (Interior Department), the Bureau of Labor Standards (Labor Department), the Bureau of Occupational Safety and Health (a HEW Department), and the Bureau of Mines (Interior Department), to name only a few. Business undoubtedly needs criticism—all organizations do. The question is how managers can attempt to cope with criticism, adverse publicity, or direct lawsuits from consumer and environmental groups.

Perhaps it is easiest to look at GM and its approach with Ralph Nader to note how *not* to proceed. GM inadvertently dug its own grave by hiring private detectives to check on Nader, leaving itself open to the possible accusation that GM would attempt to blackmail Nader into suppressing his findings. This approach resulted in such widespread publicity that the president of GM had to apologize before Congress, thereby indirectly aiding the restrictive automotive safety and pollution regulations that Congress passed. Obviously, the first lesson that can be learned is to avoid illegal activities, either in manufacturing a product or in confronting critics. Second, firms should not react in such a way that they will provide their critics with more publicity. It is more prudent in the long run to attempt to solve public criticism rather than to fight it. Even though the smelter industry may have maximized profits in the short run by fighting or ignoring pollution standards, it is entirely possible that in the long run they may have effectively forced passage of stricter pollution legislation and tighter administration of the laws. For a large firm, the long run is a more rational position to consider than the short run. A third rule for firms who expect to remain in business, then, is that a long-run approach in dealing with the public and with consumer and environmental critics is a less risky course to take to protect the firm that is open defiance or unusual intransigence, which can make front-page publicity and back-page legislation.

Much of business' difficulty in dealing with these emergent interest groups is that it was largely a new experience. Management has learned the ground rules in dealing with unions and, to a lesser degree, with government. Managers had no traditions or organizational context within which to resolve these newer criticisms and conflicts, in the same way that they had problems dealing with women and other minorities. Business found that it could be exposed to large damage suits that were quite unprecedented by merely making decisions in the way they had always been made. By the time a number of Bank of America's branches had been bombed and burned, top management realized that traditional business reactions and decisions were no longer valid. At that point, Bank of America's management arrived at top-level decisions to allocate specific amounts of loans to newer loan areas—minorities, housing, and environmental loans. Only later did they learn that they had provided no mechanism or incentive for local managers to comply with such policy. This took further rethinking and changing of banking procedures. This example points out the time lag and problems attendant with facing new problems and effecting satisfactory responses.

Obviously, if firms had been quicker to see the problems and the solutions—and thus to respond satisfactorily to consumer, environmental, and minority group critics—much legislation passed in recent years would have been unnecessary. However, reorganizations and decisions are seldom made rapidly in large corporations. The result is that when public values and public demands (even if from only a section of the population) arise suddenly, managers may be ill prepared to deal with these situations simply because they are (1) not fully informed of economic and social changes in the society, and (2) either reluctant or incapable of making rapid changes. Change at too fast a rate can destroy organizations as easily as no change at all. Management is therefore faced with the problem of knowing about a situation, deciding on a course, and then determining at what speed change can be effected.

The Press

Seldom is the press considered to be an independent interest group. However, if we link with the press reporters from financial journals (such as *The Wall Street Journal*, *Barrons*, and *Forbes*), we see that journalists have the potential power to make a company's stock go up or down, depending on what is reported and how it is reported. If one goes a step farther and includes the possibly wide distribution of security analyses from brokerage houses and bank reports about corporations—which could influence the investment decisions of millions of dollars—it becomes immediately apparent that this broad group does have the power to directly affect corporate management. Apart from the influence on stock prices, this power group has the ability to bring to public attention potentially embarrassing issues and problems about a firm. Exposing such problems can be both good and bad. Since no human organization is perfect, an investigative reporter or analyst can always find something to criticize. If such issues have either public news interest and/or potentially adverse effects on the firm's future profits, stockholders and directors can be aroused, and management may have a lot of explaining to do. Further, investigative reporting can alert governmental agencies to questionable organizational practices, which may cause them to investigate the firm and possibly result in costly lawsuits, strikes, strict enforcement of governmental regulations, or any number of consequences that management would find disturbing. The press, then, represents a major power for benefit or damage to corporate management and even to the price of the stock in the market.

Corporate management has a considerable stake in developing cordial relations with the press and in learning how to present newsworthy stories about the firm for publication. Although public relations is best done by the head of the firm, most large corporations have public relations specialists on their payrolls in addition to lawyers and advertising agencies. Often when a firm is under attack from some group, the president of the organization will appear at a press conference to read a prepared statement with his lawyer at one side and his public relations officer on the other. Managers who handle their public relations successfully tend to think in terms of "stories" that reflect well on the company, which they can tell in their various talks and press conferences. Capable public relations officers are invaluable assets to an organization. They must be able to work closely with top management to be effective, however.

Too many large firms fail to recognize that the press is looking for something newsworthy to write and sell. Firms too often develop antagonisms with some critical groups as well as with the press by failing to treat either group in a way that will win their confidence—that is, by attempting to satisfy the goals of the interest group in addition to fulfilling the firm's own goals.

Small firms find public relations programs essential at times when they may wish to issue new underwritings of stocks or bonds or when they may decide to list their stock on an exchange. Management in small firms cannot generally afford permanent staff specialists in public relations and thus must rely on aid from outside agencies, often connected with the advertising firm they use. Both small and large firms, therefore, face the press as an interest group that directly affects them, and toward whom they need to give special attention.

CASE EXAMPLES

The following cases illustrate the power of special-interest groups and their importance in managerial decision making.

Stockholders

Can a stockholder with only two shares of stock force a restructuring of a great corporation's board of directors, and perhaps oust a powerful president of fifteen years' duration? A few years ago, one might unhesitatingly have said no. Yet, recently this has happened in the Northrop Corporation, a consistently profitable aerospace company. Northrop was one of twenty-one United States companies whose executives admitted making illegal corporate gifts to the 1972 reelection campaign of Richard Nixon. A thirty-year-old lawyer owning only two shares of stock brought suit against the company. A nonprofit organization, the Center for Law in the Public Interest, took on a class-action suit against Northrop. The federal judge eventually agreed to their complaint and required a restructuring of the board of directors, bringing in from the outside four new independent directors and requiring that in the future, sixty percent of the company's directors be genuine outsiders unconnected with the company. Their responsibility will be to sort out management's financial improprieties and their possible liabilities and to prevent future political secret funds from being created and used contrary to the law. The judge ordered additionally that independent directors dominate the firm's executive, audit, and nominating committees. These committees are important because they control the use of corporate funds for proxy fights and have access to financial data generated not just by management but by outside auditors as well. Further, the judge's order requires that the outside audit firm be hired or fired at the annual meeting of stockholders. This approach hopes to ensure that management cannot make auditors entirely beholden to them.

All of these precedent-shattering requirements have been brought on the firm and its president by a stockholder with only two shares of stock in the corporation and the assistance of a free legal organization bringing a class-action suit. Apparently, the day of the small stockholder has arrived—at least as far as forcing management to be attentive to his demands and to fear the impending threat of class-action suits for illegal business practices.

Employee Interest Groups

For many years, a midwestern technical engineering firm had taken its middle-level management employees for granted. Top management negotiated salaries and grievances seriously only with its unionized hourly workers. Middle-level employees felt increasingly pinched in salary, vacations, and retirement benefits in comparison with both lower-level unionized workers and top-level management. Thus, after a particularly attractive company settlement with the union, union leaders found a ready-made opportunity to extend union membership into the middle level of the company. Top management, when apprised of the situation, refused to take the idea of middle-level unions seriously and continued to ignore middle-management complaints. A successful vote to unionize by this stratum of employees caught top management unprepared. The results were a strong decline in company morale and deep suspicion between the top and middle levels of the firm.

Further, the new terms of settlement were far costlier to the firm than if the concession had been made that would have prevented unionization. Such cases are appearing with greater frequency, suggesting that too many top managers are ignoring the fact that middle-level employees represent an interest group independent of top management and requiring recognition in its own right.

Unions

Top management faces unions as an interest group in two different situations. One is the case described above, when a group of employees is being newly recruited by a union, and the second is when a union already exists and management must engage in collective bargaining. The two cases may require quite different approaches and relationships between management and union leaders.

In the first instance, a large West Coast bank having a reputation for progressive, even aggressive, management found itself facing a vote of its clerks, tellers, and typists to join a union. Banks have not traditionally been unionized. The psychological barrier between white collar and blue collar jobs (regardless of pay) has tended to prevent people in such jobs from "stepping down in status" by becoming members of a union. What caused the crisis in this case was that the bank had just moved into lavish new headquarters. At the same time, to economize it had given raises to its nonmanagement employees that were not in keeping with the rising cost of living. This situation had antagonized its lower-level staff, who had found top management too concerned with plans for expansion to be mindful of the morale in the lower ranks. To the great surprise and chagrin of top management, its staff became the first to unionize of any bank in its region and one of the few in the nation. To many of the directors, this did not reflect well on a managerial team that had been highly vaunted as one of the best in the industry. In this case, concern with prestige, lavish buildings, and rapid expansion caused management to forget about keeping the troops at the lower levels solidly behind them.

The other, more typical, case is when the firm is involved in collective bargaining with union leaders. Unlike the above situation, in which management generally feels hostile to a new union, there are numerous cases in which union leaders and management get along—perhaps even too well to suit some stockholders. One of the most meteoric managerial careers in the steel industry may provide an interesting illustration. When this particularly successful manager was at those levels in the firm or its branches where he had to negotiate with labor leaders, his main concern appeared to be to maintain a record of good labor relations. To do so, he would frequently settle very generously with the union in order to avoid a strike and to avoid having his record besmirched as a manager who was not adept at labor relations. Although his solution was expensive for the company in the long run, his rise in the managerial hierarchy was so rapid that he was never around to face the results of his labor negotiations. However, his record continued to look great on paper and he has now made it to the top post of another steel company. This suggests, then, that there are several possible strategies with different payoffs for management, beyond the commonly taken approach of intransigence.

Minorities

A new thrust of the federal government's antidiscrimination legislation is in age discrimination. There has been a law on the books since 1967 that makes it illegal to fire, or refuse to hire, an employee simply because he or she is over forty. This law had been generally unenforced until 1973–74. Then the United States Labor Department won a judgment against Standard Oil of California, requiring them to pay $2 million in back wages to 160 employees over forty, who had been laid off between 1971 and 1973. This was a landmark case. Now the attorneys for the Labor Department are pursuing over 400 cases of age discrimination.

Perhaps the most flagrant abuses occurred during mergers, when a new company took over and decided to fire all the older executives of the merged firm. This would leave employees in their fifties and having years of experience with a particular company suddenly out of a job. It was extremely difficult for many high-level executives to move into another comparable position at their age. For many executives and their families, this has been a tragedy—coming as it did only a few years before retirement. This practice will become less common as the Labor Department begins prosecuting such cases to enforce the law against age discrimination.

Government

The savings and loan industry and the mutual savings banks have long envied commercial banks' legal ability to provide checking accounts. If savings and loans and mutual savings banks could offer checking accounts, they would be able to compete much more successfully with commercial banks. Certain types of consumer loans and more savings account business would enable the savings and loans and mutual savings banks to grow more rapidly and become more profitable. All this reasoning is behind a long-term attempt to get Congress to pass a law enabling savings and loans and mutual savings to compete with commercial banks. Commercial banks are some of the most powerful institutions in the nation. Their lobbyists are legendary, particularly at the state level, where individual state's laws affect banking very specifically. In addition, commercial banks maintain a strong voice in Congress.

The savings and loans, however, have been very shrewd despite their relative youth and size when compared to the nation's commercial banks. For some years, it has been a growing practice to make local congressmen members of the boards of savings and loan companies. As directors, congressmen are thus entitled to paid expenses to fly back to their home state for meetings, and to be able to politick and at the same time receive a fee for attending the meeting. Since congressmen are inevitably short of travel money to return home as often as possible to keep their political fires from burning out, such directorships are plums. Because of this political adroitness, savings and loan companies already have political muscle and power far beyond what might be expected merely from their financial size and numbers.

At this point, the savings and loans and mutual savings

banks have still not gotten their checking accounts. But their campaigns and arguments are gaining some ground annually. Commercial banks will have to be very clever to prevent the thus-far more politically adroit industry from eventually gaining their objective. Since such major goals are not achieved overnight, the long-run strategy presently being pursued is extremely rational and shows every sign of some day winning the additional votes necessary to restructure the commercial banking and savings and loan industries. The fact that this significant battle is a political one suggests why politics and the study of interest groups is a key to success for major industries.

Special status for particular industries is not confined to banks, however. In the Northwest, where lumber is a major industry, lumber companies have long exerted tremendous political sway with state legislators. Lobbyists for the lumber industry are among the most powerful in these states. In Washington State, for example, timber lands have a special, lower tax rate on assessed value. The rationale is that with this tax cut, the lumber companies can afford to let trees grow to their optimal size without being forced to cut them down earlier to reduce the tax burden. Although most special-interest groups can make a good case of asking for special consideration, to get legislators to accept your particular case as one meriting special treatment requires considerable skill and expense. Over the years, the lumber industry has developed skilled and powerful lobbyists in the states where they need them.

However, just because your industry has been able to get a law passed one year does not mean the law will not be repealed the next—Corporate political success requires eternal vigilance. At the state level, it is generally a good deal easier and less expensive for a group to create the political environment it desires than it would be at the national level. Thus, although state legislators may often be less politically sophisticated or experienced than their counterparts in Washington, D.C., state laws are more important than federal laws to the well-being and profits of many firms. In light of this fact, corporations and industries might well put considerable resources into state lobbying efforts, which at first glance might strike stockholders as an extravagance. In the case of the lumber industry, their favorable tax treatment at the state level is a valuable preferment.

The politics of local government may often be remote to large corporations, but to smaller businessmen, local governmental decisions can be critical. Let us consider the case of a building contractor who conceived the idea of buying a row of old houses across from a park and on top of a hill and tearing them down to build a high-rise apartment tower. He convinced a local investor to go in with him. Together they gradually and quietly assembled a whole block of houses. The zoning in existence at the time permitted them to build four-story apartments on the land. However, to build the thirty-story tower they envisioned would require a variance in the zoning ordinance. The city zoning commission gave them their variance on application. Neighborhood groups soon mobilized to stop the buildings. By registering complaints with the city council, the neighborhood groups were able to get the variance rescinded, and still later, with the use of political pressure, the community groups were able to get the city council to change the existing zoning totally so that only individual residences could be built on the land. This left the contractor with considerable expenses accrued from holding the land, paying architects' and lawyers' fees, and no hope of selling the group of houses for as much as he had paid.

Is the city liable to the contractors and investors for the money they did not make or even for the money they lost? The courts will decide. In the meantime, the contractor and the investor have had an education in community politics. Perhaps if they had gone about it differently—either by trying to solicit neighborhood support or by trying to neutralize objections—they might have been successful. On the other hand, it is difficult working with neighborhood groups, and it may be that even the most astute political advice could not have changed the outcome. There is little doubt, though, that neither the contractor nor the investor will make such an attempt again in an area where a strong neighborhood group might be able to exert political pressure on the city council.

The Military

The great cost of the American space program—launching satellites and rockets to place men on the moon—has been widely criticized, particularly when so many less costly and seemingly more immediate problems remained to be solved on earth, indeed in our own backyards. As interest in space exploits waned in the 1970s, NASA (National Aeronautics and Space Administration) found it harder and harder to acquire the money to launch each new satellite. The third ERT (Earth Resource Technology) satellite was authorized by Congress in 1973 to be built by 1977. NASA was eager to begin on the first two ERT satellites, as was General Electric, the main contractor. The problem, however, was that the Office of Management and Budget (OMB), the government spending monitor, felt that no new satellites should be built until sharper-vision cameras could be developed. In addition, the OMB was trying to cut all corners to balance the president's budget or, as it turned out, to control the deficit.

Thus for NASA and its backers, a political solution was necessary to get President Ford to overrule his own staff. How could they do it? Senator Moss of Utah, a friend of the head of NASA, wrote to every governor asking whether state agencies had made use of ERTs pictures and information. The senator received replies from twenty-five governors most of whom said the pictures had been useful. Each day Senator Moss read a new letter into the Congressional Record to help build his case. Further, he got the support of the Society of Photographic Scientists and Engineers, the American Society of Photogrammetry, the American Forestry Association, and the Association of American Geographers. At the same time, General Electric was pushing the program in all its scientific seminars. Despite all this, the OMB still turned the project down. Then NASA got *Fortune* magazine to do an article showing the OMB as a villain opposing some of the few practical benefits of our space program. Somehow, after various Republican senators had talked to President Ford, and after the head of NASA finally met with the President and the head of the OMB, the President finally overruled the OMB and permitted the new satellite to be built.

Obviously, a great deal of political pressure had to be developed and applied before NASA and GE finally got the go-ahead. This is how business is often done with government contracts. The pressure of interest groups must not be underestimated in doing business with the government.

Business versus the Environment

An operational dilemma for the citizens of Butte, Montana, and for the Anaconda Copper Company has arisen over the future of Butte and of the mine itself. Butte, Montana, is basically the company town of Anaconda Copper, the third largest copper producer in the United States. Anaconda employs one-third of Butte's work force and pays over half of the city's property taxes. But the copper pit is now miles wide and miles long and deep enough to swallow the Empire State Building. The pit has already engulfed a thousand homes and now covers twenty percent of Butte itself. The problem is that the mine will swallow the city's core, too. This fear has helped create spreading urban decay, particularly since banks refuse long-term loans for much of the city's property.

Anaconda is a force to be reckoned with. Until 1959, the company owned the seven largest newspapers in the state, including Butte's only daily paper. While Anaconda pays more than market value for the property it takes, it has invoked its right of eminent domain twice, so that with the law against them, few property owners feel they can fight. Thus their bargaining position is weak, indeed. (Eminent domain is the public power to take property provided compensation is made.)

What would you do if you were a manager of Anaconda? The mayor of Butte? Or a citizen with a home near the edge of the pit? Or a businessman whose store was in downtown Butte, directly in the path of the expanding pit? This problem has yet to be solved, but it obviously affects the whole community, including the future of Anaconda.

Security Analysts and the Press

Watergate is not the only case where investigative reporters have uncovered some well-hidden secrets (for example, the case of the break-in to Democratic headquarters in the Watergate [Hotel] in 1972). Although with less national impact (except for the stockholders and employees involved), the 1974 case of Fidelity Funding has now become a famous case, in which a security analyst began investigating the accounting and operations of a large firm and found that all was not legitimate. In this case, it appeared that Fidelity Funding, a large insurance company, was actually writing fictitious insurance policies and in turn reissuing them to other firms. In this way its growth, earnings, and stock price appeared to be going only up. In actuality, the firm was in critical shape. The security analyst and the press—not the governmental agencies, the accounting firm, the FBI, or the police—exposed this situation. Thus, the general public finds that it is often more dependent on these first groups than on official governmental agencies. Indeed, this illustrates the power of security analysts and the press to ferret out and report on the truth about corporate activities. For these reasons, management rightfully regards them as another interest group to deal with, since the price/earnings ratio of the firm, and often the future careers of the managers, are at stake.

SUMMARY

At the beginning of this chapter, we raised the central question of how management influences and works with its basic interest groups. From our discussions we see that there is no single way of working with all interest groups. To deal successfully with each group requires that management attempts to understand its unique goals, ideologies, and institutions. Further, top management must also keep an eye toward what new political interest groups may be developing, which the company should be dealing with. Thus, it is essential in studying managerial decision making to give close attention to those groups that affect the total environment in which the firm must survive. The interest groups discussed here will be of greater or lesser importance to each firm or industry. Yet, management will not last long if it fails to develop a policy toward each interest group and to effectively put such policy into operation. To do so is one of the main tasks of top management.

DISCUSSION QUESTIONS

16.1. Should businessmen be interested in politics?

16.2. Are you a member of any interest groups? Is your family?

16.3. What function does the politician perform in a democracy?

16.4. How can minorities pass laws they desire?

16.5. Do you believe the oil industry is an interest group? How is it able to exert so much influence on national and foreign policy? Discuss.

16.6. How can a businessman legally influence a legislator? What is ethical? Is there any difference?

16.7. Are business's methods of influencing government different than those of unions or other large interest groups?

16.8. Should business be allowed to bribe foreign diplomats?

16.9. Is there any difference between the ethics of American and foreign businessmen?

16.10. Is there any likelihood that large multinational firms will move their headquarters out of the United States—as some have threatened to do—because of laws governing the disclosure of political contributions paid in other countries?

APPENDIX

Posting Drug Prices

Drug pricing significantly affects a large portion of the population. The elderly, in particular, have a greater need for drugs but often have neither sufficient insurance coverage nor money for them. The average customer, no matter what his age, is at the mercy of those who price drugs, for many state laws prohibit the posting and advertising of drug prices, thereby voiding the possibility of comparison shopping for the consumer.

In one U.S. city, the Office of the Consumer Advocate decided to survey drugstore pricing practices in response to numerous complaints that different prices were charged without apparent relation to the cost of drugs but rather as a reflection of what the market would bear. A questionnaire was mailed to all pharmacists in the area requesting that they indicate the price they charged for the drugs itemized on an accompanying list. The druggists did not cooperate in the survey, especially when it became evident that the questionnaires were coded so that each respondent could be identified. The survey was recalled, and a new survey organized, with anonymity guaranteed to each respondent.

In the state where this city was located, the Ethical Practices Board of the pharmacists had been discussing the philosophy behind regulating ethical drug prices and the possibility of allowing drugstores to advertise their prices. Chain drugstores were in favor of having this right because they wanted to show their competitiveness in pricing. At first, the board rejected these modifications, but the pressure brought by the organized interest group of the retired and elderly forced the board to reconsider its previous decision. In this particular state, druggists are now required to indicate the generic name of the drug for which a brand-name prescription has been specified. In addition, after a delay of six months, drug prices were advertised in the newspapers and posted prominently in the drugstores.

1. Which interest groups do you recognize as involved in the case?
2. What is your view of the position taken by the various interest groups?
3. Why did the board delay the effective date for advertising drug prices?

SELECTED READINGS

Epstein, Edwin M. *Politics and the Corporation*. Englewood Cliffs, N.J.: Prentice-Hall, 1969.

Hunter, Floyd. *Top Leadership, U.S.A.* Chapel Hill, N.C.: University of North Carolina Press, 1959.

Key, V. O. *Politics, Parties, and Pressure Groups*. New York: Thomas Crowell, 1947.

Monsen, R. Joseph, Jr. *Modern American Capitalism*. Boston: Houghton Mifflin, 1963.

Monsen, R. Joseph, Jr., and Cannon, M. W. *The Makers of Public Policy*. New York: McGraw-Hill, 1965.

Reich, Charles A. *The Greening of America*. New York: Bantam Books, 1970.

Votaw, Dow, and Sethi, S. P. *The Corporate Dilemma*. Englewood Cliffs, N.J.: Prentice-Hall, 1973.

Chapter 17 Environmental Issues and Business

Our environment is the physical world in which we live—a world of trees and lakes, automobiles and buildings, parks and dams. In short, the environment is the physical world of natural and man-altered materials. Most people believe that pollution damages the environment. It is extremely important, therefore, to understand exactly what pollution is and how it affects our surroundings.

POLLUTION DEFINED

The pollution problem is easier to understand if we picture the production and consumption of goods as a flow process. Imagine people grouped between two giant baskets, one labelled "resources" and the other labelled "wastes." From the basket of resources, people obtain the raw materials needed to produce goods. As these materials are transformed into consumer products, a portion of the materials is wasted, and is placed in a giant wastebasket. This waste can occur in several stages: Some occurs during the extraction of resources from their natural state; other waste occurs during the production or transformation phase, for example, when coal is burned to produce heat and the residues of soot, smoke and gases become waste products. Still other waste develops while goods are being distributed and consumed. Goods are discarded when they become worn out, broken, or obsolete. Kenneth Boulding, the well-known economist, has called this process "throughput,"[1] that is, when resources are extracted, used, and then discarded as waste. All this means that it is important to consider the total cycle of man's relation to the environment—the system that includes the input of resources, their transformation, and the output. The word *pollution* generally refers to the output of wastes, or more specifically to the generation and disposal of wastes.

Pollution has always existed to some degree. In 61 A.D., Seneca complained about smoky air coming from the many chimneys of Rome. The smoke-laden air of London brought citizen pro-

See R. Joseph Monsen, *Business and The Changing Environment* (New York: McGraw-Hill, 1973), Chapter 1, for a further discussion of this topic.

[1]Kenneth E. Boulding, "The Economics of the Coming Spaceship Earth," in Henry Jarrett, ed., *Environmental Quality in a Growing Economy* (Baltimore, Md.: Johns Hopkins Press, 1966).

tests nearly a hundred years ago. In many parts of the world, open sewers running down the street were the primary method of waste disposal for centuries. Thus, pollution is not a new phenomenon; we cannot even say with certainty that it is worse now than during some previous periods. It is true, however, that organized programs to combat pollution have been increasing in recent years. Concern over the environment resulted in some action during earlier periods. For example, under President Theodore Roosevelt's administration, laws were passed to develop and to conserve natural resources, and the national parks and forests were established. The conservationists of these early times were primarily interested in saving resources, whereas today's environmentalists are also concerned about the proper disposal of wastes.

Nature, fortunately, possesses the inherent capability to recycle its own waste material. New growth will hide the scar left by a clear-cut timber area, air will be purified with the help of wind and water, and fouled brooks will clear as the water runs over gravel and sand. Damage to the environment results from an imbalance, or disequilibrium. It occurs when the *pollution rate is greater than the natural capacity of the environment to assimilate and process the wastes.*

Pollution as a Serious Problem

If pollution exceeds nature's and society's ability to reprocess waste, and thus impairs the overall quality of life, it becomes a serious problem. The problem of waste disposal can no longer be ignored; when society has to struggle to accommodate the increasing accumulation of waste, priorities must be re-evaluated and reordered. Difficult decisions must be made about the nature and quality of life desired, and what price will be paid to achieve it. For example, is it realistic to assume that mass transit or even the bicycle would replace the car as a means of commuting to work? If such a decision were made, either by law or voluntarily by citizens, the gain for clean air would be tremendous.

A tradeoff in both financial and human terms is involved in every pollution analysis. The cost of pollution must be weighed against potential benefits—as we compare, for example, the pollution of cars, electric utilities, and paper mills with the convenience of transportation, electricity as a source of power, and an industrial plant as a place of employment. If a product or service produced by business causes severe problems, such as foul air, mercury poisoning, or devastation of vegetation, society must reconsider whether or not to continue production. According to the National Environmental Act of 1969, public policy shold be "to achieve a balance between population and resource use which will permit high standards of living and a wide sharing of life's amenities." The act requires that society establish a goal of equilibrium between man and his environment.

An example of this continuous review of costs and benefits occurred in the case of the planned jet aviation port in the Florida Everglades. Construction plans were dropped after the National

Audubon Society pointed out that a jetport would permanently damage the natural wildlife habitat, and subsequently the ecology. Business managers today face the unique challenge of maximizing the organization's interests while considering the new costs, such as river effluents, foul air, and the gradual deterioration of vegetation. Managers did not always concern themselves with costs occurring outside the enterprise. Today, few communities ignore environmental quality standards, although their responses differ widely. Just as individuals may have to buy small cars, industry may have to forego some large profits. The passage of federal laws on clean water and pure air indicates society's desire to redirect governmental efforts and funds to preserve a healthy environment.

The principle of *a common good*, which has served as the basis for legislation, is increasingly used as the rationale for environmental measures that interfere with or limit managerial decision making. This principle reflects an ordering of values in cases in which the public interest, as reflected in a majority vote, overrules the will of a minority. Clearly, this idea is fraught with danger if misapplied. However, the nature of modern, urban, industrial life forces society to alternatives and decisions that might have been avoided in the past.

FORMS OF POLLUTION

Each person has his own idea of what types of pollution or waste are most objectionable. One's perspective may be comprehensive and logical, or it may tend to be somewhat superficial. Therefore, it is important to analyze the basic primary forms of pollution—solid waste, water pollution, and air pollution. Other forms that may not be as pervasive or as common, such as noise pollution, are also important.

Solid Waste

Where do we hide our garbage—or at least our solid waste? The collection points for society's **solid waste** are usually hidden behind the scenes—sanitary landfills, waste treatment plants, incinerators, and others. The largest concentration of waste can be found in cities whose motto for solid waste disposal seemed to be "out of sight, out of mind." Since it was first collected, municipal refuse generally has been transported from high-population areas and dumped in low-density areas, on the outskirts of the city, or out to sea. However, the rapid growth of cities and suburbs increases the difficulty in locating landfill areas suitable for the disposal of garbage. If the decision is to choose a reclaimed area available for new uses such as parks, housing, and golf ranges, those who live near these areas raise serious objections. The neighborhood may be made unpleasant by the steady traffic of large trucks, the dust clouds, the odor, and other features of a live dump.

The most commonly used method for solid waste disposal is the sanitary landfill, which involves continuous layering of garbage covered and packed with dirt and sand. But many experts consider this method obsolete. In addition, areas suitable for sanitary landfill are becoming scarce, and dumping solid waste ignores the potential for recovery of valuable byproducts and recycling opportunities.

San Francisco has experienced such a land shortage. The city is located at the tip of a narrow peninsula, surrounded by the San Francisco Bay and the Pacific Ocean. As the region's population increased, the pressure for more shoreland led to dumping garbage, sewage, and industrial waste to fill in the bay. The San Francisco and Oakland international airports, waterfront, housing developments, marinas, refineries, and other industries were relying on fill for new land. In fact, the Army Corps of Engineers had shown that two-thirds of the bay was less than twelve-feet deep and that much of the underwater land rights were privately owned.

A group of environmentalists and conservationists concerned about the bay's future formed the "Save the San Francisco Bay Association." Water area in the bay had been reduced by one-third by 1960, due to diking and filling. The group's report, published a few years later, showed the harmful effects of the reclamation on the city's environment, including a decrease in annual rainfall because of the decrease in water surface. This decrease in rainfall aggravated the air pollution because pollutants were being washed from the air less frequently. In addition, the tidal flux was reduced, diminishing the oxygenation of the water.

The conservationists were successful in gaining passage of a state law creating a regional Conservation and Development Commission with power to plan for the maximum usefulness of San Francisco Bay while preserving its recreational and scenic values. Its report, in 1969, declared that any fill was harmful to developments in the bay. The grassroot support for protecting the bay convinced those with interests in bay properties that the public interest had prevailed. The area probably has the world's largest concentration of conservationists' groups, ranging from the Audubon Society and the Sierra Club, to local park groups to ecology centers and governmental regulatory agencies. All are concerned with the environmental ecology of the bay area.

Business representatives now cooperate with these groups through the Environmental Information Clearinghouse, a nonpartisan, nonsectarian, and nonprofit group. The Clearinghouse does not lobby or take sides; it functions as a forum for information and discussions about environmental issues between businessmen and environmentalists and conservationists. In the final analysis, these discussions center on the impact of environmental decisions on the quality of life of the area's citizens.

And what is San Francisco going to do with its garbage? New ways of disposing of solid waste are being studied. The current plan is to compact the refuse, load it into railroad cars, and haul it several hundred miles to the barren land of Lassen County in northeastern California.

People-Related and Industrial Refuse

Unfortunately engineers, managers, and legislators do not agree on how to dispose of solid waste. The 350 million tons of waste generated annually in the United States are disposed of in a wide variety of ways. This high amount of waste presents an enormous problem to society. In 1920, the estimated figure was about 2.7 pounds of waste per person per day; in the 1960s, it was about 5.3 pounds per person per day. This figure could reach 8 pounds per person per day by 1980 for a population of approximately 235 million, if the trend continues. The growth in population will cause nearly half of the predicted increase, while the increased per-capita production of waste will bring about the other half.

These figures do not include solid waste handled by business organizations themselves. The mining and metal-processing industries generate more than 1 billion tons of mine tailings and smelter slag every year. Slag and mineral wastes from mining operations have rendered an estimated 5 million acres of the United States useless for agriculture.

Farm wastes were once considered a real asset, and manure and vegetable wastes were tilled under as fertilizer. However, large factorylike farms and feedlots have changed this practice because chemical fertilizers are easier and more economical to use. Operators of both feedlots and chicken farms now have a surplus of manure, and a number of manure-processing plants have been forced to close because of the dust and odors created by the process of reducing manure to pellets. Individual, industrial, and agricultural solid wastes total roughly 4 billion tons a year. To this we must add the 7 million cars, trucks, and buses that are junked annually.

Personal solid waste is the most critical, even though it is only a small part of the total, because everyone is affected by the filling up of our "waste reservoirs." Cities and municipalities are under pressure to find new disposal methods because of new legal requirements. There is a rising interest in recovery processes due to the rising costs of traditional waste disposal methods, as well as the disappearance of landfill areas. The largest portion of urban waste (close to half) is paper and paper products, followed by lesser quantities of organic waste (food scraps and the like), grass and dirt, glass, metals, wood, plastic, leather, rubber, and textile materials. According to the Council on Environmental Quality, the cost of the nation's solid waste disposal systems increased from $5.7 billion in 1970 to $7.8 billion in 1975.

Packaging and Recycling

What is solid waste? Although there is some disagreement on the exact composition of solid waste, it is certain that paper and paper products form an important component. However, packaging of all types—paperboard, cans, and bottles—is no more than thirteen percent of all solid waste. (Again, keep in mind that municipal waste is only a portion of the waste generated in the nation.)[2]

The problem of cans and bottles does not go away, as can be seen along any highway, in the park, or on the beach. The landscape from Portland, Oregon to Portland, Maine is disfigured because of the public preference for single-portion packs, throwaway bottles, and other packaged foods. Environmentalists are becoming more insistent that the United States not accept the pollution caused by broken bottles, aluminum and tin cans, crushed cartons, and antipackaging bills are being considered in states throughout the country. A model law was enacted in Oregon prohibiting nonreturnable bottles and enforcing a payment and recycling system for both cans and bottles, which are the most critical litter items because they do not decay or decompose as paper products do. One returnable bottle that can be filled nineteen times can replace a like number of throwaway bottles, and thus reduce the quantity of packaging refuse.

Recycling programs have been sponsored by both bottling companies and aluminum can companies, and these have been partially successful. Frequently, the recycling chain is based on citizen volunteers who collect, sort, and haul the cans to collection points without charge. The technology has not been perfected for all types of container refuse. The successful efforts by the Coca Cola Bottling Company and Reynolds Metals depend on a deliberate policy decision to design economically feasible recycling efforts. The Container Corporation of America started a program in which paper container producers use a recycling symbol to identify recyclable materials. But the supply of waste packaging materials is still much greater than the demand.

This nation has emphasized a high standard of living, including the availability of material goods, and packaging materials have increased along with the standard of living. As a nation, the United States has not cared much about the disposal of trash, garbage, and refuse, in constrast to Japan and Germany. Consequently, disposal systems aimed at **recycling** and recovery have not received much public support. Although business has often been supportive of such efforts, rarely has it taken the lead in consumer education or in reclamation and recycling. However, the changing emphasis on the quality of life means that effective and useful disposal of solid waste has become one of the most important environmental and technological objectives of society.

Current Approaches to Solid Waste Disposal

Dumping is one simple approach to handling solid waste; another is selectively reclaiming and recycling valuable elements. Municipalities and cities around the country are now adopting new solid waste disposal methods. One such method involves shredding,

[2]William N. Gunn, "Packagers and the Environmental Challenge," *Harvard Business Review* (July–August 1972), pp. 103–111.

which reduces the bulk of solid waste and sets the stage for separation equipment, which can then reclaim nonferrous metals and iron and steel for the local scrap market. Shredding can also return plastic to the raw material cycle. Further, plastics can be pelletized and sold to secondary processors, who manufacture items such as toys, irrigation tiles, and pipes.

Recovery efforts are being directed at paper, glass, and aluminum. Technological snags have troubled paper recovery, however. Fibers from pilot recovery processes are not yet good enough to find a market under normal price conditions, and virgin wood fibers are still preferred. Color separation problems are still occurring with the so-called glass cullet, but some progress has been made. Successful experiments have been conducted using glass particles in road pavement surfaces, but as in the case of paper fibers, this has not led to permanent demand. However, progress has been made in the recovery of aluminum, which is nonmagnetic, probably because of its relatively high value and permanence.

In many communities, the use of incinerator operations merely transfers land pollution to air pollution. However, research is being done on superheat furnaces, which could receive solid waste mixed with oil. The heat from such furnaces could be used to generate electricity. Such pyrolysis (chemical decomposition by heat) processes could also be used to manufacture gas and oil. These furnaces differ from conventional incinerators because they can consume plastics and rubber, but at present, only pilot plants are operating.

Innovative ways to use solid waste have been developed in some cities. For example, Chicago deliberately piled up its landfill refuse to a height of 125 feet and built an artificial mountain, appropriately called Mt. Trashmore. It is used as a winter ski and recreational area in a location without natural hills or valleys.

Solid Waste Research

Several research projects on waste disposal technology have been sponsored by the federal government. Pilot plants are being built to demonstrate how a "garbage factory" could produce marketable by-products, thereby recovering some of the processing costs. These pilot plants demonstrate the recovery of pulp, ferrous metals, glass and fibers, as well as metal-, glass-, and steam-generated power. Other plants concentrate on incorporating metal, fiber, glass, and plastics into building products and materials.

Alternative packaging methods using biodegradable materials have been investigated by manufacturers. For example, in Sweden there are reports that a layered bottle, developed from resins, completely decomposes over time through exposure outdoors. Another example is Polaroid's development of a new camera that uses a totally integrated process, thus eliminating chemicals and potentially poisonous wastepaper from instant picture taking.

The cooperative efforts of government—federal, state, and municipal—and sanitary engineers and industry representatives should eventually be successful in modernizing solid waste processing and recycling operations. Consumers may also be required to cooperate by selectively discarding refuse by pregrading wet garbage, different color glass, ferrous and nonferrous metals, and so on. And, business must show a new attitude toward industry's solid and liquid wastes or effluents.

Water Pollution

Water pollution is usually less visible than solid waste pollution. But the presence of thick industrial effluents may be a health hazard for those living near or using polluted waterways. Fishing runs have been destroyed and recreation areas for swimming and boating have been ruined by water pollution. The Cuyahoga River in Cleveland was the scene of one of the more spectacular incidents of river pollution: In the summer of 1969, the river's oil content had risen so high that it caught fire! The blaze almost destroyed two bridges. What more vivid example of oil pollution could one imagine than to have the water so polluted that it could actually burn!

As another example, many believe that Lake Erie is rapidly dying. The sludge worm population increased and provided a biological indicator of the deterioration of the water. Other forms of aquatic life had long ago disappeared because of the pollution with industrial effluents. Before the fire, the Cuyahoga River had become so thick with effluents that even the sludge worm had given up. Whether Lake Erie can ever be revived is still an open question.

Thousands of different substances make up the pollutants in our waters; pollution is not the result of a single pollutant. Sources such as domestic sewers, industry discharges, agricultural flowoff, mining, and others contribute to the problem. Industrial and municipal wastes, as they mix, may regularly react with each other to jeopardize the healthy functioning of water sewage and treatment plants, because of the sudden presence of high chemical concentrates and their effect on the decomposition process.

Oxygen and Water

The oxygen content of water is its life-giving element. Problems occur when the demand for oxygen exhausts the available supply faster than the rate of natural replenishment from the atmosphere, life processes of aquatic plants, or artificial and natural aeration. Oxygen facilitates the decomposition of common organic matter such as sewage or dead plants and animals, which are attacked by *aerobic* bacteria. These bacteria rely on the presence of oxygen in the water to carry on the decomposition process. Oxygen can be replenished naturally as long as the waste load is reasonable. But, with large amounts of organic sewage, oxygen is used faster than it can be replaced, and the balance can be changed. At the same time, bacteria multiply, consume the oxygen in the sewage-saturated water, and cause organic debris to accumulate. Fish and other aquatic life cannot survive when the oxygen content falls below a certain level. Eventually, normal bacterial action ceases and *anaerobic* bacteria (microorganisms that do not need dissolved oxygen) continue to decompose the organic solids. The decomposing process of anaerobic bacteria produces bubbles of methane and foul-smelling hydrogen sulfide gas. Without extensive treatment, the water is no longer useful for many purposes.

Degradable and Nondegradable Pollutants

Awareness of the variously composed detergents used in the United States has brought about an understanding of the important differences between degradable and nondegradable pollutants. *Degradable pollutants* are those subject to biological decomposition through bacterial action requiring oxygen. Certain other pollutants, which are *nondegradable* still consume oxygen in the process of becoming mineral forms. These include phosphates, which come from such sources as detergents, fertilizers, and human waste. Phosphates bring about a process known as **eutrophication**, which is the mineral enrichment of water, and are excellent fertilizers for algae and other microscopic plants. As these plants overgrow and rot, they use up the oxygen in the water, causing rivers (such as those in Illinois) to lose their ability to renew themselves. Industrial waste discharges play an important part in eutrophication, but even more important is the damage caused by farms. Nitrogen, used as fertilizer to increase crop yields, later drains into the rivers, which are choked by the decaying algae.

Eutrophication was literally choking Seattle's Lake Washington to death during the early 1960s. The communities surrounding the lake had been pouring raw and treated sewage into the lake for years. Even treated sewage usually has a high phosphate content; and pollution increased as rapidly as the population grew. The situation was obvious when a white disc with an eight-inch diameter that could be seen at a depth of twelve feet in 1950 could hardly be seen at a depth of three feet in 1962. With concerted citizen action, a large bond issue to finance a regional sewage treatment utility was approved by voters in communities around the lake. A network of underground pipelines was installed to divert and redirect municipal sewage to a central treatment plant. After purification, the residue was discharged into saltwater Puget Sound. Because saltwater is already rich in minerals, the effect of treated sewage is much less. The result of this procedure is a dramatic change in Lake Washington, one of the largest freshwater bodies in the Northwest. The lake is now purer than it has been for many years, and it is still improving.

Thermal Pollutants

It is seldom realized that heat itself is a form of water pollution. Water can also be deprived of its oxygen-carrying ability by heat. The best estimates are that ninety percent of all water, excluding that used for irrigation, is used to absorb heat. Among the industries that employ water as a coolant, the electric power industry is the largest user. Millions of gallons of water are used each day by one large steam-powered generating plant. Water temperatures of more than 105° F have been registered for the Mahoning River in West Virginia, which is located near industrial areas where generating plants operate. Another example is the Dow Chemical plant at Tittabawassee, Michigan, situated on a tiny river with a flow of water of only 200 cubic feet per second, compared to other rivers such as the Detroit River, which has a flow several thousand times faster. In 1970, the waste discharges heated the water 17° F; in 1972, 10° F. In agreement with the Michigan Water Resources Commission, Dow Chemical committed itself to meet strict federal-state standards by 1975, with the expectation that the heat load would be down to 5° F in 1974.[3]

What harmful effects does **thermal pollution** have on a stream? Specifically, water has less ability to absorb and retain oxygen as the temperature rises. High oxygen levels are required by many valuable game fish such as salmon and trout. When streams are artificially heated, the desirable fish are lost and replaced by less valuable fish such as carp and suckers. Aerobic bacterial action is diminished as water loses oxygen, causing the heated stream to lose some of its ability to cleanse itself of organic pollutants.

Cooling towers can be used to control heat pollution by lowering the temperature of the water before it is returned to streams and lakes. However, utilities have been reluctant to spend the additional money because these towers add an average of about ten percent of the cost of a power plant. Of course, the utility can preserve its profitability by shifting the entire cost for pollution solving to the consumer in the form of higher utility bills. Another possibility might be for the cost to be shared by utilities and their customers.

Minerals

The oxygen level of water is not affected by most mineral elements, except for phosphates, which are plant nutrients. However, minerals may do damage in other ways. Soil erosion caused by mining, or industrial manufacturing processes may generate large particles that form silt deposits. Such deposits decrease the clarity of the water, and they may also settle to the bottom, gradually filling lakes and stream beds, smothering aquatic life, and covering the gravel beds used by spawning fish.

Salts and some other mineral compounds dissolve in water rather than settling. They create *hard water*, that is, water having a high mineral content that may taste unpleasant, make laundering difficult, and cause rust and scaling in pipes. Water becomes useless for almost all purposes when the mineral content is very high. Two methods can be used to make hard water usable again, water softening and desalinization (removing the mineral salts),

[3] "Dow Cleans Up Pollution at No Net Cost," *Business Week* (January 1, 1972), p. 33.

although the latter has not yet become economically feasible on a large scale.

Minerals are not removed by the current techniques for purifying water and treating sewage. In some densely populated areas, the entire flow of certain rivers is recirculated and reused several times before it reaches the sea. The more frequently our water supply is reused, the greater the concentrations of minerals and other pollutants that cannot be removed either by nature or man-made processes.

Metals and Chemicals in Water that Poison

Increasingly, we learn that certain metals and chemicals in waters are poisonous to both human and sea life. We have indicated that some pollutants are becoming more concentrated in water. An even larger danger occurs from concentration in the bodies of animals, including man, of various heavy metals such as mercury and lead, which are toxic even in low concentrations. Although the body can rid itself of most foreign substances, it tends to keep many of these metals until a lethal level is accumulated. Further, there is no known medical cure for advanced mercury or lead poisoning.

Synthetic organic or man-made carbon-based chemicals are accumulated in much the same way as poisonous metals. Certain of these chemicals, such as the pesticide DDT, are called "long-lived" because they resist degradation, remain unchanged, and eventually become part of the (animals on which other animals feed) *food chain*. When these chemicals are introduced into water supplies, they are absorbed by plankton and other small organisms. Other animals that eat the plankton, in turn, absorb most of the pesticide. This process continues with each link in the food chain, each predator retaining most of the pesticide that its prey had accumulated originally. This effect was illustrated dramatically in a study conducted at Clear Lake, California. From a once large population, the grebes (a small water bird related to the loon) were rapidly disappearing. The area had been heavily sprayed for several years with DDD (a derivative of DDT) to control gnats. This was the resulting food chain: Grebes ate sunfish, which ate frogs, which ate insects and plankton. An analysis showed that the plankton had accumulated concentrations of the insecticide 250 times stronger than the original application. Frogs had accumulated 2,000 times the original dose, sunfish 12,000 times, and grebes up to 80,000 times the original concentration. The outcome was that the level of DDD in the Clear Lake grebes was high enough to interfere with their ability to reproduce and led to their becoming nearly extinct.

Incidents of mercury poisoning have been discovered in fishing populations, who are accustomed to eating fish that had deposits of mercury in their bodies as a result of paper mill effluents in a river. Lead poisoning used to occur in children who chewed on painted railings or toys covered with lead paints. Other lead poisoning incidents result from fallouts from refining and smelter operations, where the pollutants spread over a large geographic area. As these illustrations indicate, man becomes subject to various forms of harmful substances as a result of his position in the food chain, direct consumption, or air pollution.

Management, increasingly, must also pay attention to the nature of its manufacturing operations and their possible impact on human life. Vinyl chloride, which forms an important compound in plastics materials, was declared potentially cancer-causing as a result of studies done for industries involved. Recently, studies financed by the Dry Color Manufacturers' Association alerted its members and others to the possibility that chrome pigments might cause cancer, though further studies are still in process. Potentially from 100,000 to 200,000 workers might be involved in some way in industries that fabricate the pigments into other products—plastics, paints, printing ink, and others. The largest U.S. manufacturers of chrome pigment include American Cyanamid, Hercules, DuPont, Borden, Reichhold, as well as several manufacturers in Canada, Europe, and Japan.[4]

Man is at the top of his food chain, just as the grebe is. The average American would be ruled unfit for human consumption under federal food purity standards, having accumulated twelve parts per million of DDT. Even though the average human intake of DDT is only a fraction of the lethal dose, some scientists fear that long-term exposure may be injurious even when low amounts are involved. The long-term effects of DDT and many other man-made organic pesticide compounds are not certain, because these compounds are relatively new. Further, every year chemists are creating hundreds of new chemical compounds, and there is no way to determine any adverse effects.

Treatment of Sewage and Water Wastes

Most U.S. cities have water treatment plants; many industries also provide some form of treatment for their waste water. As might be expected, not all treatment plants and techniques are equally effective. Three levels of treatment plants are commonly in use: primary, secondary, and tertiary.

The simplest and least effective method is *primary*, or mechanical treatment. First, sewage is filtered through screens that remove large floating objects. Then the sewage is held in ponds where smaller particles settle. These primary treatment techniques remove about sixty percent of the suspended solids. Unfortunately, this method is less effective on the degradable pollutants, which are reduced by only thirty-five percent. Although this method is generally considered inadequate, about thirty percent of the U.S. population live in communities using only this primary treatment.

Secondary treatment techniques are more successful and are in widespread use. These include the first step of the primary

[4] "Chrome-Pigment Workers Are Told of Cancer Dangers," *The Wall Street Journal* (July 3, 1975), p. 2.

treatment. Then, the effluent is subjected to a further biological treatment, which is an accelerated version of the natural bacterial degradation process. Sewage is mixed with a bacteria culture. It is then aerated to provide oxygen. Thus decay process that would take weeks under natural conditions occurs in hours. This process reduces suspended solids and degradable organic matter by ninety percent. Further, if the effluent is treated with chlorine, as much as ninety-nine percent of the disease-causing bacteria can be eliminated.

Some pollutants, however, still pass unchanged despite the secondary treatment. Neither primary nor secondary treatment removes phosphates, salts, complex organic chemicals such as DDT, and viruses—all common components of municipal and industrial waste water. More advanced treatment can remove these substances, but no single technique will remove them all. Rather, a collection of processes generally called *tertiary*, or advanced treatment, has been developed to treat particular classes of pollutants. Because they are new and expensive, only a few large-scale advanced treatment facilities exist. However, water is reused more often because of our growing population, and therefore some form of advanced treatment will be necessary.

Air Pollution

Polluted water has one advantage over polluted air—it can be treated and made relatively safe before use. Water pollution is often considered an aesthetic or economic issue rather than a health issue. However, air pollution is a different problem because it is quite difficult to avoid contact with air when it is polluted. We notice the problem when we see or smell the smoke and foul odors emanating from factories, cars, and burning garbage. Unfortunately there are some colorless, odorless gases even more dangerous.

Gases as Pollutants

Certain gases are toxic even if they cannot be seen adding smog to the skies. Chief among these odorless, colorless gases is sulfur dioxide, which is released in great quantities by the fuel oil and soft coal containing sulfur burned to generate heat. Sulfur dioxide pollution becomes serious when as few as six parts per million are present in the air, irritating the nose and throat. Both man and animal can become seriously ill after exposure to this concentration for more than a few hours. It also corrodes limestone, iron, steel, mortar; and is particularly destructive to synthetic fabrics—a strong concentration will melt the nylon stockings off a woman's legs. Green plants are even more sensitive to it than animals are. Sulfur dioxide, extremely dangerous by itself, can combine with oxygen and moisture in the air to form droplets of sulfuric acid, and this combination is lethal when carried into, or formed in, the lungs of animals and man.

Despite extensive publicity about pollution exposure, even the citizens in the grimiest cities are exposed to a much lower concentration than that proven harmful in the laboratory. But prolonged exposure to even very low levels of sulfur dioxide may cause chronic respiratory disease. Few of us will pay much attention to low lifetime exposure, except for medical researchers and scientists, who have determined that people living in polluted areas tend to have more respiratory ailments (and a generally higher death rate as was the case in the Ohio River valley), but they have not yet determined how much of this illness is caused by sulfur dioxide and how much may come from other pollutants or from cigarette smoking. We do know that those who are already weak—the young, the old, and people with respiratory and heart problems—are the first to suffer long before the rest of the population is affected.

Because air, like water, is constantly purifying itself, air pollution may not be a daily problem. The pollution is dispersed by air currents, and eventually cleared by rain and chemical reactions. Occasionally, however, weather patterns may increase the danger to human life. Cities such as Los Angeles may become locked into a stagnant air mass, causing the pollution level to climb until the air is again able to move upwards across the mountain ranges.

One day in October 1948, such a temperature inversion (cooler ground air trapped by warmer air above) occurred in Donora, Pennsylvania. This small industrial town has a steel mill, a zinc plant, and a sulfuric acid plant and is located in a valley along the Monongahela River. The inversion layer acted as a lid and prevented the cooler air from escaping by a normal upward flow. Because the walls of the valley kept the air from dispersing horizontally, Donora became a closed atmospheric system. Sulfur dioxide and other pollutants produced by the factories added to the valley's stagnant air. Before the inversion lifted, nearly half of the town's population had become ill and twenty people had died.

Steubenville, Ohio, has been considered one of the dirtiest cities in this country.[5] Now there is some evidence that it also might be one of the *deadliest*. A recent survey indicated that the death rate in this community is seventeen percent greater than the national average. For every 100,000 people living in this eight-county region, 1,158 die each year—compared to the national death rate of 940 per 100,000. Although no one can definitely attribute this difference to pollution, there is growing speculation that this is the reason. A chief radiologist in one of the local hospitals said, "We see enough chronic bronchitis on x-rays that we virtually consider it a normal finding." Within a 100-mile area, 12 electric power utilities, 8 steel mills, and numerous industrial plants and chemical factories are the source of nearly 1,500 industrial pollutants. The coal dust, silica, sulfur dioxides, metallic sulfides, carbon dioxides, and other solid particles are inhaled daily by the dwellers in the Ohio River Valley. As elsewhere, fluorides from aluminum

[5]"Air Pollution—Ohio City's Deaths 17% Over U.S. Rate," *Seattle Times* (February 18, 1975), p. 5.

processing plants have turned the grass blue and have caused the teeth of cows to soften and crumble—the animals die of starvation. Automobiles rust and corrode, unleaded paint peels, and leaded paint turns black—imagine what it must do to the inside of the human body!

Great amounts of sulfur dioxide are emitted by steel, zinc, sulfuric acid, and pulp plants, and the petroleum refineries also are known polluters. Nationally, however, the burning of fossil fuels rich in sulfur, such as coal and low-grade fuel oil, produces most sulfur dioxide. The electric-power-generating plants are the largest users of coal, and they currently account for more than half of the total emissions of sulfur dioxide. By the year 2000, it is estimated that the demand for electricity will require burning 3.5 times the amount of coal now used.

Sulfur dioxide is a critical pollution factor because it is hard to control, and also because the energy crisis may increase the use of low-grade coals and fuel oils. Power companies must solve the problem as required by new antipollution laws. Some companies have converted to natural gas or low-sulfur oil, but these more expensive fuels are probably not a long-term solution because their supply is limited. It is possible to install devices that either remove the sulfur from the coal before it is burned or remove the sulfur dioxide from the exhaust gases afterward, but most of these devices are experimental and expensive, adding ten to twenty percent of the cost of producing electricity. However, companies can offset part of these costs by selling the sulfur removed by the process. Nonetheless, this will undoubtedly force the cost of power up, and add to the general inflationary pressures on our economy.

Automobile as Pollutor

Many people consider the automobile the major destroyer of clean air. About sixty percent of the gross annual tonnage of air pollutants is contributed by automobiles, and in Los Angeles the figure is about eighty percent. The various sources of air pollution are summarized in Table 17.1. Note that the "transportation" category

Table 17.1 Proportion of Nationwide Emissions from Major Sources (in percentages)

	Carbon Monoxide	Particulates	Sulfur Oxides	Hydrocarbons	Nitrogen Oxides
Transportation	76.0	5.6	1.3	53.0	47.0
Fuel Combustion from Stationary Sources	2.0	28.0	72.7	3.0	39.1
Industrial Process Losses	8.2	27.4	23.7	14.0	1.6
Solid Waste Disposal	4.7	5.6	0.3	5.0	4.1
Miscellaneous	9.1	33.4	2.0	25.0	8.2
Total Emissions (in millions of tons per year)	94.0	21.5	30.5	25.9	17.0

Source: U.S. Department of Health, Education and Welfare, *Air Quality Data*, 1967.

(primarily automobiles) is chiefly responsible for three of the five major pollutants. One of the most dangerous of the automotive pollutants is carbon monoxide—colorless, odorless, tasteless, and deadly. It is produced when carbon fuels—coal, gasoline, or oil—are not completely burned, as occurs in the internal combustion engine used in most automobiles.

Laboratory experiments indicate that concentrations of approximately 120 parts per million (p.p.m.) in humans for one hour or 30 p.p.m. for eight hours may cause dizziness and blurring of vision. Lethal concentrations are several times greater. The average level of city air is 6 to 8 p.p.m., with 15 to 20 p.p.m. during peak traffic hours. Because the body can rid itself of carbon monoxide, most U.S. health officials reason that long-term, low-level exposure leaves no permanent effect. Other European experts, however, claim that exposure to only a few p.p.m. of carbon monoxide over a period of several years has caused cases of permanent brain damage. Therefore, the standard of 2 p.p.m. has been set by the Soviet Union, while California has a lower standard of 30 p.p.m. for its "serious" level.

Two other pollutants—oxides of nitrogen and hydrocarbons—are also produced in large quantities by the automobile's internal combustion engine. These are identified primarily with causing **smog**, a combination of *smoke* and *fog*, which appropriately describes the grey pall that used to hang over industrial towns. However, man has learned that modern urban smog is little related to smoke or fog. The famous Los Angeles smog, now so common across the land, is something else again.

Even in the 1940s, metropolitan Los Angeles had a serious smog problem. By the 1950s, the city was finally forced to take action to try to save their famous sunshine. Most people blamed the factories, oil refineries, and incinerators for the problem. However, when a vigorous abatement program did not improve the "smog" (in fact, it continued to grow worse) by the 1960s, people had to admit what a few experts had been arguing all along that the real cause of the smog was the automobile.

Today there is general agreement that the automobile is the source of most smog. Many large cities such as New York now have emergency plans that call for banning or severely restricting automobile traffic when pollution reaches critical levels. In addition, national emissions standards have been created for all new cars and are continually being made more rigorous. New cars are now required to have pollution control devices. According to the requirements of the Clean Air Act of 1970, emission standards for automobiles must eliminate ninety percent of all hydrocarbons and carbon monoxide within five years, and ninety percent of nitrogen dioxides by the following year—with a year of grace, if necessary. It is interesting to note that when the automobile industry suffered substantial economic reversals during the inflation and recession of 1974–75, many of these standards and controls were relaxed.

Particulates

The most common forms of **particulate** pollution, smoke, soot, and ash, are created from burning coal, wood, oil, and refuse. Since the beginning of the industrial revolution smoke, soot, and fly ash have been common and have been the cause of much of the pollution in industrial cities. These particulates, or particles of solid matter in the air, soil clothes and the exteriors and interiors of buildings. The result is that people must spend more money for cleaning or resign themselves to living in a dingy environment. Those who live in heavily polluted areas also face a health hazard similar to the "black lung disease," an illness common among coal miners. Even cancer may be linked to the lodging of small particles in the lungs.

In addition to smoke, soot, and chemical dusts, humans face particular danger from particles of metal in the air, such as lead, cadmium, and others. For example, chronic kidney and cerebrovascular disorders can be caused by accumulations of lead, most of which come from leaded gasoline. The lead concentration in the atmosphere today is more than one thousand times greater than it was before the industrial revolution. The main solution seems to be to use unleaded gasoline.

Asbestos, beryllium, cadmium, iron and iron oxides are other minerals that are dangerous particulate pollutants. Cadmium is released into the air in several industrial processes including paint manufacturing and electroplating. Only recently has cadmium been recognized as a health hazard. A few years ago, a young girl working in a Japanese cadmium smelter committed suicide to escape the severe pains that accompany advanced cadmium poisoning—contracted from exposure to cadmium particulates while at work. Because medical science has no known cure, it is imperative that the emission of such particulates be controlled. Unfortunately, regulations and testing methods for particulates in the air have a hard time keeping pace with progress in science. Some firms have installed controls to stop heavy particulate matter, because weight was thought to be a critical factor. As size and chemical composition become the critical factors, companies may need to acquire additional equipment to collect the fine particles.

Other Air Pollution Problems

Only a few of the major air pollution problems have been mentioned here. Other less common pollutant substances, such as arsenic, beryllium, hydrogen, sulfate, chlorine, and fluorides, may still pose serious hazards in certain locations. Let us consider the example of fluorides, which are given off by aluminum smelters and phosphate fertilizer plants.

Florida is known for its orange groves, beef production, and large deposits of phosphate rock. During the 1950s, when the demand for phosphate fertilizers in Polk County increased, the cattlemen and grove owners became disturbed about the general deterioration of their animals and trees. A study showed that most green plants, including orange trees, are quite sensitive to airborne fluoride poisoning. The quality and quantity of the fruit yield was affected by severe leaf damage to the orchards; and cattle were poisoned by eating grass that had absorbed fluorides. This situation is not new, nor is it limited to Polk County; it repeats itself across the country wherever aluminum or fertilizer factories are placed near or on agricultural lands. Some factories have installed emission controls, but these are costly and raise the price of fertilizers and aluminum.

Controlling Air Pollution

While the supply of air remains constant, every year the populations of people, cars, power plants, and factories increase. The result must inevitably be a deterioration in the quality of air and ultimately in the quality of life, and possibly even the length of life. The American Chemical Society has suggested some alternatives for controlling air pollution.

1. *Remove the pollutant before it is released into the atmosphere.* Filters and electrostatic precipitators are typical equipment used.
2. *Select energy and manufacturing inputs that do not contain major pollutants.* One solution, such as switching from coal to gas or oil, usually means higher costs. Present inputs have usually been chosen because they are the cheapest.
3. *Remove the pollutant during the process of use.* For example, certain factories can desulfurize heavy fuel oil as it is burned. Thus, less sulfur dioxide is produced.
4. *Operate the process so as to minimize creation of pollutants.* For instance, if every motorist had the engine of his car tuned regularly, automobile emissions could be reduced measurably.
5. *Replace pollution-generating processes with nonpollution-generating processes.* For example, use electric or turbine-driven cars (which so far are impractical or expensive) or atomic power plants (which create thermal pollution, however).

BUSINESS AND POLLUTION

Is business responsible for pollution in the United States? American industry administers the production processes and manufactures many of the products that pollute the environment. Millions of pounds of pollutants pour into the water from industrial pollution; in fact, one large paper mill alone emits as much pollution as a medium-sized city. Even though domestic sewage is a definite problem, forty percent of all wastes processed by municipal sewage plants originate from industry. Further, American industry contributes to consumer pollution through its policies of planned obsolescence, unnecessary packaging, and the convenient, throwaway container.

In the past, businessmen too often have appeared unconcerned about the effects of pollution on our environment. When smoke from the steel mills kept the city of Pittsburgh, Pennsylvania in perpetual twilight, the mill owners had a stock reply to any complaints: "Smoke means jobs." The accepted line of reasoning was that industry provided jobs and industry could not operate without pollution; so pollution was tolerated.

Pollution today is no longer ignored or considered a mere nuisance. It is recognized as a menace that threatens a sounder quality of life and health for this generation and the next, and may even endanger our very lives. Increasingly, the public, aware that technology can sometimes control pollution, is voicing strong disapproval when environmental abuses continue. As a result, business has become more sensitive to the impact of pollution. Public relations men outline their firms' records of abatement efforts, often with the statement, "We were spending millions on pollution control long before it was a popular issue." Such a claim may well be true for some organizations but many other firms are stalling. Some experts predict that the majority of expenditures for pollution control must continue to come from the private sector. Table 17.2 supports this viewpoint.

It is not easy to prevent firms from passing the costs of pollution control on to the consumer by charging higher prices when pollution-control standards are applied across the board; but the issue is complex. The best explanation is an example. Three firms—I, II, and III—producing beer are located along a stream from which they take the water necessary to make their brew; pollution control applied to all three firms could have quite different effects. If Firm I is located not far above Firm II, and Firm I pollutes the river, Firm II suffers because it must purify the water before it can produce its beer, thus paying for Firm I's waste while Firm I gets its pure water free. This increases Firm II's costs, making it impossible to produce beer as cheaply as Firm I can. If the river can purify itself of Firm II's pollutants before it reaches Firm III farther downstream, then Firm III, like Firm I, has no cost for cleaning its water. Firm II's operations would therefore be

Table 17.2 Distribution of Total Pollution Control Expenditures, 1972–81

Type	Private Expenditures	Public Expenditures
Air	35.6%	3.1%
Water	16.2%	28.1%
Solid Waste	10.2%	5.9%
Other	——	2.0%
Total*	63% or $172.7 billion	37% or $101.5 billion

*Figures do not total due to rounding.

Note: Nearly two-thirds of estimated total environmental expenditures during the decade 1972–81 will be in the private sector. This compares with a private share of somewhat less than half in 1973. Of an estimated $173 billion in cumulative private spending over the decade, some $150 billion will be over and above what would have been spent by business and industry in the absence of recent federal legislation.

Source: Adapted from Committee on Economic Development. *More Effective Programs for a Cleaner Environment* (New York: Committee for Economic Development, 477 Madison Avenue, New York 10022, 1974), p. 7.

less profitable, which might eventually force it out of business unless the public would pay more for Firm II's product.

Enacting a single water-pollution standard that would require all three firms to purify their wastes would raise the cost of manufacturing beer equally—that is to the high level of Firm II. However, because the river naturally purifies itself between Firm II and Firm III (and because Firm III dumps its wastes into the ocean with no apparent problems), consumers might argue that such pollution controls are unnecessary and would increase all beer costs and prices. Thus, the consumer with the same income could afford to buy less beer after uniform pollution standards were required. Because the pollution in the river would not affect all consumers directly, consumers' total welfare, by this reasoning, might be hurt rather than benefited by a single pollution control. In many cases, however, the opposite would be true—the public and the consumer benefit from pollution control that improves the air they breathe or the water they drink.

Some companies have virtually been dragged into the new ecological responsibility. For example, the Environmental Protection Agency recently issued a deadline for the Union Carbide ferroalloy plant in Marietta, Ohio, to reduce its particulate and sulfur dioxide emissions. The company's response was that only by reducing operations could it reduce emissions. It also stated that given a few more years, something might be worked out; otherwise, they would have to lay off 625 employees in order to comply with the directive. Eventually Union Carbide was given more time. This tactic has often been successful for firms wishing to get around environmental restrictions. Another instance occurred in 1970, when the State of Vermont enacted a series of strict new antipollution laws. Spruce Tissue Mills of New York had planned to build a new paper plant in depressed Bennington County, and said that it would abandon its plants unless Vermont lowered its antipollution standards. The state gave in and lowered its requirements rather than lose the promised jobs.

Many firms have been much more cooperative than the instances described above. For example, Georgia Pacific used a system for removing and recycling mercury long before the effect of mercury pollution became a public issue. Recently, Dow Chemical pledged that it would build only *closed-system* plants in the future, in order to recycle and remove all pollutants. Under this system, waste products would be reused and various catalytic chemicals would be recovered and reused with only minimal losses to be disposed of as industrial waste. Dow announced that "no plant will be built if there is any waste effluent or air pollution."

Managerial Perspective on Pollution Control

The previous discussion indicates that business and industry have become subject to close public scrutiny for their pollution practices. The prevalent legislative climate favors the protection of the environment and management cannot avoid or ignore the need to preserve environmental quality—in contrast to past practice when costs to be carried by another generation were generally ignored by management.

Antipollution measures represent a challenge to management. They do not need to be seen solely as additional costs that the market may or may not be willing to absorb through higher prices. For many companies, they may represent opportunities to expand into new fields of manufacturing pollution abatement equipment or new waste recovery systems. Previously, a manufacturer's cost-benefit analysis only allowed for recovery of waste material insofar as its value merited the processes involved. With the added demands to install new equipment to maintain clean air and pure water, the recovery or recycling processes may seem much more advantageous. By eliminating pollution at the source, fine-tuning production processes, and by recycling waste streams, Dow Chemical has been able to reap a profit from pollution control. It recovered raw materials, boosted product yield, and was able to keep down its investment in waste treatment equipment.

Dow Chemical has integrated its approach to pollution control with its general management organization. Thus, a twelve man ecology council, including the company president and three members of the Board of Directors, was established in 1969 to formulate guidelines. Subcouncils were also established in both domestic and foreign divisions to plan each plant's approach to eliminate pollution, to offset costs incurred, and to decentralize responsibility down to the level of plant foremen. In addition, the ecology council established an Environmental Research Laboratory for waste prevention, or waste conversion. An Environmental Testing Advisory Board studies new products for possible ecological hazards. Finally, Dow's Environmental Control Systems Group advises not only the parent organization but also other clients on tricky pollution control problems, using its technical skills and sale of chemicals and filters.

It is obvious that handling the situation alone is a problem which has plagued every plant manager. Personally he might favor pollution control, but costs have to be competitive in a market situation where other suppliers may decide to go slow. In order to eliminate or at least minimize the need for further legislation and its potentially stringent requirements, voluntary pollution control can be assumed by a whole industry, as in the case of the Dry Color Manufacturers' Association and its concern with worker health. Another avenue is represented by joint efforts with government agencies—federal, state, and municipal—to attempt to solve contamination or pollution problems. One example of this involved companies and municipalities along the Delaware River. After the Delaware River Basin Commission gained broad powers over the polluted waterway from the adjoining states, Delaware, New York, Pennsylvania, and the federal government, the companies and municipalities set up a pilot plant on secondary-treatment of effluents. Companies such as DuPont, Mobil Oil and Scott Paper expect to be part of the financial pooling arrangement on ultimately handling the waste discharge, while a public bond issue provides financing. The plant would be operated by the Commission, and an appropriate fee based on waste handled would be charged to municipalities and companies using the treatment plant.

Pollution Control Enforcement

With both the public and Congress focusing their attention on the environment, many firms have tempered their stands against pollution control in order to avoid appearing out of step with public values. A former commissioner of the Federal Water Pollution Control Agency pointed out this desire to avoid adverse publicity, when he admitted that federal enforcement powers are not strong, and that any firm could delay an abatement order for years through lengthy legal appeals. "Frankly, he said, "our enforcement actions have been effective largely because they have been tried in the Court of Public Opinion. Our enforcement conferences are open to the public and are covered by the media."

Obviously, firms and industries that sell directly to the consumer are most sensitive to public opinion. An official of the Washington State Department of Ecology compared two case histories to illustrate this point. Safeway Stores had a practice of disposing of old cardboard boxes in incinerators located behind their grocery stores. These sources of air pollution were small but still bothersome. The state official made a single telephone call to the company, and a week later, Safeway stopped using its incinerators. In contrast, a smelter in Tacoma, Washington has been using every weapon at its command to fight a sulfur dioxide abatement order. Part of this difference obviously stems from the fact that Safeway deals directly with the public and may be hurt by negative publicity, whereas the Tacoma smelter deals only with industrial customers, and its managers believe, rightly or wrongly, that they can be largely indifferent to public demands.

People have suggested that businesses that curb pollution on their own initiative do so not because of public interest, but rather in order to discourage antipollution legislation. In recent years, major corporations have spent millions on advertising their pollution-control programs. After a tour of the headquarters of a major detergent manufacturer, a professor reported that company officials repeatedly demonstrated the company's concern for the environment. At the end of the tour, the professor asked when the firm planned to market a nonpolluting detergent. The answer was, "When we're forced to!" From an environmental standpoint, this response is not encouraging. Yet, how far should a business go in considering the public and potential future issues? After all, businesses are trying to make a profit, and money spent on pollution control means less money paid for dividends to stockholders, for higher salaries, or for new capital equipment. In addition, pollution control expenditures probably will increase product prices.

In spite of some uncertainty with regard to the scientific basis on which to pursue enforcement of pollution control, federal and state agencies are flexing their muscles as they prepare to embark on a stronger approach to enforcement of new laws and regulations. Thus, the Clean Air Act of 1970 has produced a number of enforceable regulations on maximum allowable particulate concentration levels, particularly for carbon monoxide and nitrogen dioxides caused by car traffic. Similarly, the 1972 water law makes it possible to issue legally binding water-pollution control permits. It requires that industry attain a level of "best practicable treatment" by 1977, and "best available treatment" by 1983, for realizing potentially in the future "no discharge of pollutants" as the standard rule.[6]

Some people have suggested that the profit motive must be eliminated before the problem of environmental pollution can be solved. The form of economic system may not be the critical factor, however. For instance, gigantic new pulp and paper plants in the Soviet Union are now endangering the future of Siberia's Lake Baikal, one of the natural wonders of the world and the home of some species of fish and wildlife found nowhere else in the world. Now the forest cover around the lake is being stripped and soil is eroding into the lake. Millions of pounds of wastes, untreated or partially treated, are being poured into Lake Baikal. And this pollution is occurring in a communistic and nonprofit system. It appears that the profit motive is not the only factor in the pollution picture.

SUMMARY

Pollution, defined as the generation and disposal of waste, becomes a problem only when the waste flow is greater than environment's ability to cleanse itself. A single car in the middle of Iowa creates no air pollution problem because the atmosphere can render its pollution harmless. The same car in industrialized and congested Los Angeles creates a serious air pollution problem.

If pollution may eventually affect the very existence of man, what can be done? We have referred to various types of pollution and ways in which they can be controlled. An important issue emerges regarding who will pay the cost of pollution abatement. In the long run, consumers will be picking up the tab through higher prices or lower quality of goods purchased, or if the government pays, through taxes or subsidies levied against the public. If the pollution control remains minimal, we will all pay through the deterioration of the environment and perhaps the destruction of our health.

Some conservationists argue for pollution control regardless of whether the pollution has been proved to have any adverse effects on humans or animals. Such an approach is highly expensive for the public. However, in a society that decides to send men to the moon, the choice to use its resources to eliminate pollution is probably just as rational. How the public chooses to spend its resources involves values in general rather than just economics.

[6]Robert W. Fri, "Facing Up to Pollution Controls," *Harvard Business Review* (March–April 1974), pp. 26–34, 150–52.

Management faces a challenge in reversing its impact on the nation's pollution problem. Business and industry must accept the fact that society today wants to preserve and improve the quality of the environment. People will no longer tolerate discharges of wastes into the atmosphere or rivers; they expect business and industry to contribute to the cleanup process. For companies, pollution control may not necessarily mean higher costs and thus reduction in profits, or disappearance of markets. Where an industry voluntarily assumes responsibility for pollution control, or federal and state laws require compliance with minimum standards, the total effect on cost structure will leave it about the same. Managerial leadership and creative attention to pollution control may reveal opportunities for expanded markets for systems, processes, and equipment. They may also reveal new opportunities for profitable recovery and recycling of wastes, as the value of the materials and the potential expenditures for pollution abatement equipment may be nipped at the source.

DISCUSSION QUESTIONS

17.1. Compare the pollution problem of cities today with those of fifty years ago.

17.2. What is your view of the relative difficulty in controlling air versus water pollution?

17.3. What are some approaches used in persuading business to voluntarily limit its pollution?

17.4. Provide an example of the costs and the benefits from both the business point of view and the public point of view in attempting to solve a pollution problem.

17.5. Do you have any evidence to suggest that the Soviet Union would be more concerned about pollution than the United States is? Discuss.

APPENDIX
High-Tension Right-of-Way

Mr. Claude Hagerty, President of the Schnook Electric Power Company, found himself in the midst of a mushrooming controversy. The High-Tension Wire Division had been assigned the task of having a new high-tension wire system built on Peninsula Island. This would improve the distribution of electricity to customers in the system who frequently found themselves subject to electrical outages due to snow, rain, water, and fallen branches.

From irate customers who did not hesitate to call him at home at any hour and also from angry letters, Mr. Hagerty learned that public resistance to the new high-tension wires was rising to a feverish pitch. Because Peninsula Island included the summer homes of a large number of influential people in the community, the matter had become more serious than he had initially expected. The truth of the situation was that heavy equipment had already been brought in by the High-Tension Wire Division, and workmen had already proceeded to clear a broad swath through virgin forest to provide for the high-tension transmission towers and wires. In the process it had affected some of the streams through silting that were in turn harming fishing runs. This coincided with a court suit brought against the power company by a land developer claiming destruction of irreplaceable scenic and aesthetic values.

Mr. Hagerty was doubly disturbed by the incident, since he had distributed a statement throughout the organization to the effect that extreme caution and sensitivity should be exercised to avoid any accusation of environmental polluion or destruction of environmental values. Yet he was now being accused of aesthetic pollution.

1. What different types of pollution might be discussed generally and specifically for this case?
2. How might the power company have approached the problem of new high-tension wires?
3. Why do you suppose that Mr. Hagerty's directive had not been followed?

SELECTED READINGS

Anderson, Walt, ed. *Politics and Environment*. Pacific Palisades, Calif.: Goodyear, 1970.

Engler, Robert. *The Politics of Oil*. New York: Macmillan, 1961.

Hays, S. P. *Conservation and the Gospel of Efficiency*. Cambridge, Mass.: Harvard University Press, 1959.

Kneese, A. *Environmental Quality*. Baltimore, Md.: Johns Hopkins Press, 1972.

Ridgeway, James. *The Politics of Ecology*. New York: Dutton, 1970.

Ridker, R. *The Economics of Air Pollution*. New York: Praeger Publishers, 1967.

Schultze, Charles L. *Pollution, Prices, and Public Policy*. Washington, D.C.: The Brookings Institution, 1975.

Chapter 18
Management and Minorities

Minority groups have been an issue in American society since before the Civil War. With the passage of the Civil Rights Act of 1964 and its subsequent amendments in 1968 and 1972, the treatment of minorities in business became a special concern for managers. This legislation generally prohibits discrimination in employment based on race, color, religion, sex, or national origin in industries involved in interstate commerce, in state and local governments, and in educational institutions.

There are two broad categories of minorities in the United States: nonwhites (blacks, Asians, Mexican-Americans, and American Indians) and women. Nonwhite women are included as members of both groups. The employment problems of nonwhites are directly associated with their social and economic history in this country as a result of the effects of racial prejudice, whereas the employment problems of women are based on overcoming traditional attitudes toward the different roles for the two sexes. There are, of course, other minority groups that come under the protection of the law, but nonwhites and women are the most prominent groups and will be the basis of our discussion.

The interpretation and application of antidiscrimination statutes have resulted in court judgments against business firms and the cancellation or suspension of government contracts during the past decade, costing employers millions of dollars. Another consequence of discrimination in the long run may be economic losses for society due to underutilizing the talents and skills of minority employees. While not all minority persons (nor all white persons of European descent) are equally employable, there are, of course, many well-qualified or qualifiable individuals who are systematically excluded from appropriate employment through the effects of racial prejudice or stereotyped ideas of behavior on the part of those who make personnel decisions in organizations.

In this chapter, we will outline the important dimensions of the minority issue, discuss some of its practical implications for managers, and indicate effective ways managers can cope with these problems in organizations.

See R. Joseph Monsen, *Business and The Changing Environment* (New York: McGraw-Hill, 1973), Chapter 2, for a further discussion of this topic.

NONWHITE MINORITIES

Black Americans make up the largest group of nonwhites in the United States. Ever since blacks were brought to this country as slaves, they have been struggling to achieve social acceptance and economic success. Until recently, the opportunity for blacks to secure an education adequate to compete in the job market was extremely difficult. The kinds of jobs they have held and the resultant low level of economic success reflected this condition.

Some Perspectives on Black Employment in the United States

By the turn of this century, 87 percent of the black population were still employed in agriculture or as servants. By 1972, however, a major change had occurred, and only 3.6 percent were still in agriculture and only 8 percent in domestic service. Of the rest, 5 percent worked in construction, 5 percent in transportation and public utilities, 13.8 percent in wholesale and retail trade, 16.4 percent in nondomestic services, 3.2 percent in financial, insurance, and real estate, and 22.6 percent were employed in government.[1] The presence of blacks in the mainstream of trade is still far less than proportional, but the switch over three-quarters of a century from mainly agricultural jobs to urban employment has been encouraging.

Blacks are mainly employed in industries where the demand for unskilled labor is high and where they have not been victimized by the deliberate antiblack policies of strong unions. The volunteer military service has attracted large numbers of blacks. More recently, civilian hiring by government has been a source of employment. There has also been a rapid increase in the numbers of black professionals engaged in teaching and nursing, where most hiring is done by public agencies.

In 1942, an historic milestone was reached when President Roosevelt ordered that discrimination in employment in the United States must end. His order came, however, only after a threat of a black people's march on Washington, D.C., during wartime. Nevertheless, the decree was an important historical event. It marked the first time the American government had ordered that employers, and therefore management, become actively involved in putting an end to discrimination in employment.

Between 1940 and 1944, the number of blacks in civilian jobs increased by nearly a million—not including the 700,000 in the military. Very little progress was made, however, in clerical, sales, professional, or managerial employment until the Korean War expansion created a major improvement in the blacks' economic position. An even greater relative economic gain for blacks compared with whites was achieved in the 1960s and early 1970s, when the moral force of the civil rights movement and favorable Supreme Court decisions contributed to improving the blacks' economic situation and focused nationwide attention on the problems of all nonwhite minorities.

[1]Andrew F. Brimmer, *Employment and Income in the Black Community* (Washington, D.C.: Federal Reserve System, 1973), p. 51.

The long-term departure of blacks from agriculture and from the South accelerated during the post–World War II period. By 1972, the number of blacks in manufacturing and construction (2,381,328) almost quadrupled compared with 1940. Those in wholesale and retail trade increased from fewer than 300,000 in 1940 to more than 1,000,000 in 1972. In addition, during this period, substantial gains in teaching, nursing, and the lower levels of public administration were made.[2] Government remained the largest employer of blacks until about 1972. Beginning in that year, manufacturing equalled government as the major employer of blacks. Today, jobs are opening up at the top of the occupational ladder for trained and educated individuals.

Education

A key to economic improvement for blacks and other minorities is the amount and the quality of the education they receive. Forty-five percent of all black applicants have failed examinations for the U.S. military service in the past. This exam tests aptitude and general knowledge at an eighth-grade level. In some regions, for instance in the southeastern states, seventy-five percent of black applicants have failed to pass. This indicates that there is a considerable lag in the educational program for blacks, particularly in the southeast corner of the country.

The educational opportunities in our central urban cities may not be good enough to prepare blacks to compete in the job market of an increasingly technological society, but it does appear to be relatively better than in many rural areas. The recent high unemployment rate among young black females has been attributed largely to lack of training in using a typewriter. Because many jobs for young women in this country require typing skills, girls without such training are at a serious disadvantage when applying for a job.

The data in Table 18.1 show a dramatic rise in the level of education by both whites and blacks in the United States since 1940. Whereas the percentage of whites who complete four years of high school and some college has almost doubled since 1940, this increase among blacks has been dramatic—nearly a 500-percent jump in those who complete four years of high school and some college! The data for those who complete four years or more of college show an even greater increase: For whites the increase in the past thirty years has been nearly 300 percent; for blacks it has been 600 percent! This indicates a major change in our society and a tremendous total increase in educational attainment by both blacks and whites.

Advances in education for blacks represent a positive sign for their future employability. As we have said, part of the employment problem for blacks has been their lack of education.

[2]Ibid ; U.S. Bureau of the Census, *1940 Census*, "The Labor Force," vol. 3, pt. 1, U.S. Summary, (Washington, D.C.), Table 76, p. 188.

Year	All Races		White		Black and Other Races	
	4 Years of High School or More	4 Years of College or More	4 Years of High School or More	4 Years of College or More	4 Years of High School or More	4 Years of College or More
1970	75.4%	16.4%	77.8%	17.3%	58.4%	10.0%
1969	74.7	16.0	77.0	17.0	57.5	9.1
1968	73.2	14.7	75.3	15.6	57.6	7.9
1967	72.5	14.6	74.8	15.5	55.7	8.3
1964	69.2	12.8	72.1	13.6	48.0	7.0
1960	60.7	11.1	63.7	11.8	38.6	5.4
1950	51.7	7.7	55.2	8.1	23.4	2.8
1940	37.8	5.8	41.2	6.4	12.1	1.6

Source: *The Chronicle of Higher Education* (February 15, 1971).

Table 18.1 Educational Attainment by Race and Year in the United States

This lack has restricted the number of jobs for which they could qualify and, as a consequence, has forced blacks into the lowest skill categories, where the number of applicants frequently exceeds the availability of jobs. Table 18.2 gives the percentages of black and white employment for the entire U.S. population in 1973. The *Monthly Labor Review* (February 1971) has explained the data in Table 18.2 in the following way:

At least four factors may have contributed to the relatively slower rise in Negro joblessness and the consequent narrowing of the relative gap between the white and the Negro unemployment rate: (1) the rise in the education and skill level of Negroes, which enabled many of them to enter occupations where joblessness tends to be low; (2) the impact of government manpower programs, which have enrolled a relatively large proportion of Negro participants; (3) some lessening of discrimination by employers in the hiring and retention of Negro workers; and (4) the relatively smaller proportion of Negroes employed in industries experiencing the sharpest reduction in employment.

It is quite clear that Negroes have continued to make significant occupational progress despite the economic slowdown. The number of Negroes employed in white collar occupations rose by 150,000, or seven percent. Even more significantly, there was a measurable increase in Negro employment in the professional, technical, and managerial fields.

But education is only part of the problem. Much more needs to be done to change societal and managerial attitudes, which may work to discount the value of educational advances of blacks and other minorities.

Poverty

Many blacks are trapped in the environment of the central city. Between one-quarter and one-third of all black families in the

	1973	1972	1971	1970	1969	1968	1967	1966
White Jobless Rate	4.3	5.0	5.0	4.5	3.1	3.2	3.4	3.3
Black Jobless Rate	8.9	9.9	10.0	8.2	6.4	6.7	7.4	7.3
Black/White Jobless Rate Ratio	2.0	1.8	2.0	1.8	2.1	2.1	2.2	2.2

Source: U.S. Bureau of Labor Statistics, *Monthly Labor Review* (February 1971); U.S. Bureau of the Census, *Statistical Abstract of the United States: 1973.* Table 358, p. 224; U.S. Department of Labor, *Black News Digest* (Washington, D.C.: Office of Information and Reports. January 21, 1974).

Table 18.2 Unemployed by Race (percent of population)

United States live in poverty. In 1971, over twenty-four million people had incomes below the $3,745 poverty level, a decrease of fifteen million from 1959.[3] Of these people having incomes below the poverty level, sixteen million (or roughly two-thirds) were white and eight million (one-third) were nonwhite. When viewed in relation to the total population, however, only nine percent of the total white population was at the poverty level, whereas thirty-three percent of the nonwhite population earned incomes below that amount.[4] This indicated an improvement from the 1956 level, when fifty-six percent of nonwhites had incomes below the poverty level.

Other Nonwhite Minorities

The two other large groups of nonwhite minorities in the United States are Mexican-Americans and American Indians. The former have higher average incomes than black Americans but lower than white Americans. Among the Indian population, the average family income is about $2,000 per year, far below the minimum poverty level of $3,745. About ninety percent of Mexican-Americans live in the southwestern states. The majority are agricultural workers who, until recently, were unorganized and were paid minimum wages by American farmers. In 1970, the Farm Workers' Union was organized by Cesar Chavez, who has worked unceasingly to improve both the income and the standards of housing for large numbers of itinerant Mexican-American farm workers. Nonfarming Mexican-Americans who live in the cities are faced with the same problems of poverty and employment as are other inner-city minorities. Their median family income is about two-thirds of the median family income of whites. From the standpoint of education, Mexican-Americans have generally less education than blacks and far less than most whites.

About seventy percent of the American Indians in the United States (total population of approximately 670,000) live on government reservations in substandard housing. Nearly one of every two is unemployed. About forty percent of Indian children drop out of school before completing high school, and fewer than sixty percent have completed the equivalent of an eighth-grade

[3]U.S. Bureau of the Census, *Current Population Reports*, Series P–60, no. 88 (Washington, D.C., 1973), Table 1, p. 3.

[4]Ibid., Table 3, p. 5.

education. Three-quarters of the American Indians have incomes substantially below the poverty level. Indians who have migrated to the cities have found themselves disadvantaged in both education and background. Lack of education and skills qualifies them for only low-paying, menial tasks, and they must make difficult social adjustments because life on the reservation contrasts sharply with life in urban areas.

WOMEN

The traditional role of women as wives and mothers who stay at home to care for house and children is rapidly changing. Just prior to World War II, only fifteen percent of American women worked outside the home. By 1972, this had increased to forty-two percent.[5] Further, the number of families in which both husband and wife work is increasing, as the role of women in the economy changes. Economic bias against women, however, remains strong. According to a Bureau of the Census report to the president, "Sex bias takes a greater economic toll than racial bias." For example, in 1972 the median income of white men was $10,918; of black men, $7,373; of white women, $6,172; and of black women, $5,280.[6] Women, both white and black, with some college education earned less than black men with an elementary school education.

When men and women are employed for the same kind of work, men are generally paid from forty to sixty percent more. In Civil Service clerical jobs, women are usually graded as Class B and men, as the higher-paid Class A accounting clerks. Similar differences are found among tabulating machine operators, in which women typically occupy lower-level jobs. Women are found also in the lower-paying draftsmen and engineering categories. In the cotton mills, women are usually the battery hands, spinners, and yarn winders (the lowest paying jobs), whereas men are loom fixers, maintenance machinists, and card grinders. In the managerial ranks, women are often found at the small retail store levels, whereas men manage the manufacturing plant or the wholesale outlet. In institutions of higher learning, women are much less likely than are men to be ranked as associate or full professors. Data for 1972 from the Bureau of Census covering all occupations illustrate, as in Table 18.3, that overall differences in wages for men and women are substantial.

Note from Table 18.4 that in recent years, differentials between men and women college graduates seem to be narrowing. Starting salaries reported by about 200 companies for June graduates showed slight gains for women in 1974 over 1970 and, except in accounting, the differentials between men and women for 1974 were not great. Salaries reported by 160 companies for June 1975 college graduates showed that in all categories starting

[5]U.S. Bureau of the Census, *Current Population Reports*, Series P-60, no. 87 (Washington, D.C., 1973), Table 353, p. 222.
[6]Ibid., Table 5, p. 5.

Table 18.3 Median Income of Workers by Sex and Occupation, 1973

Occupation	Median Income: Women	Median Income: Men	Women's Income As Percent of Men's
Professional, Technical, Kindred Workers	$ 9,095	$ 13,945	65%
Managers, Officials, Proprietors	7,998	14,737	54%
Clerical and Kindred Workers	6,458	10,619	61%
Salespeople	4,674	12,031	39%
Operatives and Kindred Workers	5,420	9,481	57%
Service Workers (except private household)	4,745	8,112	58%

Source: U.S. Bureau of the Census, *Current Population Reports*, Series P-60, no. 97, Table 69.

salaries for both men and women were practically the same, and in some instances, the larger companies are offering women slightly higher entering salaries.

Thus, women, who are actually the majority of the population in this country, have until just recently been accorded the status of a minority in relation to employment and pay. Today, as the data point out much of the discrimination in salary has been wiped out. Whereas, in 1970 (as Table 18.4 shows) women were paid less than men in all the listed fields. In only five years, by 1975, women actually led in starting salaries in half of the listed fields. Thus, discrimination in this particular form seems to be ending rapidly. However, beyond the starting salary issue lie the more difficult ones for managers—of promotion and equal opportunity for best qualified—that imply neither reverse discrimination nor minority discrimination.

Table 18.4 Salaries for June 1970, June 1974, and June 1975 College Graduates

Field	1970 Women	1970 Men	1974 Women	1974 Men	1975 Women	1975 Men
Accounting	$ 747	$ 832	$ 868	$ 920	$ 986	$ 990
Chemistry or Science	765	806	878	890	950	992
Economics, Finance, Business Administration	700	718	802	806	840	814
Engineering	844	872	952	963	1075	1062
Liberal Arts	631	688	752	741	776	784
Mathematics, Statistics	746	773	838	848	918	915

(Average Monthly Salary)

Source: F. S. Endicott. *The Endicott Report—Trends in Employment of College and University Graduates in Business and Industry.* Copyright 1974, 1975, by Northwestern University, Evanston, Illinois.

BARRIERS TO HIRING MINORITIES

There are several barriers in addition to education and cultural differences associated with hiring minority persons in the United States. Some of these are based on social attitudes whereas others appear to be the consequence of overreaction by minority job seekers and possibly the courts, leading either to aggressive exploitation or to alleged reverse discrimination. Each of these barriers will be explored separately.

Prejudice

Prejudice is an adverse opinion that has no grounding or support in fact. It is often applied unconsciously by one group to members of another—even when those demonstrating prejudice make conscious efforts to eliminate it. Laurens Van der Post illustrates this fact in his book *The Dark Eye in Africa*, when he recalls the surprise and consternation of the Dutch in Java when they were faced with the realization after World War II that the Indonesians wanted them to "go home." Stricken Dutch leaders pointed out that they had given the natives everything—roads, railroads, hospitals, schools, and a strong economy. "Why do they want us to go?" they cried. Van der Post's response was, "Because you've never had the right look in your eye when you spoke to them."[7]

Women are also the subject of a form of prejudice known as "sex stereotyping," a phenomenon discussed below. In relation to nonwhites, stereotyped characterizations tend to become self-fulfilling prophecies. Blacks, Indians, Mexicans, Puerto Ricans, and others are characterized in the white man's opinion as lazy, shiftless, and irresponsible. The resultant deprivation from employment, as a consequence of this negative portrayal, has alienated them from the mainstream of American life and forced them into joblessness, poverty, and an anomie that manifests itself in rejection of the white man's work ethic. Thus, these groups come to fulfill the prejudicial prophecy.[8]

Sex Stereotyping

Managers must be alert to various forms of sex stereotyping that affect decisions involving women in their employ. For example, one of our societal beliefs about men compared to women is that men will more often give priority to their job over their family, whereas women should always put their family's interests above job demands. These beliefs affect managerial decisions involving women employees in a variety of subtle ways. Married women executives, for example, are less likely than men to be sent on out-of-town trips such as sales meetings, training workshops, and so forth. In the long run, this form of discrimination works to the disadvantage of women executives because it deprives them of contacts, visibility, and the benefits of training programs.

Managers also tend to value male employees more than females. Some of the reasons for this have to do with assumptions

[7]John D. Glover, Ralph M. Hower, and Renato Tagiuri, *The Administrator*, 5th ed. (Homewood, Ill.: Richard D. Irwin, 1973), p. 628.

[8]Michael Harrington, *The Other America* (Baltimore, Md.: Penguin, 1971), p. 82.

about male versus female characteristics. Men are thought of as being strong, aggressive, logical, career oriented, and less likely to allow emotions or feelings to interfere with their work. Women, on the other hand, are more likely to be thought of as weak, compassionate, intuitive, and family and home oriented. Since the male "character" is more compatible with traditional organizational values, men and women tend to be treated differently when it comes to disciplinary issues or decisions affecting salary, promotion, and retention.

Joan E. Crowley, Teresa E. Levitin, and Robert P. Quinn at the Institute for Social Research, University of Michigan, recently completed an interesting study on the "Facts and Fictions about the American Working Woman." They collected information from a sample of 539 working women and 993 working men.[9] In the study they set up eight stereotypical notions about working women.

1. "*Women in the U.S. work only for pin money.*" Instead, the study found that one-third of the women were the sole wage earners in their households; further, an additional eight percent of the women said they provided the major share of the family's income.
2. "*Women work because they have to for economic reasons.*" Only slightly more than one-half of the women indicated that this statement was true.
3. "*Women are more satisfied than men to hold jobs which are intellectually undemanding.*" The study found that fifty-five percent of the men indicated that their jobs were intellectually demanding, but only thirty-seven percent of the women felt their jobs were intellectually demanding.
4. "*Women are less concerned than men with getting ahead in the job.*" The researchers found no difference between men and women in regard to the importance placed on promotion. A significantly higher percentage of women, however, did indicate that they did not want to be promoted. This might have been connected with the women's feeling that they had no chance for promotion.
5. "*Women prefer to be followers on the job.*" The investigation showed that women had less desire to do their own decision making than men.
6. "*Women are more concerned with the hygiene of job context aspects than men.*" Indeed this was true. Women did attach more importance than did men to good working hours, convenient commuting to work, and pleasant physical surroundings.
7. "*Women are more involved in the socioemotional aspects of the job.*" This appeared to be somewhat true but was related by the researchers to the fact that women held jobs where interpersonal relations were important.

[9]"Facts and Fictions about Working Women Explored; Several Stereotypes Prove False in National Study," reviewed in *ISR Newsletter* (Autumn 1972), pp. 4–5.

8. *"Women are less concerned than men with obtaining self-actualizing jobs."* The study concluded that women are just as concerned as men with an opportunity to develop their special abilities through work, being involved in interesting work, and doing things that they are best prepared to do.

If American managers are to take advantage of the pool of skills and talents inherent in the female population and comply with antidiscrimination laws, such stereotyped thinking as outlined above must be carefully analyzed and eliminated from managerial decision making. Family values are changing in the United States from exclusive male dominance and sharply differentiated sex roles toward a power-equalized, sharing relationship between men and women. Thus, men and women in the organization must be treated more as equals and less as "different" members of the human species. In addition, the tendency for managers to expect that "women must change" should be modified by the idea that the organization's attitudes toward women may also need to change. Although many organizations are concerned with these issues, their management needs to acknowledge and discuss them openly in executive and supervisory training programs.

Tokenism

The problem of **tokenism** emerges as a set of attitudes and behavioral patterns on the part of management that is perceived by members of minority groups as patronizing, paternalistic, and even ingratiating. Minority persons are made to feel that their jobs, promotions, and special training are favors bestowed on "inferiors" by "superiors" who want to demonstrate a desire to compensate them for historic injustices. For example, a black draftsman was hired and added to a previously all-white design team. Other applicants, who were white, had had better qualifications. The draftsman later learned from a counselor that the only reason he was hired was that the president of the company had passed the word: "We need to do our part for equal opportunity—hire some blacks." As a result, the black employee began to sense that he was needed for his minority classification rather than for his ability. His resulting psychological reaction affected his work and led eventually to his leaving the organization. In another example, a female supervisor was promoted to a position sought after by her and several men. She eventually found out that she was not promoted on *merit* but simply because the organization wanted to improve its male-female managerial ratio in response to feminist pressure. Or, a Mexican-American professor is chosen over several better-qualified white applicants because the university needs a higher minority ratio for the purpose of obtaining a government contract.

Tokenism can take a number of forms, including prominent displays of minority employees on company literature or the use of a token minority actor in a television commercial. Although such practices may often be legitimate and indicative of a genuine interest in employing qualified minority individuals, they are more often just the opposite—an ill-concealed attempt to manipulate employees or the public. Managers must use great care in practicing affirmative action and avoid tokenism because of its adverse effects on the psychology, motivation, and productivity of minority employees—and indeed of all employees.

Reverse Discrimination

Government regulations based on civil rights statutes not only outlaw discrimination but also require "affirmative action" by government contractors and departments of the federal government. In effect, these regulations require those involved to provide concrete evidence that there are tangible efforts to have minority employees represented *proportionally* in the organization's ranks. The penalty for noncompliance is severe—loss of government funding.

Affirmative action programs have sometimes led to *reverse discrimination*—that is, hiring minority applicants for jobs in place of better-qualified whites, or in the case of women applicants, in place of better-qualified males. In a famous case at the University of Washington, several black applicants were admitted to the law school at the expense of better-qualified white applicants. One of the nonadmitted whites sued on the basis that this was discrimination in reverse. The U.S. Supreme Court finally sidestepped the issue by declaring it "moot" since the white student was allowed to enter law school and graduated before the Court made its decision. Thus the courts seem to prefer to avoid facing such difficult questions. Many business organizations that work on government contracts have felt the effect of these governmental regulations through increased personnel costs and sometimes through losing government contracts for failure to present an acceptable affirmative action program.[10]

RESPONSE TO THE PROBLEM OF DISCRIMINATION

Some business firms actively hire qualified minorities and have adjusted their standards to accept those who are only potentially qualifiable. A qualified minority person today may have an advantage in getting a job. Business' response to hiring minorities, although varying from firm to firm, has become increasingly positive. Recently the *Harvard Business Review* indicated that about one-sixth of the largest 500 firms in the nation were actively hiring the hard-core unemployed. This can be viewed positively as a major improvement in a relatively few years, or negatively as only partial involvement by big business in the problem. In any case, there is room for a greater commitment on the part of all organizations.

Manufacturers have generally avoided locating their plants in the central city. As business locates outside the city, those who can afford to, move to the suburbs. The result of these moves is that fewer dollars from business and property taxes go to the city,

[10]For a discussion of some of the legal ramifications of affirmative action as well as of the Civil Rights Act, see Antonia H. Chayes, "Make Your Equal Opportunity Program Court-Proof," *Harvard Business Review* 52 (September–October 1974), pp. 81–89.

making it more difficult for the city to purchase needed goods and services. The ultimate effect is a vicious circle resulting in a further decline in the central city's appeal to businesses. Businesses avoid cities largely because problems of crime and violence increase costs, such as insurance. An acceptable environment for business is characterized by social stability and reasonable assurance that the risks of locating in the inner city are not unduly high relative to the gains. However, the minorities' attitude toward traveling to jobs far from home can be easily understood, especially when they may often see business discrimination as a source of many of their problems.

Managers have not been inclined to cooperate spontaneously with governmental programs, such as retraining, except when required by law. This response may be unfortunate, but it is typical of business-governmental relations in this country. The long-run solution to the problem of minority hiring in the United States will come about only through cooperation at all levels.

Some Further Suggestions

Although the preceding discussions have been pessimistic in many respects, there are some reasons for optimism. Eli Ginzberg, a noted authority on the subject, gives five reasons business can and will help integrate minorities into their operations:[11]

1. *Long-run self-interest.* In the long run, it is best for business and for the country to create a congenial business environment in terms of jobs and income that will include all segments of the working and consuming public.
2. *Business acumen.* Business is goal oriented. Managers prefer to set goals that have a time limit. Minorities also want evidence of improvement in their position over a specific time period. Inasmuch as business is adept at setting and meeting goals, if such plans include minorities, business is likely to be more successful than most governmental programs in achieving specific targets for minorities.
3. *Profit orientation.* In some businesses that are both pragmatic and profit oriented, management historically has tended to be open to all classes. In recent years, this trend toward openness is becoming more accentuated.
4. *Power.* Firms are interested in market power. As business sees more market power among minorities, it will respond both to their consumer demands and to their job demands.
5. *Cooperation.* If business and government together are understanding and sympathetic to the needs of minorities, the basic economic problems of minorities in this country should be solved. Neither government nor business can solve minority hiring alone, Ginzberg argues. Both groups hold too much power to be ignored. If business and government can work together, however, almost all economic projects can be

[11]Eli Ginzberg, "The Dynamics of the Problem," in Eli Ginzberg, ed, *Business Leadership and the Negro Crisis* (New York: McGraw-Hill, 1968), pp. 19–20.

realized effectively. Indeed, it was a combination of business and governmental efforts that put the first man on the moon.

Government's Response

In addition to enforcement of the Civil Rights Act, the federal government has developed a number of manpower programs. These programs have been designed to aid the hard-core unemployed. Nearly $3.2 billion was spent in fiscal year 1971 for such programs as training, placement, and job creation; a substantial increase over the past few years. The federal government specifically has spent substantial money for vocational rehabilitation, employment services, and neighborhood youth corps: $2.7 billion in 1972, in which minority groups' participation ranged from twenty-seven to seventy-one percent of total enrollment. The economic slowdown in 1974–75 made further progress in hiring the hard-core unemployed more difficult than was the case when total unemployment rates were lower. Progress in minority hiring in general is adversely affected by downturns in the economy.

SUMMARY

Nonwhites and women constitute the two largest minority groups in the United States. Nonwhites—that is, blacks, Mexican-Americans, and American Indians—have traditionally suffered from race prejudice; and women have suffered from sex-role stereotyping. Since 1942 a variety of antidiscrimination statutes have been passed by Congress for the purpose of protecting minority groups from the disadvantages of unequal economic and educational opportunities. Managers of business firms and heads of educational and governmental organizations have been required to adjust their hiring practices and internal policies in order to follow the legal standard of equal opportunity for all.

Black Americans constitute the largest nonwhite minority group. Because of a variety of historical circumstances and past discriminatory practices, many blacks find themselves struggling to survive at the lower end of the economic and educational scales. The problems of blacks have been compounded by the (only recently ended) movement of black families from the rural South into crowded city slums where the effects of poverty are more keenly felt than in rural areas. Education is often better in the cities, however, and is a positive force toward improved economic status. Of the other nonwhite minorities, Mexican-Americans are only slightly better off than blacks, while the American Indian is the worst off of all major minority groups. Increasingly, problems of these latter two groups have come to public attention and attempts are being made to cope with the problems of these minorities as well.

Women, while still underemployed and generally underpaid, seem to be progressing more rapidly toward equality than nonwhites. Most women, however, occupy lower status, nonmanagerial jobs and are held back because many employers feel that they have emotional and physical characteristics and life role demands that are incompatible with a working life, particularly in the ranks of management.

In addition to prejudice and sex stereotyping, tokenism aggravates the minority problem in the United States. Tokenism—or the hiring of members of minority groups simply for public display—is resented by most minorities. Likewise, the problems of reverse discrimination, or the hiring of members of minority groups who are less qualified than majority applicants is causing bitter reactions by some whites. Thus the problem of hiring and promoting are fraught with difficulties for today's managers.

The conclusion that emerges from this brief discussion of the role of minorities is that managers must play a major part in any solution. Their cooperation and initiative are needed to help minorities make important contributions to organizational life. As management reflects the values of the country in general, it must respond to the public mandate of equality and fair play (increasingly enacted into law) in hiring, salary, promotions, and in its relation to community development.

DISCUSSION QUESTIONS

18.1. Why does black income distribution have two peaks—one below the poverty level and another way above it?

18.2. Why are women considered a minority group even though they represent a majority of the population?

18.3. Which major minority group has the lowest average income level in the nation?

18.4. Should management play a major role in solving minority problems in this country? If so, what specifically can it do?

18.5. Do you think the minority problem of blacks is really different from the problem faced by women?

18.6. Affirmative action seems to be aimed at improving the lot of minorities and women at the expense of the white male worker. Do you agree? Is this fair? Discuss.

APPENDIX
The Trainee

In a reorganization of the Cooper Company, two departments were merged. The surviving department located in a neighborhood of the city, where professional employees lived decided to retain Lucille Mitchell in her position. She was a young black who had been operating key-punching equipment. She was intelligent and quick. Her efficiency had impressed management so she was kept on, although other employees were sent to other departments or released.

Later, management promoted Lucille Mitchell to become the supervisor of the key-punch operators, though she was younger than many of those she supervised. The notion of promoting minority employees was well established in this branch department. Mexican-Americans and Japanese made up close to twenty-five percent of the total number of employees working in this department.

The company announced that it would start a management training program and Lucille Mitchell was one of those selected to prepare for supervisory and departmental managerial assignments in the company. After the formal part of the training had been completed, the plan was to rotate the trainees through a number of departments located in various parts of the city.

When the time came to place the trainees, all of them were picked for assignments except Lucille Mitchell. In fact, not a single departmental manager would accept her as a trainee. It was alternately reported that she had snubbed customers, that a customer had called and complained, or that she had lost an account for them.

Lucille returned to her old department, still a trainee. Normally she would have been released from employment in the company, but management did not think this was proper, given their stated policy of supporting the advancement of members of minority groups. Nevertheless, she was considered as surplus in the department.

1. What do you think about the promotion and advancement policies of the company and of the department that was home for Lucille Mitchell?
2. Do you think that the company handled her training assignment properly?
3. Was it appropriate for the other departmental managers to refuse to employ Lucille?
4. What would you do about the career of Lucille Mitchell?

SELECTED READINGS

Banfield, E.C. *The Unheavenly City*. Boston: Little, Brown, 1970.

Cross, T.L. *Black Capitalism*. New York: Atheneum, 1969.

Ginzberg, Eli, and Eichner, A.S. *The Troublesome Presence*. New York: Free Press and Mentor Books, 1964.

Sethi, S.P. *Up Against the Corporate Wall*. Englewood Cliffs, N.J.: Prentice-Hall, 1971.

——. *Business Corporations and the Black Man*. Scranton, Pa.: Chandler, 1970.

Chapter 19 Consumerism and Social Responsibilities

Americans have one of the highest standards of living in the world. They have accumulated the greatest abundance of consumer goods known to any people in history. This general affluence has made it possible for people to evaluate our society and its institutions from a new perspective and, interestingly, to become much more critical of our social institutions. Mass communications and a better-educated population provide the basis for the new standards and expectations about business behavior, ethics, and consumer products. Governmental and business organizations run the risk of not being responsive to these changes, which may result in new legislation that tends to restrict business operations. This trend is demonstrated by the new laws of the 1960s and 1970s protecting consumer interests.

Consumer protection groups and their accompanying political power represent a countervailing force to the decision making of business in society. It is probable that for the first time this nation will have an effective and probably lasting consumer movement that will assume responsibility for protecting the public interest. We will refer to this movement as *consumerism*.

RISE OF CONSUMER AWARENESS

In the eighteenth century, Adam Smith wrote in *The Wealth of Nations* that consumption is the focal interest of production; that is, production is rational or justifiable only to the extent that the consumer's interest is promoted. This attitude was in contrast to that of the age of mercantilism, which had focused on the producer. Thus we can see the emergence of an organized, influential consumer movement as the continuation of Adam Smith's new economic system—**capitalism**—and the growing demands for individual political rights, which served as a basis for the American Revolution. However, the difficulty of maintaining the balance prescribed by Smith, in view of the strong producers and weak consumers in the marketplace, has led to repeated redress on behalf of the consumer through intervention of protective legislation or groups of consumers banding together.

See R. Joseph Monsen, *Business and The Changing Environment* (New York: McGraw-Hill, 1973), Chapters 3 and 4, for a further discussion of this topic.

The Era of Criticism

An early and significant piece of consumer protection legislation occurred in 1872, when the federal government passed a law prohibiting the use of the mails to defraud. Then in 1890, the Interstate Commerce Commission was created to regulate the railroads in the public interest. The National Bureau of Standards was created in 1901 to fix common measures for use by business and government. Between 1879 and 1905, more than one hundred bills were introduced in Congress to regulate interstate production and sale of food and drugs, but because of congressional and public apathy and strong business opposition, no action was taken on any of these bills. However, a major landmark for consumer protection occurred in 1906 with the passage of the Pure Food and Drug Law, which outlawed misbranding food and drugs. Upton Sinclair's novel, *The Jungle*, published in 1906, exposed the unsanitary conditions in meat-packing plants and influenced the passage of the Meat Inspection Act of 1907. This law authorized the Department of Agriculture to inspect slaughtering, packing, and canning plants, filling in some gaps in the Pure Food and Drug Law of the previous year. Besides Upton Sinclair, other writers were critical of industry's efforts on behalf of consumers and the public generally—they became "muckrakers" or critical commentators of events of the day.

In 1914, The Clayton Act created the Federal Trade Commission to protect consumers from unfair trade practices. In the 1920s, the Federal Power Commission was established to protect the consumer from public utility monopolies. However, many governmental agencies became the "captives" of the very groups they were intended to regulate—for example, power companies, airlines, unions, and agriculture—because special-interest groups were able to lobby successfully to get "sympathetic" individuals appointed to these regulatory agencies. (This concept was discussed in detail in Chapter 16.)

The Consumer Movement Era

Under President Franklin D. Roosevelt, the New Deal of the 1930s brought with it a surge of consumer legislation. This interest in consumers began in the late 1920s with Stuart Chase's bestseller, *Your Money's Worth*, which pictures the consumer as an "Alice in Wonderland" amid conflicting product claims and promises. This book urged the establishment of impartial product-testing agencies and, along with the subsequent protest literature, aroused consumer demands for more protection than the Food and Drug Act of 1906 had provided. The most significant step in the interest of consumers was the passage of the Food, Drug, and Cosmetic Act of 1938, which strengthened the definitions of adulteration and misbranding and authorized the government to inspect factories producing food, drugs, and cosmetics. Manufacturers were blamed for mixing less valuable ingredients into products thus reaping higher profits through adulteration. Similarly, goods were sold under a familiar label though the mix or quality were different. At times competitors were using a label almost identical to another product. Despite all this, labels were not yet required to show contents, net weight, or manufacturer's identity; thus, information disclosures as well as safety requirements for many products still did not exist.

Although the 1930s introduced experiments to have the consumer represented in government (notably with the presence of consumer councils in various government agencies), such groups had little power or influence. Consumer protection efforts during this period did not succeed in bringing about any lasting changes or any means for hearing and rectifying consumer complaints. The New Deal politicians were unsure whether to give consumer representation equal voice with other pressure groups, such as business and labor. This lack of conviction severely damaged the consumer movement during this period.

Only piecemeal consumer legislation was passed during the 1950s. The Flammable Fabrics Act of 1953, for example, was passed because of publicity reporting the deaths of children dressed in flammable materials. The act was designed to prohibit the manufacture of apparel made from flammable fabrics. However, some consumer groups claimed that this act was not administered conscientiously and that flammable clothing materials were still being sold in this country despite the legislation. The 1950s were not a decade of major consumer concern either by Congress or the public.

CURRENT AGE OF CONSUMERISM

To the post–World War II generation, born into an affluent society, the materialism of the older generation has seemed obscene. The clogging of city streets and highways with cars, the pollution of water and air, and the seemingly unfettered emphasis on growth and expansion of the economy appeared to predict doomsday. A concerted public outcry by the conservationists and environmentalists, believed the post–World War II generations, would be necessary to influence the management of business and industry to exercise some self-restraint or at least to consider the total cost of their activities to society. In the same way as Upton Sinclair's *The Jungle* represented the nation's muckraking phase at the turn of the century, Ralph Nader and his associates have galvanized consumer sentiments into new activism. Ralph Nader has become the key consumer advocate of today's generation and has given dramatic perspective to the powerlessness of unorganized consumers in the marketplace.

Other circumstances also have contributed to the increase in consumer awareness, including the distribution of defective, unsafe, or impure products; the poor safety records of some makes of automobiles; apparent deceptive grading and pricing of automobile tires; the flammability of children's clothes; worthless automobile and appliance repairs; warranties and guarantees that are not honored; and merchandising promotional schemes that border on unethical and frequently victimize the buyer. What is more, the consumers have had their health jeopardized by products that have not been tested adequately, such as cosmetics, television sets, kitchen ranges, and microwave ovens, among others. These examples represent applications of technology to consumer products without a complete understanding of their potential danger to the users. Information about the hazards of such products has become available through the press and more widely, through television newscasting. Cases of unsafe products have caused great concern and indignation among the public.

Deceptive Packaging and Promotion

Housewives participating in one test were asked to choose in their view the best buy among various products carried by a supermarket. Because it was difficult for them to compare the prices of like products in packages of different sizes and weights not marked with unit pricing, the housewives proved to be wrong forty percent of the time. In viewing product quantities in relation to price, to get the most for your money, any shopper will find this a difficult task, and in the past manufacturers and merchants have not made this task any easier. The differently shaped bottles in which such liquids as shampoos, cooking oils, and syrups are packaged make it nearly impossible to determine visually how much is contained in each unit. If you can determine the quantity, you still face the problem of deciding whether nine ounces for thirty-seven cents is a better buy than seventeen ounces for sixty-eight cents. This is the problem faced by shoppers in stores without unit pricing. Obviously, as most people do not take the time to make the calculation very few make the best decision.

Promotional contests have recently received much criticism for alleged deception. One congressional committee reported that in many contests there were few winners for the big prizes—automobiles, Caribbean cruises, yachts, color televisions, $10,000 in cash, or $100 a week for life, among others. In fact, the odds are so poor that the contestant can only justify the time spent entering these contests by any personal satisfaction gained. Contests generally fall under the supervision of the Federal Trade Commission (FTC), which found that in some contests the chance of winning (even if an award were to be made) was 1 in 1.2 million for the prizes in the $1,000 category.

The Problem with Guarantees

In 1969, the Presidential Task Force on Appliance Warranties and Service[1] concluded that, "In some instances the exclusions, disclaimers, and exceptions so diminished the obligations of the manufacturer that it was deceptive to designate the document as a warranty, because the remaining obligations were lacking in substance." Indeed, of the 200 warranties examined by the task force, 34 had some exceptions, disclaimers, and exclusions, several contained all 3, and in some the same exceptions were repeated several times. In other words, if you buy a product because of its warranty, you may be disappointed later.

[1]*White House Report of the Task Force on Appliance Warranties and Service* (Washington, D.C.: The Committee on Commerce of the U.S. Senate, January 8, 1969).

Indeed, getting a guarantee activated is often frustrating. Large hidden costs may be involved. In some instances, the guarantee covers parts but not labor, and the labor costs alone would be prohibitive. Manufacturers, of course, tend to estimate when repairs on a product will be necessary, and therefore they write their warranties with an expiration date that will avoid the cost of fulfilling a warranty. Sometimes the manufacturer offers an unreliable and expensive repair service. In such instances, there is little the consumer can do.

The Presidential Task Force on Warranties found that many manufacturers do not respond to consumers who ask them to honor warranties. In such cases, the consumer is often asked to first contact the distributor, which may lead to what is described as the "run around"—a considerable exchange of correspondence, broken appointments, and no satisfaction for the consumer. The manufacturer, distributor, and retailer all disclaim any responsibility for, or ability to solve, the problem. Although the consumer has the legal recourse of filing suit against the manufacturer, the cost is not likely to be worth the trouble. Because some manufacturers and retailers know that consumers lack effective recourse, they view consumers as easy victims. Better Business Bureaus and *the courts* may be the only avenues of recourse open. The Better Business Bureau records complaints but was seldom able to solve problems for the consumer in the past, until it introduced a consumer arbitration system whereby disputes are submitted to arbitration binding on business.

Lack of Repair Service

Consumers often find themselves at the mercy of mechanics or repairmen. Surveys of repair services have produced dramatic examples of disreputable practices. One recent study of television repairmen indicated that 17 out of 20 who were called to replace a burned-out tube charged up to $30 for a job that should have cost about $9. Increasingly, reputable manufacturers are setting up their own training programs for mechanics to service company products in an effort to preserve customer goodwill in their brand.

Consumer discontent from experiences such as those discussed above has helped make consumer advocate Ralph Nader one of the best-known figures in this country. Through such sensational and well-publicized consumer cases as that involving General Motors and the Corvair, business has been its own worst enemy in arousing suspicions about its practices. Consumerism and the pressure for legislation to protect the consumer are the natural consequences.

CURRENT CONSUMER LEGISLATION

Although consumer protection depends ultimately on businessmen's attitudes and honesty rather than on the passage of laws, legislation over the years has attempted to protect the consumer and protective legislation will undoubtedly increase in this current "age of consumerism." The 1960s were a period of unprecedented concern for the consumer. The Kefauver-Harris Drug

Amendment of 1962 required the Food and Drug Administration to pretest drugs for safety and effectiveness and to label drugs by their generic names. The availability of drugs under the generic name in addition to the same drug labelled with a brand name suggested lower drug costs for the public, as physicians were not restricted to prescribing only the higher-priced brand-name drugs.

Between 1966 and 1973, more than a score of major legislative acts covering a wide range of products and activities were passed to protect the consumer. Some examples include: the Fair Packaging and Labeling Act (Truth-in-Packaging) of 1966, which took five years to pass over the opposition of business; the Consumer Credit Protection Act (Truth-in-Lending) of 1968, which requires full disclosure of terms and conditions of finance charges to borrowers (its passage was delayed for several years in committees by senators who represented special-interest groups); and the Securities Investor Protection Act of 1970, which provides protection for customers of registered brokers and dealers and members of national securities exchanges. Legislation also was passed in the following areas: hazardous toys and articles, automotive safety, hazards of cigarette smoking, meat and poultry inspection, toxic insecticides, radiation control, unsolicited credit cards, air quality, and boating safety.

One of the most important recent legislative measures was the Consumer Product Safety Act of 1973, which established a Consumer Product Safety Commission. Through the commission's offices, some of the prior legislation on product safety is being monitored. In addition, any manufacturer who discovers that an unsafe product has been distributed in the market must inform the commission within twenty-four hours. This has become one of the most effective devices for increasing standards of product safety. It seems that Congress finally discovered that consumer issues could attract votes. However, as an editorial in the *Wall Street Journal* cynically commented, "Since consumer legislation doesn't cost much it . . . [is] an attractive alternative to big new spending programs that are being deferred because of budget problems."[2]

The lowly matchbook is an interesting illustration of the far-reaching efforts of governmental agencies to provide product safety in areas in which company management has been slow to respond. It is commonly accepted that certain care must be exercised so that a matchbook does not catch fire when a match is struck—an estimated 9,000 injuries per year have been attributed to matchbooks bursting into flame. Now the Consumer Product Safety Commission proposes that every matchbook have its striking surface on the back of the cover to prevent the remaining matches from catching fire. In addition it suggests that each matchbook carry a safety message on its cover. Finally, it urges matchbook manufacturers to adopt a standard requiring a chemical coating on the match head that will reduce the burning time

[2]*Wall Street Journal* (January 13, 1968), p. 1.

from 30 seconds to 15 seconds. The standards for this proposal were worked out by the American Society for Testing and Materials.

Many companies use matchbooks as an advertising medium for their products. As a result, the price to the consumer of the matchbook has been very low, even gratis. The loss of advertising space, the need for the development of a new chemical coating, and the redesign of the cover could mean a price increase to the consumer. However, the gain in return would be a safer product. Again, management's delay in uniformly dealing with this present fire hazard may have brought on legislative demands that are more restrictive than a voluntary industry effort to self-regulation, or for that matter credibility for having taken the leadership in product safety.

The Consumer Protection Law

In general, business' response to consumerism has ranged from indifference, to critical, to defensive. Many business executives have tended to take a narrow view of the issues, relying on their lawyers to tell them simply whether a given package design, label, or advertisement is legal. If informed that it is legal, management frequently assumes that it has demonstrated adequate responsibility toward the consumer. This attitude has helped to stimulate the passage of even more consumer legislation in most states, notably, a consumer protection law that cracks down on deceptive sales practices. This law goes beyond supervising technical matters such as weights and measures or degree of cleanliness, and tries to protect the consumer from his own ignorance. Through such legislation, a state attorney general's office can bring a mass restitution suit or a class-action suit on behalf of all of the victims who have been defrauded.

Consumer Forces

Currently, the consumer organizations that attempt to speak for the consumer are as follows: the relatively new Consumer Federation of America; the National Consumer League, which was founded in New York City in 1891; and the Consumers Union, an independent testing agency that publishes *Consumers Report*. The trade unions and the credit unions also have maintained an active interest in consumer protection measures. Ralph Nader and the various groups he sponsors remain perhaps the most effective lobbying effort on behalf of consumer interests. There is not yet a permanent, powerful, political lobby to keep the consumers' interests in balance with those of business and labor.

These consumer problems suggest that *self-regulation* on the part of business has been unsuccessful in protecting consumer interests. As we have mentioned repeatedly, the regulatory agencies have too often been captured by the groups they are set up to regulate. The federal role in regulation has continued to grow, with an estimated 1,000 federal consumer programs administered by 57 agencies at an annual cost of more than $3 billion.[3] The conclusion

[3]"Consumerism's New Twist," *The Morgan Guaranty Survey* (October 1972), p. 5.

we might reach is that in an economy such as ours, with many groups representing various economic interests, unless a group is *organized effectively and permanently*, it will have little long-run influence. Our economic system requires counterbalancing organizations to represent often conflicting interest groups—not just consumers alone but also minorities and environmentalists.

The past decade has seen an increase in protection for a much-neglected sector of the American economy. For this protective legislation to be meaningful and enforceable, consumer involvement needs political power to counterbalance the demands of the other power groups in the economy, such as business, labor, and agriculture. The future of consumerism lies in the ability of its supporters to maintain themselves as a politically permanent and stable force.

THE SOCIAL RESPONSIBILITY OF BUSINESS

Frequent charges of corporate irresponsibility are publicized in the news media because the public has begun to question and criticize decisions made by many leading corporations—decisions that critics argue destroy the environment or even endanger life. The public often believes that the business community has yet to respond satisfactorily to complaints regarding environmental issues and the quality of life. The public seems unwilling to accept the explanation that pollution is part of the price of economic growth. Thus, the change in public values toward business decisions has not been balanced by a comparable change in the business community's values. We must consider why business often seems out of step with public values and what exactly its responsibility to the public is.

Three Approaches to Social Responsibility

Three different approaches to the social responsibility of business seem to exist. The first is the traditional view that management's only responsibility is to maximize the profits of its shareholders. This is the classical view held by such noted economists as Milton Friedman, an economic advisor to Barry Goldwater in Goldwater's 1962 bid for the presidency, an unofficial advisor to President Nixon in the first years of his administration, and the main pillar of the so-called Chicago school of economics. This classical approach maintains that social problems are not the direct concern of the businessman and that many such problems can be resolved through the impersonal market mechanism. *If* the market mechanism *cannot* satisfactorily resolve social problems, then the solution should be sought through legislation. According to this philosophy, minorities should be hired for jobs according to their skills and should be paid at the wages prevailing for those jobs. If this system does not work, and no employment is available at any wage, however small, then the situation should be corrected through legislation.

The second view is that corporate social responsibility reflects the personal value system of management. That is, a firm's social involvement is directly related to the interests and involvements of its executives, who are also its decision makers. Thus, their values and preferences prevail. In contrast, the third view argues that corporate social responsibility means that a corporation acts on a social problem in response to demand by consumers and/or public opinion. In this case, demands from outside the firm force management to make decisions that it may otherwise have been unwilling to make.

These two latter approaches to social responsibility probably lead to business decisions that result in lower short-run profits. In the past, such decisions were difficult for many businessmen to justify. But attitudes toward social responsibility and its economic justification are changing, even among the older generation of corporate executives. As the result of a shift in values among a major portion of our population—from a profit motive to a growing concern about pollution, product safety, and minorities' rights—profits and economic growth are sometimes assuming a lower priority than the quality of life and the rights of individuals. However, stockholders, who through direct or indirect ownership of stock currently represent one-third to one-half of the U.S. population, still demand that they receive the highest possible dividends on their investment. But as members of society who are affected by business' decisions, they demand also that business demonstrate a sense of responsibility to the public interest. This responsibility extends to a large number of groups outside the corporation. As a consequence, emphasis has clearly shifted away from the impersonal, unrestricted decision-making process in which the motives of the firm are considered more important than the needs of society.

Why Managers Often Ignore Societal Values

We might ask why managers often ignore public interests and demands. In the following analysis, Votaw and Sethi suggest that although some managers may recognize their obligations to society, they choose nonetheless to increase their profits by ignoring the public's interests and to let their successors worry about the consequences.[4] There are additional reasons for business' noninvolvement in social responsibilities, as described below.

Changes in Social and Physical Environment

First, the social and physical environment of the United States changed rapidly after World War II, resulting in new social attitudes and values. Yet many managers' static views of society still do not reflect these attitudes. For instance, the management of a firm still might not hire a worker precisely *because of* his race or sex, particularly if the firm has the possibility of hiring another worker.

The Role of the Corporation in Society

Another reason that many firms seem out of step with public values is that managers seldom give much thought to the corpora-

[4]Dow Votaw and S.P. Sethi, "Do We Need a New Corporate Response to a Changing Social Environment?" *California Management Review* (Fall 1969).

tion's role in society because they are busy "meeting payrolls" or balancing demands between the traditional interest groups—labor and stockholders. In the large corporations, however, ownership and management have become increasingly separated. Further, the shareholders, who elect the board of directors, are largely disassociated from each other. On a second level, a more subtle separation exists. In most large corporations, management tends to perpetuate itself by selecting its successors and paying only casual attention to its board of directors. As a result, the boards of directors have lost control over management, since the ownership has become so diffused that no one group possesses much power, and management is seldom challenged. In fact, many large firms have boards of directors composed solely of managers. Under these circumstances, management can usually weather any crisis. Even if mismanagement is apparent, managers are seldom replaced by such an "inside" board of directors. Thus, managers may feel little pressure to evaluate their traditional approaches or to consider what the role and function of the firm should be in a rapidly changing society.

The Difficulty of Assessing Society's Needs

Another, and perhaps a more compelling reason that managers fail to act socially responsible is that it is often difficult to anticipate what behavior the public will consider socially responsible. In retrospect, it is simple to say that automobile manufacturers should have acted more responsibly by manufacturing nonpolluting automobiles—but society has persistently emphasized the production of automobiles at the lowest possible cost. Managers must anticipate and continually assess what the public's priorities may be in the near future.

Many managers may actually have failed to notice the fundamental changes in values that society has undergone. When business fails to see how public demands are changing, it can hardly satisfy them. For the sake of analogy, just as Ford Motor Company misread consumer demands for the Edsel, so many managers seem to be misreading public demands regarding the social role of the corporation. The result is a moral "Edsel" crisis regarding current social issues. In this country's largest industry—the automobile industry—and also in many other large firms, the lack of responsiveness to the issues of safety, pollution, the economy of small cars, and discrimination in hiring have been grossly apparent.

Lack of Holistic Thinking

A final reason for business' noninvolvement with society is that managers are accustomed to solving problems as separate entities, as subsystems unrelated to the rest of the corporation or to society. This tradition in Western thought of subdividing issues or problems and dealing with them individually often results in conflicting solutions. Managers must recognize that a business subsystem is meaningful only in its total context; that is, they must adopt

holistic thinking. The total system, and not the subsystems, should be optimized. For instance, in its simplest terms, if corporations in our country are making huge profits but many people are hungry, the total system is not functioning well. Since the systems approach to problems (which includes both economic and noneconomic considerations) is relatively new, managers who are unaccustomed to thinking in this manner cannot be criticized if they continue to approach problems in a more traditional way. Yet managers are more often being pressured to change their thinking and to become aware of the broadest implications and effects of their decisions. The ideal result of changed viewpoints would be a situation in which the profit motive of business and the values of society will become compatible.

The history of business repeatedly demonstrates that the failure of business firms to conform voluntarily to societal values is likely to result in social legislation to compel such compliance. In our society, voluntary actions are usually preferable to legal compulsion. Waiting until the government or the public demands action only weakens the whole economic system of free enterprise by creating a lack of trust in business' motives. Further, legislation imposes more restrictions on business in the process. Managers who want to operate in the most open and liberal environment must develop an approach that considers both business and society.

Management's Responsibilities

As we consider the framework of legislation within which today's managers have to operate, we recognize that a market economy, as in the United States, pays a price in its effort to reward individualism and success under competitive conditions. So far society has not been sensitive to the need to create incentives that would cause business behavior to consider the public interest equal with that of the enterprise and its managers and owners.

A very perceptive business executive, Mr. Irwin Miller, Chairman and Chief Executive of Cummins Engine Company, has written that managers in business and industry have been too close to the ends of their enterprises. They now find themselves being attacked by everybody—workers, customers, government, educators, the church, racial minorities, and even by the managers' own children. He felt very strongly that executives of business and industry need to assume a position of leadership instead of abdicating it.[5] This means that instead of just reacting, managers need to accept a changed society and a changed environment. They have to include in their decisions consideration of the rights of the consumer, and the expectations of the public. This means adopting a systems view that incorporates the organization, the nation, and even the world to the effect of decisions on pollution, product safety, employment opportunities, and the health of urban and industrial life.

[5]Irwin Miller, "Business has a War to Win," *Harvard Business Review* (March–April 1969), pp. 44–52, 164, 168.

The responsibility of business introduces a new dimension to executive decision making. Business has been accustomed to considering only efficiency, in terms of measurable costs and output, in judging the success of the organization. There is now increasing evidence of a surge of interest in the ethical dimensions of business. Universities, such as Columbia, Washington (Seattle), and California (Berkeley), have offered courses on business and its environment, including discussions of social responsibilities and ethics. The new environment of business and the interests of the new generation of students have created expanded offerings in all business schools. Using real life case material, students consider the new demands on today's managers.

This environment is becoming very complex. For instance, a federal law effective July 4, 1975, defines product warranty, which, as we have discussed previously, has been a bane with consumers. This law makes a distinction between "full" and "limited" warranty. Full warranty implies that the buyer will have repairs "within a reasonable time and without charge," including defects and replacements. If these efforts fail, the buyer must be given full refund. Although the FTC issued nineteen pages of instructions on this new law—for instance, it is unclear as to what is a reasonable number of attempts to repair—its implications for companies are still vague. Only one thing is clear—the intent of the law is to strengthen the hand of the consumer. The consequences may well be that prices will have to rise. It may also be interpreted as the public's countermove to the problem, because it perceives that management has approached the problem until now only in terms of what is absolutely required.

SUMMARY

This chapter takes a critical look at the relations between the consumer and business—that is, to see what business is *not* doing right. The writings of Upton Sinclair in the early part of the century and Ralph Nader's and others' recent work have aroused the public to the need for greater consumer protection. Unfortunately for the public, the interest in consumer problems has not persisted after each flurry of legislation. Throughout our history, consumers have been at a serious disadvantage in achieving their rights.

The current consumer movement and the legislation it has spawned are derived both from a rising level of public expectation regarding business' behavior and also from the past failures of business to protect the consumers of its products against fraud and harm.

The recent development of consumerism is the result of public indignation about unfair treatment from business. This movement has produced the greatest wave of protective legislation in American history. The publicity and resultant legislation have often served to create a negative image of business.

In recent times, consumerism appears to have gained a stronger base because of the current generation's general concern with the conservation of natural resources and with the environment. New legislation on problems of unsafe products, deceptive packaging and promotion, worthless guarantees, incompetent repairs, and hazards to the environment have created a public demand for change, which must be seen as expressing the public opinion that business has not succeeded in achieving acceptable practices through self-regulation. The formation of consumer protection organizations suggests the embryonic stages of a political lobby on behalf of consumer interests.

The socially responsible business firm can be defined as one that anticipates what the public will expect from it and attempts to meet these demands before public criticism is reflected in legislation to bring about the desired results. Since values in contemporary society are changing rapidly, the definition of business' responsibility is changing also. The areas in which a business must be socially responsible are determined by the values of the total society in which managers live and work. The issues currently facing business are concerns with conservation of resources, pollution of the environment (water, air, and others), and the inclusion of members of minority groups in the work force.

Some managers still view the role of the firm and its social responsibility in the context of the past, when social demands were fewer. Today, managers who are unwilling to update their views, approaches, and values, and to adjust and readjust continually to newer markets and changing social demands are likely to go out of business. The speed with which managers adjust to change is also critical. In the long run, few large corporations can escape public pressures because such pressures eventually create laws if business fails to respond adequately or rapidly enough. Yet the issues society raises today will cost firms considerable sums of money. In most cases, the cost will be passed on to the consumer. Consequently, society is making a choice between having a safer and more pollution-free physical environment and a probably increased cost of living.

DISCUSSION QUESTIONS

19.1. How do you define consumerism?

19.2. Are there advantages to business in the new consumer movement?

19.3. In what ways can consumers influence business decision making?

19.4. What problem is caused for consumers by the customary practice of lobbying for legislative change?

19.5. What appears to have been the changing meaning of "socially responsible behavior" on the part of a business enterprise?

19.6. In most cases, management assumes increased costs when it responds to a demand for "social responsibility." Can you reconcile the conflict that appears to exist between "social responsibility" and "maximization of profits"?

19.7. What causes the feeling that business decisions seemingly take into account only what it gets away with?

19.8. Can a firm be socially responsible and still be a major cause of pollution? Describe a hypothetical situation in which this might occur.

APPENDIX The National Heating Corporation

Mr. Howard Grimes, President of National Heating, a large manufacturer of home heating equipment, was concerned about recurring consumer reactions to faulty products marketed by American manufacturers. Recently, the company had been sued by a customer following a furnace explosion in the customer's home. The suit charged that the company's product was unsafe, did not have adequate installation and operating instructions, and lacked effective protective devices. Mr. Grimes believed that National's furnaces were as safe as those of any of his competitors. The furnace in question had been installed by a distributor who was also the authorized service agency for the company's products. Grimes did not see how this furnace model, the company's biggest seller, could be made any safer without increasing production costs. This would automatically increase its price, reduce its sales volume, and reduce the company's profit as well.

Although Grimes was not sure that the customer could substantiate his charges, he nevertheless wondered if the company had gone far enough in its concern with quality and safety in its products. As he thought about it, he believed that his company was caught in the middle of a conflict: Customers seemed to be demanding a safer product, but he was not sure they would pay the higher prices for a safer product while standard models made by Furnco, Home Heat, and other competitors remained available at the old price. He also felt that he must protect the interests of stockholders, who naturally wanted a satisfactory return on their investment in the company. Finally, he was not unaware of the political implications of consumerism, a movement that seemed to be gathering strength in American society and that implied further federal controls on business.

Mr. Grimes believed that his firm must somehow find a way to cope with adverse public reaction, whether based on real or imagined defects in his product. He wondered if an officer of the firm should take on "social responsibility" as part of his job. As he thought about it, however, he was not sure from a practical standpoint just what a person who had this responsibility should or could do. Should he become involved in public relations? Should he consult with other manufacturers about their plans? Should he become involved in actual manufacturing decisions as a sort of "consumer representative"? Was this a job that would require someone who had special talents?

1. How would you handle the problem of consumerism in this firm if you were Mr. Grimes? Be specific.
2. Do you agree with Mr. Grimes' analysis of competition in his field?

3. If Mr. Grimes would have been told by a member of his company that his product was unsafe, should he have discontinued its sales?
4. What do you see as the factors that should govern Mr. Grimes' behavior as a responsible business executive?

SELECTED READINGS

Fritschler, A.L. *Smoking and Politics*. New York: Appleton-Century-Crofts, 1969.

McKie, James W., ed. *Social Responsibility and the Business Predicament*. Washington, D.C.: The Brookings Institution, 1975.

Nader, Ralph. *Unsafe at Any Speed*. New York: Pocket Books, 1966.

Sethi, S.P. *Up Against the Corporate Wall*, 2nd ed. Englewood Cliffs, N.J.: Prentice-Hall, 1974.

Steiner, George. *Business and Society*. New York: Random House, 1971.

Votaw, Dow, and Sethi, S.P. *The Corporate Dilemma*. Englewood Cliffs, N.J.: Prentice-Hall, 1973.

Walton, C.C., ed. *Business and Social Progress*. New York: Praeger, 1970.

Chapter 20
The Urban Crisis

In its original meaning, the word *civilization* related to cities. Indeed, the heart of Western civilization is still located in our urban centers. If our major cities were destroyed, we would lose most of our great art, music, science, architecture, libraries, and organizations. Few people, given a moment to think about it, would question the relationship between civilization and life in our cities. It is all the more curious, therefore, that at the same time the United States has become the leading edge of Western civilization, it is also letting its cities collapse and become progressively unpleasant, dangerous, and uninhabitable for the large middle class.

Over the past quarter-century, the United States has been busy creating a great suburbia around most of its cities. The cities are now being left to the few very rich who can afford to live in protected compounds and to the aged, poor, and minorities who cannot flee the crime, noise, and ugliness associated with these centers. The majority of our middle-class population has moved out and into a more serene, if duller, suburbia. In the process, the middle class has deprived itself of easy access to the high points of our culture. Yet they believe that living in a neighborhood with less crime and better schools is worth it. Given the necessities of life, they are not wrong. But the present process of suburbanization resembles all too strongly the decay of the Roman cities and the fleeing of their citizenry, which was a prelude to the death of the Roman empire. For at least the last fifty years, some scholars have been predicting that Western civilization is dying. That is has not yet happened does not mean that these predictions are necessarily wrong, for the decline of a civilization is often a slow process.

There are other thinkers who argue that the United States represents the reconstituting or reorganizing of Western civilization. After the elite and aristocratic traditions of European civilizations, some suggest that this country is currently expanding the concept of civilization to a broader and more democratic level. Thus, pop art is an expression of democratization in which common, ordinary objects—coke bottles, soup cans, detergent boxes—are glorified on canvas by notable contemporary artists.

Whether it is possible to build a great civilization based on the common man has never been tried before; and whether it can succeed is open to debate. Using the philosophy of democratization, it is possible to discard our old cities just as we discard our old automobiles and with no fewer consequences—except that both litter up the landscape. Of course, there is a great difference between cities and automobiles, and it is not clear how the former can be left behind so easily. Although we cannot answer this question, we can note that the problem of our cities is complex and involves several separate, but interrelated, issues.

THE DISINTEGRATION OF OUR CITIES

Why are we permitting our major cities to disintegrate? As we have said, a large number of the major American cities are becoming unsafe, ugly, and unpleasant to live in. This is happening even though more and more of our population is living in the megalopoli (for example, cities and suburbs) that stretch from Boston to Washington, D.C., through parts of the Midwest, and along growing segments of the West Coast.

The second major question is why are we "parking" more and more of our poor, aged, and minority groups in the cities? This segregation has made it possible for large numbers of the society to lead relatively homogeneous lives in suburbs with relatively few poor, elderly, or racial minorities. In a sense, the cities seem to have been relegated as a place to put those who would disrupt the homogeneity of the white, middle-class suburbs. The result is that the middle class must now commute long distances to work (as more businesses leave the city for the suburbs this will change) and is becoming culturally deprived, being removed from the center of cultural institutions. Ironically, the poor who live in the cities cannot afford to use the cultural centers or have not been taught to do so, even though they may live nearby.

The trend toward more minorities living in city centers seems to be continuing. A number of American cities are already more than fifty percent black. It is estimated that by 1985, Chicago, Philadelphia, St.Louis, Detroit, Cleveland, Oakland, Baltimore, New Orleans, Richmond (Virginia), and Jacksonville will also be dominantly black cities. Nonwhites are now in the majority in the schools of many of these cities. By 1985, according to some estimates, the percentage of nonwhites in the schools of these cities may approach ninety percent—unless busing becomes a major force in redistributing the student population.

With federal subsidies for highways and the financing of suburban tract homes following World War II, the great boom of the suburbs began. This was inevitable in view of the burgeoning population and the preference for single-story houses. Although there have been efforts to renew and redevelop the cities, such attempts have not been massive enough nor sufficiently well planned to help house all classes and types of citizens in the city or to prevent the cities from becoming the province of only the very rich and the very poor. There is no single solution to these complex problems. A solution will require a concerted effort on the part of both government and business and probably even a change in attitudes and values by much of the public to alter the current trend and save the cities.

The Trickle-Down Process

To understand why many of our large central cities are in serious decline, we may begin by noting how we have traditionally housed our population in this country. The **trickle-down process** most aptly describes the situation. In this process, we build new high-quality housing at the edge of the metropolitan areas, where land is available for new construction and is relatively inexpensive. But because of the cost of new housing, only those in the upper half of the income distribution in this country can afford to buy or rent them. This is due in large part to the fact that we rigorously enforce high standards of quality in these new areas, thereby raising the costs of construction. As this housing ages and newer units are built farther out of the metropolitan center, successive groups of lower-income individuals occupy these houses. Eventually, the housing deteriorates because the occupants cannot afford to meet rising maintenance costs. This change, then, from new subdivisions to old low-income housing (and even to slums, in the worst cases) takes about forty to sixty years in most areas. As each area ages, building-code enforcement becomes less stringent and is finally almost ignored in the oldest slums. Thus, we have a double standard of housing-law enforcement, which is unavoidable in this system because the standards of quality required by law are too costly for many poor households to maintain.

In the major metropolitan areas, then, the trickle-down process describes the way in which people are housed. This process is not new: It has dominated urban growth in the United States for decades. Earlier, code enforcement was equally lax in all parts of our urban areas, so that the least costly housing or even slums were not concentrated in the older city center.

A new trend toward a significant slowdown in central-city population pressure seems to be developing today. This trend is the result of much lower "in-migration" (mainly of blacks) from rural areas to central-city slums and a forty-percent drop in U.S. fertility rates since 1957. Hence, there is a much smaller wave of newcomers to the central city. The reduced immigration into the United States also affects this movement. Unless the high level of migration from rural, and particularly southern rural areas, into our large cities lasts, there will be no one to take over the oldest and worst housing at the bottom of the trickle-down process, when its former occupants move to better areas. This drop in population pressure is a key reason for the present abandonment of many older buildings in some old central-city areas.

Economic Growth of the Suburbs

Another critical factor that is changing the cities is the shift of population and economic growth and jobs to the suburbs. In the 15 largest metropolitan areas in the United States, the central cities lost seven percent of their jobs between 1960 and 1970; while in the suburbs, an increase of three million jobs occurred, or a gain of forty-four percent. By now, the suburbs have more total jobs than the central cities. Indeed, this analysis becomes even more surprising if we look at all metropolitan areas. On a national scale, we find that the suburbs have captured over seventy percent of the nation's population growth in the decade 1960 to 1970. Central cities, on the other hand, received only about one-fifth as much growth. Further, 53 of the 156 cities with over 100,000 people in 1970 *lost* population during the past decade.

In addition to losing inhabitants, central cities are losing to the suburbs many of their basic economic functions; that is, manufacturing is increasingly moving from the central city to the suburbs. If we ask ourselves what indispensable functions still occur in the central cities, it is surprising how few we can name. For instance, it used to be that the railroad terminals were the key to transportation and were located in the heart of the city. Today, airlines generally have located airports in the suburbs. Indeed, three-fourths of all the workers who now live in the suburbs are also working there—at least in the big metropolitan areas. This is an extreme reversal; we have normally perceived the suburbs as merely "bedrooms" for workers who spend their working hours in the city.

If we analyze the functions and activities that are occurring in the cities and suburbs, it becomes apparent that there is a rise of suburban nodes that rival the central business district. That is, complexes of office buildings, banks, hotels, theaters, and other commercial units are now grouping around shopping centers to form a minicenter (a node in the suburbs that often rivals the downtown functions). This development makes the old central area just one among many comparable focal points. Further, the city center is often at a disadvantage because of the congestion of automobiles, costlier parking, less personal security, and few affluent residents. Recognizing these changes is important for understanding the dilemma now posed for the central cities and their downtowns.

Crime and the Lack of Security

The rise of crime and the lack of security also are key factors in the deterioration of central cities. No other single factor is so significant in causing households to withdraw from the central cities, avoid downtowns at night, move from older residential areas, withdraw from central-city schools, or move business locations to the suburbs. Banfield argues that this aura of crime that permeates so many central cities is due not to blacks, or for that matter to any race or ethnic group but to the lower-class patterns of life that are exhibited, particularly by those at the lower end of the social scale.[1] In many city centers today, there are large black populations, just as there were Italians, Irish, and Jews a few years ago. Banfield argues that it is lower class behavior—exhibited by

[1]E.C. Banfield, *The Unheavenly City: The Nature and Future of Our Urban Crisis* (Boston: Little, Brown, 1970), pp. 158–84.

those living in the lower class housing, usually the slums in the central cities—that is the cause of crime, not race.

The criminal justice system is an almost total failure in deterring crime in low-income areas and even in more affluent areas. Fear is causing blacks in the central city to abandon whole blocks of housing in "crisis ghetto" areas. Finding some way to counteract such insecurity should be society's top urban priority, according to urban economist Anthony Downs.[2] Otherwise, he sees little hope for reversing the present trend.

Housing

Despite a general belief that there is a national housing shortage, this is not true in most areas. In fact, in many parts of the nation there is an actual market surplus of units because of the back-to-back record production of new housing in 1971 and 1972, when more than 2.6 million units were built each year. In contrast, well under 1.5 million new households were formed during each of these two years; under 700,000 units were actually removed from the housing inventory. These figures suggest that under the trickle-down process, there may not be enough demand for older houses, given the huge increments of new units built largely on the edge of the suburbs. Although such market situations can change in a few years, the interesting point is the effect such new units can have by creating a strong "pull" and accelerating the abandonment of housing in the central cities, where they may remain abandoned. In other words, the cause of much of the growing number of vacant and abandoned houses in the central cities is the massive new housing built in recent years on the urban fringes.

Continued concentration of the poor in the cities is a direct result of the trickle-down process, indirectly caused by the poor's exclusion from the suburban areas. This concentration of the poor leads to a "reverberation of problems" in such areas and creates an environment that drives out those households that can afford to live elsewhere. By avoiding "forced economic integration" in the suburbs, some authorities argue that we create "forced concentration of poverty" in the central city. Others argue that in a society that does not want to be reminded of the problems of old age or poverty, the central city is the "sandbox" in which we park the old and the poor.[3] Efforts to solve problems that do not face us directly are generally given a low priority in our society. Consequently, in many ways, the movement of more and more of the population to the suburbs and away from the city's problems may suggest that major efforts will not be made to solve the urban crisis.

Taxes and Services

Some urban economists argue that there has been a constant diversion of the central city's limited tax resources into a growing

[2]Anthony Downs, *Urban Problems and Prospects* (Chicago: Markham, 1970), pp. 27–74.

[3]George Sternlieb, "The City as Sandbox," *Public Interest*, no. 25 (Fall 1971), pp. 14–21.

city bureaucracy which, coupled with rising wages for city employees, has deflected increased tax revenues away from improving city services. As public employees such as garbage collectors, police, firemen, and teachers have unionized, they have been able to increase their pay during the past decade without significantly increasing their service outputs. Hence, this has been mainly a redistribution of income from central-city taxpayers to city employees. City taxes are regressive and have made the city more and more expensive to live in, even though services have not improved. Not surprisingly, the rising city taxes and low level of services tend to further the desire of the middle class to move to the suburbs—not because of lower taxes in the suburbs but because of better service per dollar of taxes spent. The poor also would like to move to the suburbs but are prevented from doing so by zoning laws and building codes that keep most suburban housing costs high.

Transportation

Transportation changes—such as the decline in importance of railroads and the rise in the importance of cars and airplanes—have deemphasized the advantage of central locations for living or working. The constantly expanding auto ownership and usage is a key factor underlying this change and is largely responsible for "urban sprawl." The mobility permitted by cars means work opportunities can be located everywhere, rather than concentrating them in central areas. Traffic congestion in older central areas not built for present transportation patterns discourages locating new facilities there. In contrast, new "belt" expressways linking different parts of the urban periphery influence suburban locations. Finally, airports located far outside the city have replaced downtowns as "model transfer points" where many activities naturally cluster. All these factors reinforce the loss of basic functions formerly located in high-density central areas. A permanent energy crisis in the supply of gasoline, however, may result in a fundamentally new perspective on the transportation problem in the city and the suburb.

REVITALIZING THE CITIES

We may ask, "Why don't whites move back into the central cities if property values are declining relatively there and a central location (for some jobs, at least) is more convenient?" Much of the answer lies in the continued racial segregation discouraging white "reentry" into deteriorated central areas. In spite of the rising affluence of middle-class blacks and laws against housing segregation, residential segregation is not declining in the United States. The suburbs as a whole are still ninety-five percent white, central cities as a whole are now over twenty percent black. The refusal of whites to move into mainly black areas may be due to the law of racial dominance, which suggests that in this country, middle-class whites will abandon and not reenter any area that is no longer dominantly white. Their reluctance may also be due to the problem of personal security. These reasons have discouraged cities from "recapturing" the white middle class through gradually renewing deteriorating central areas. Renewal of bigger chunks of the central cities is politically resisted by those who would be displaced, which includes those who would have no place to live because they would be unable to afford the higher rents, and includes also minority group politicians who see their political bases eroding if integration occurs. Yet the only way to change the atmosphere of deterioration and crime in the city is by renovating large enough areas so that a real difference occurs.

These factors indicate a long-range trend toward abandonment of the most deteriorated central-city areas. So far, this has occurred mainly in black areas. In essence, the housing process outlined here is the final result of the trickle-down process, when that process is accelerated by massive construction at the urban fringe. Moreover, the situation described above represents a powerful set of well-entrenched forces and attitudes in America that benefit particularly the middle-class majority. These forces will not go away spontaneously; nor could the government alter them without great difficulty, if at all. The accelerated trickle-down process is producing good housing and housing improvements for many white, middle-class Americans. The poor living in concentrated poverty areas, people trapped in transition areas undergoing "trickling down," and central-city taxpayers are those being badly injured by these trends—along with the owners of abandoned properties and their neighbors.

Naturally, this long-range process has an adverse effect on property values in many central cities. If abandonment and less drastic forms of decay spread outward from "crisis ghettos," the affected areas will increase in size and number, eventually including many older suburban neighborhoods when decay spills over central-city borders. Fortunately, however, not all central cities or all central-city areas have been hit by this trend of decay and abandonment. Many areas anchored by desirable attractive features like parks, lakes, and universities have *risen* in value recently. But except for unusual local amenities, time is on the side of deterioration—that is, if we keep accelerating the process by duplicating or exceeding the housing production years of 1971 and 1972, when housing was built mainly on the suburban edges.

This whole trend could be reversed only through major changes in both present governmental policies and public attitudes. It would require large-scale subsidies of (1) central-city renewal, so that a middle-class environment could again be created in parts of the central cities, and (2) suburban housing for low- and moderate-income households, so they are not forced to cluster only in the central cities. Although private investors can help by maintaining and creating viable downtown areas, to solve the problem will require both public and private cooperation—and government programs to create an incentive for such cooperation.

The decay of the cities could be reversed also by the altered cost of transportation (higher costs of automobiles and fuel) and/or a change in tastes to live within the cities again. Such radical changes in both economic facts and public tastes may well affect the city in the coming decade, although neither is likely to alter the urban ghetto appreciably. Governmental legislation along the lines suggested above is probably the only force in our society powerful enough to create such major changes in the urban core of our older cities. Whether we shall decide to save our cities or discard them is a decision that our country has not yet made.

SUMMARY

Although the heart of Western civilization is its cities, the United States today is allowing them to collapse and become more and more unpleasant and uninhabitable. As suburbia grows, and the middle class moves out seeking less crime and better schools, the cities are left to the few very rich who are protected, and to the aged, poor, and minorities who cannot leave.

The development of suburbs and the decay of cities has several results: many middle class people lead relatively homogeneous lives in suburbs with few poor, elderly, or racial minorities. The middle class must travel long distances to work and to enjoy cultural institutions. A number of American cities are already more than fifty percent black, and this trend is increasing.

One major factor in this decline is the trickle-down process, by which high-quality housing is built at the edge of the metropolitan areas, and only those with higher incomes can afford to live there. As the housing ages and newer units are built farther out, lower-income groups move in, and the housing deteriorates, thus producing low-income housing. This process takes about forty to sixty years. Another important factor affecting the cities is the shift of population and jobs to the suburbs. Indeed, many of the basic economic functions of central cities are moving to the suburbs rivaling the central business district. But society in general does not consider these high-priority problems, because it is not directly affected by them.

Taxes are another problem in the central city. Some economists argue that the tax resources are moving more and more into a growing city bureaucracy with constantly increasing wages for city employees. This means a rise in taxes with no increase in services. Changes in transportation patterns have also deemphasized the importance of central cities, particularly with the new expressways linking suburbs, and the location of airports far outside the centers.

In considering revitalizing the cities, we must ask why whites do not move back into the central cities. The primary reason is the continued racial segregation—whites are unwilling to reenter an area no longer dominantly white. Another reason is personal security. Governmental legislation and massive subsidy programs of central city renewal, coupled with suburban housing for low and moderate-income households, are the only forces strong enough to save the cities. But this would require public and private cooperation, and the nation has not yet decided whether to preserve the cities or to discard them.

What do urban problems have to do with business? If you were deciding where to locate your new store, office, or plant, where would you locate? This is one of the most important and common decisions businesses must make. Yet it is also a critical decision for individuals for it determines where jobs will be; it is critical for future land values; and it is critical for the quality of life of the total population. Ultimately the decision on where to locate your business is part of a chain of decisions that involves both the government and the public. Thus it is essential to analyze this issue using a systems approach. Urban environmental changes are part of the total arena in which business operates, in addition to affecting who the work force will be, who the customers may be, what the overhead cost of rent and taxes will be, and finally, what the profit-and-loss picture of the firm may be. To the business firm, the urban problem ultimately involves profits; and for the general public, the urban crisis affects the society and its physical, social, economic, and aesthetic environments.

DISCUSSION QUESTIONS

20.1. What effect may the energy shortage have on the trickle-down process? Explain.

20.2. What would your ideal city be like?

20.3. What urban and suburban areas seem most ideal to you? Why? How could more such physical environments be created in other cities?

20.4. What are the basic reasons Americans tend to let many of their cities disintegrate?

20.5. Is there a direct relationship between taxes and the quality of services to the public?

20.6. Can crime be more effectively controlled by the special design of our communities? Of our buildings?

20.7. How would you forecast the future of your city? Your suburb? Are you satisfied with the direction in which it is developing? How can you have any influence on this direction?

APPENDIX
The Parkhill Development

Sheffield, England, like many European cities, had a large complement of apartment houses. In the 1950s, the city fathers, who represented a well-financed Labor party, decided to tear down the shacks and low-grade housing that had occupied the slope above Sheffield and replace that crime-infested community by building a new housing development, Parkhill. There were approximately 1,000 apartments in this development, which was built of concrete and brick and stretched for a quarter of a mile diagonally across a rising slope some fourteen stories above a valley. Sheffield had built a number of city-owned, high-rise apartment buildings before, and Parkhill was neither innovative nor unique. Parkhill soon became sooty and black from the acid smog rising from Sheffield's industries.

The city architect had briefed the people who moved into this new housing from urban areas so that they would become part of the new community as soon as possible. It was not his intention to ask for suggestions, but the indoctrination became rather informal and very informative. In fact, from one of these briefing sessions came a modification in the design of Parkhill that was incorporated into succeeding units: The architects had forgotten that many people worked at night and slept during the day. In their original design, the bedrooms were placed below the street level, and the sounds from large numbers of children running back and forth from school made it difficult to sleep. When the architect accepted the suggestion to move the bedrooms upstairs, as well as other suggestions, a real community spirit developed in Parkhill.

In the 1970s, the same architects who had been involved in the fundamental design of Parkhill had a chance to utilize their experience in another housing complex, Hyde Park. This project was designed to discourage vandalism by providing the possibility of surveillance from many points in each building. In spite of these design features, vandalism in Hyde Park has been high—doors have been ripped off, bottles are thrown, graffiti are everywhere. There is a general belief by the tenants who live in Hyde Park that this is a poor place to live.

1. Does it appear relevant to draw on English experiences in city life to apply to planning U.S. cities?
2. What would you visualize life to be like in a stone and brick building with 1,000 apartments and up to fourteen stories high?
3. What might have kept Parkhill from experiencing vandalism of the type cited for Hyde Park?
4. What types of housing developments are you aware of in your own city or neighborhood?

SELECTED READINGS

Banfield, E.C. *The Unheavenly City: The Nature and Future of Our Urban Crisis*. Boston: Little, Brown, 1970.

Bish, R.L. *The Public Economy of Metropolitan Areas*. Chicago: Markham, 1970.

Downs, Anthony. *Urban Problems and Prospects*. Chicago: Markham, 1970.

Faltermayer, E.K. *Redoing America: A Nationwide Report on How to Make Our Cities and Suburbs Liveable*. New York: Collier Books, 1968.

Highway Research Board. *Indirect and Sociological Effects of Highway Location and Improvement*. Highway Research Record no. 75, Washington, D.C., 1965.

Jacobs, Jane. *The Life and Death of Great American Cities*. New York: Random House, 1961.

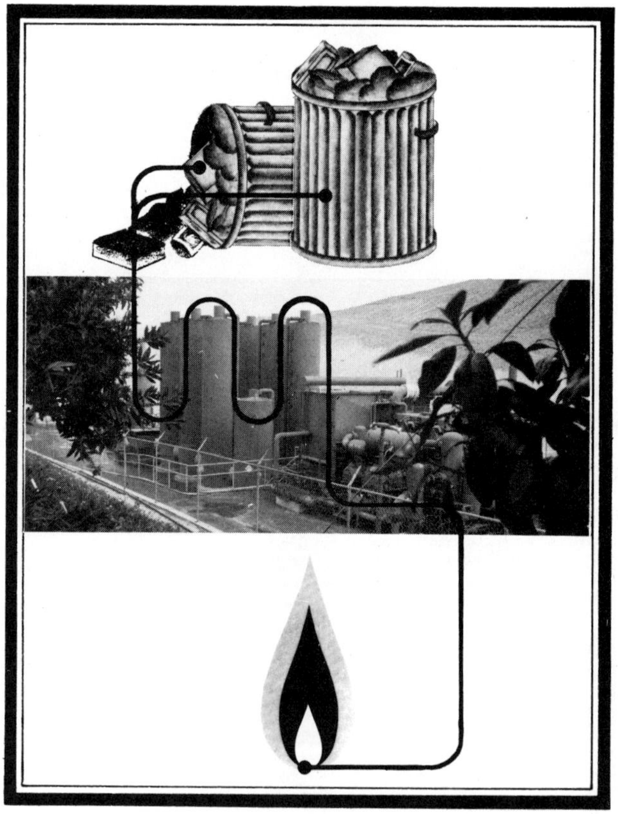

Chapter 21
Applying the Systems Approach to Public Issues

Man tends to influence his future by what he thinks will happen and then shapes his destiny by what he does. Significant progress occurs when there is a clear vision of what should be done, followed by a sincere dedication toward achieving that goal, and coupled with a professional capability to manage the task properly.

The pressures of an expanding population, social unrest, rising human aspirations and needs, and the growing interest by governmental units at all levels dramatize the challenge for social planners. Traffic congestion, inadequate housing, crime, pollution, water shortages, and the minority groups' struggle for economic and social equality are issues that have been discussed at length in the preceding chapters. Problems of such magnitude defy traditional approaches and solutions, and although the economic wealth of the United States has grown to record levels, we still face an absolute shortage of resources to do all that must be done. Consequently, we must organize and establish priorities for using our available resources to create a better economic world within a framework of peace and social adjustment.

Some believe the problems that confront society are impossible to solve. Such people are frustrated by the complexity and magnitude of what has to be done and the difficulty of getting people to change their ways of life, particularly when it may mean giving up some individual freedom. If we perceive and operate projects as systems and improve the quality of decision making by adopting the appropriate techniques and tools, we will have the possibility of making the kind of progress that will help solve our social problems. In this chapter we will illustrate the relevance of using an integrative philosophy—the systems approach—to study problems and issues related to economic, social, and political organizations.

Whereas Chapter 5 presented general concepts related to managerial planning and decision making, this chapter illustrates the application of decision making to public issues, or problems that are normally beyond the scope of a single organization. Since a manager's chief function is to make decisions, we can hardly overemphasize the need to understand the decision-making process.

THE TOTAL SYSTEM

In Chapter 4, we stated that the total system is established by definition; that is, by defining the meaningful boundaries of the system in question. In most systems organized for profit, these boundaries fall within the limits where direct decision making occurs, even though the environmental influences of other systems are also considered. In economic, political, and social systems, the scope of analysis can be much greater, and the criteria for defining the system might be to measure the degree of influence rather than to measure the decision-making capability. In every instance, the definition of the total system should be made in full awareness of the system's relationship to its environment. The first task, then, is to define the total system to determine relationships within the system and to assess the effects of the many variables—but this task is very difficult. Looking at the inputs and outputs of the defined system may often provide the basis for a meaningful analysis.

Inputs to Operations

A system operates on the inputs it receives. The operations on these inputs may include a number of possibilities—transforming, servicing, moving, holding, mixing, or aggregating. Secondary inputs, such as direct and indirect labor, capital goods, supplies, and other services also come into play. Primary inputs may be unplanned or they may be regulated to fit costs, profits, or other objectives. Garbage is an instance of an unplanned input; all of it must be accepted by the garbage disposal system. In contrast, the number of persons accepted to continue through the higher levels of a public educational system represents a regulated type of input.

We can get some insight into the significant difference between fixed and regulated inputs by considering the design of a public transportation system in an urban area. If we assume that the system has been designed to meet current and predicted demands, we can expect it to respond to changing input requirements. However, that same system can serve as an instrument of public policy by determining the direction of future growth. For example, real estate development is more likely to occur adjacent to an established transportation system, and so the inputs to the system are determined by the system itself.

Some latitude is permissible in selecting secondary inputs. One may substitute capital for labor, for example, or one type of capital equipment for another and still maintain the system's capabilities. A close relationship exists between inputs to system design and system operation. System design is established by capital expenditures for buildings, equipment, and machines, which may be substituted for direct energy inputs (labor) in the operation of the system.

Outputs of Operations

Depending on the nature of the operating system, the objective of the system may be to provide goods at a certain place at a given time, or to provide a product of a specified quality for a given number of dollars. Since the operation of a system is rarely perfect, the statement of the desired primary outputs should include a description of how much variation will be tolerated. Random elements in almost all operating systems combine to produce a distribution of outputs. Understanding the inherent variability of the system output is necessary before we can make a proper interpretation of the observed results.

In some cases, available inputs determine what the outputs from the system will be. In this case, the objective becomes one of selecting and producing outputs that maximize return or benefits within the resource constraints. For example, an oil refinery serving an oil field with an input of crude oil having specific chemical characteristics would have an output limited to certain petroleum products.

Another way to analyze the output of a system is to determine the tradeoffs that can be made between outputs that are obtained from the system and factors such as cost, resources used, and method of operation. In this case, the study is directed toward determining the cost-benefit tradeoff rather than maximizing output from a given input or minimizing those inputs necessary for a given output. For example, a fire protection system may be analyzed to determine the tradeoff curve between cost and the average time required to place equipment at a fire. When there are more fire stations and more equipment, cost is higher but response time is lower. Conversely, with fewer stations and less equipment, cost is lower but response time increases. Figure 21.1 illustrates the shape of the curve.

Figure 21.1 Tradeoff Curve for Fire Protection System

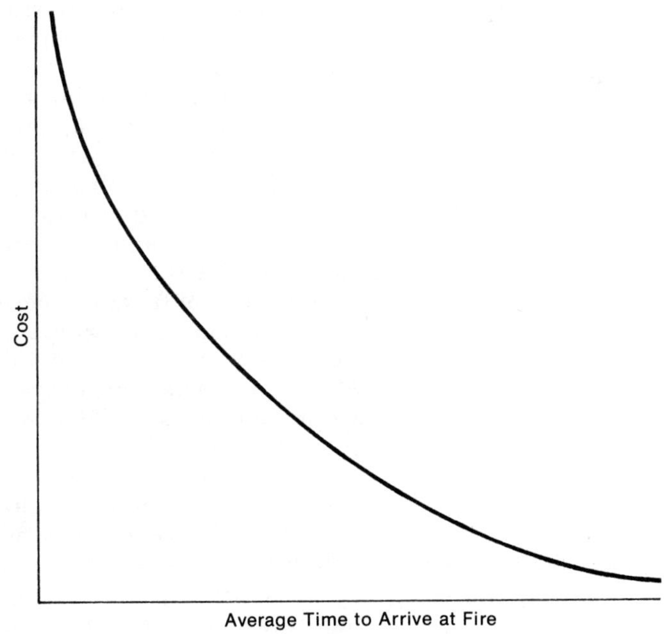

From such an analysis, we can assess the benefits derived from incremental investments in the fire protection system. If we could relate the average losses due to fires to the average time taken for the equipment to arrive, it might be possible to determine the optimal level of fire protection, where the marginal cost of protection is equal to the marginal gain achieved from reducing losses. However, functional relationships of this nature may be impossible to obtain, and in many cases, the analysis would be directed simply at obtaining the tradeoff curve.

Secondary outputs of operating systems include waste materials or other undesired by-products of operations. At one time, secondary outputs from operations were largely disregarded because of their failure to contribute to economic gain. Wastes often were dumped on the adjacent land, into the water, or into the air—with little direct cost to the system that produced them but at considerable cost to the surrounding environment and to the public at large. The point of view now being reflected in recent regulatory legislation is that such secondary outputs must be eliminated or made harmless. If this is not possible, the operating system that is responsible for producing these secondary outputs must bear the ecological cost that previously had been inequitably borne by the public. Regulating the effects of strip-mining activities, of dumping oil from ships, and of the disposal of wastes in rivers and streams illustrate secondary outputs from some operations that are now receiving increased public attention and concern. An analysis of the operations involved must consider the consequences of these secondary outputs and the possibilities for regulating them.

Transformation

At any point where transformation or service is performed on the input, and where energy, information, and/or materials are consumed in the process, we may identify a specific subsystem of the operation. Identifying a subsystem depends, however, on (1) also being able to identify inputs and outputs, and (2) determining whether the boundaries that are identified have meaning and feasibility for managerial supervision.

The principal factors that are of significance to transformation units are capabilities, capacity, and the cost characteristics. A unit may have the capability to perform only a single function or a great variety of functions perhaps on a number of different inputs or transactions flowing through the system. Specialized units are more likely to be suited to continuous processing operations, whereas units having general-purpose transformation capabilities are more often associated with varying inputs and outputs. For example, the crude-oil refinery is a special-purpose processing facility designed for the continuous transformation of crude oil, with limited capability for altering the mix of inputs and outputs. This system's particular network of activities includes collecting, refining, and distributing petroleum products. Because the network has tight physical coupling, balancing the capacity of the elements and the flow through the system is critical.

In contrast, a hospital must have the capability for servicing a wide variety of patients having different needs. There are accident victims, maternity cases, contagious diseases, and patients who need surgery. Establishing a facility to provide such a network of activities requires great analysis: the number and location of subsystems, the functions to be performed at each location, the parties responsible for the work in each instance, the mode of transportion between subsystems, and the information flow required to support the desired rate of output at minimum cost.

Measuring the capacity of a transformation subsystem requires consideration of the basic operating rate and the length of time that the facility operates. Some facilities such as electrical power generating plants are designed for continuous operation, but most are not utilized continuously. To determine the size or number of facilities to be incorporated in the system, the analyst must compare the marginal costs of higher utilization rates with the costs of providing additional facilities to be operated at lower utilization rates.

SYSTEMS ANALYSIS

We have emphasized the value of managing a problem, a profit, a program, or an organization within the context of the total system. In addition, we have stressed the need to organize the analysis relative to inputs, outputs, and transformation. At this time, we would like to point out the logic of organizing the entire decision-making process in an orderly and comprehensive fashion.

In Chapter 5, we presented a model of analysis that highlights objectives, constraints, analysis, decision criteria, tradeoffs, and synthesis. The model features an orderly process, the development of feasible alternatives, model development and improvement, planned feedback, and evaluation of the final decision. This model of analysis includes many of the significant characteristics of a manager's job—the task of outlining specific objectives, considering internal and external constraints to decision making, conducting comprehensive analyses, establishing meaningful priorities, weighing alternative decisions, instituting new programs, and reviewing and evaluating established programs in relation to the consequences of previous decisions.

Although it is essential to understand the model first, the process of systems analysis has more meaning when a person is actually involved in such an exercise. To help the reader become more familiar with the process, we will apply the model to one of the more interesting, vital, and controversial issues facing the world today.

Applying Systems Analysis to Zero Population Growth

A great debate is shaping on whether the United States should attempt to control the growth of its population, and if so, how to do it. To achieve **zero population growth (ZPG)**, the birth and immigration figures cannot exceed the number of deaths in any given year. Until recently, concern over excessive population was focused on the underdeveloped nations with their masses of poverty-stricken people. In this country, early attention to population growth also pertained to the poor or to those people on governmental economic assistance programs. Lately, the concern over population growth has been related to all segments of society. In 1970, there were 3.6 million births, and nearly 400,000 immigrants came to this country, the sum of which was nearly double the 1.9 million deaths that occurred.

There are many arguments over the moral, ethical, and legal ramifications of steps the government might take in pursuing a no-growth population policy. Should there be an effort to curtail immigration? Should government support and finance free abortions? Provide incentives for small families and/or late marriages? Tax "excessive" childbirth? Perhaps an intensive campaign emphasizing the potential dangers of population growth and making the small family or the childless couple "fashionable" might result in lowering population growth and make direct governmental intervention unnecessary.

Even if achieving zero growth were possible by whatever means, it would not happen immediately. Figure 21.2 shows the highest projection (about 325 million people) by the year 2000, which compares with approximately 285 million at the present rate of growth. If the growth were to slow below the present rate, it would still be year 2037 before zero population growth were

Figure 21.2 U.S. Population Growth

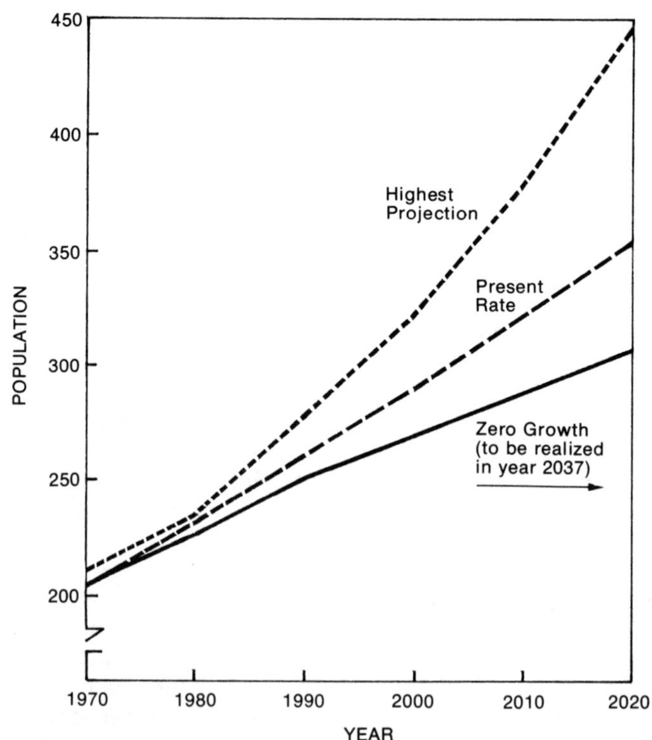

Source: U.S. Bureau of the Census.

achieved. A two-baby average (statistically, 2.11 babies) per woman would mean a gradual decline in population growth until it levelled off in about seventy years. The present fertility rate is 2.45 babies per woman, which is down from the 3.6 rate of a decade ago.

A stable population would be older, and the median age would rise about ten years. There would be fewer children, and the need for new schools and those things associated with new family formations would go down. It surely would be a different country. Figure 21.3 illustrates the interrelationship between population and resources, food, living space, and pollution. Assume that the earth is a large spaceship, completely dependent on the resources contained therein. Therefore any action that would waste these scarce resources or pollute the environment of the spaceship would tend to destroy the livability for everyone stationed on this spaceship. Likewise, an increase in the number of people would reduce the available resources per capita, as well as increase the possibility of pollution.

Figure 21.4 illustrates how the model of the systems approach can be applied. The analysis is meant only to illustrate each stage of the process. In an exhaustive analysis, translation of objectives should specify some time (for example, year 2037) when the goal could be reached. It should also include a statement of some arbitrary target level around which the balance between births and deaths as influenced by both normal and unpredictable factors would revolve. Another example of how this model could be developed to be more comprehensive would be to suggest other alternatives, such as encouraging emigration while reducing immigration toward zero. Assuming that there were common agreement

Figure 21.3 The Relation of Population to the Environment

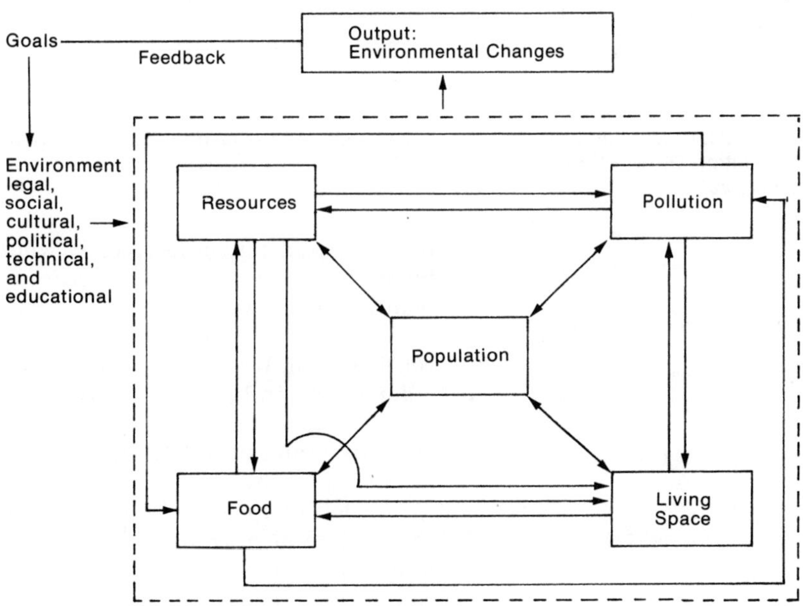

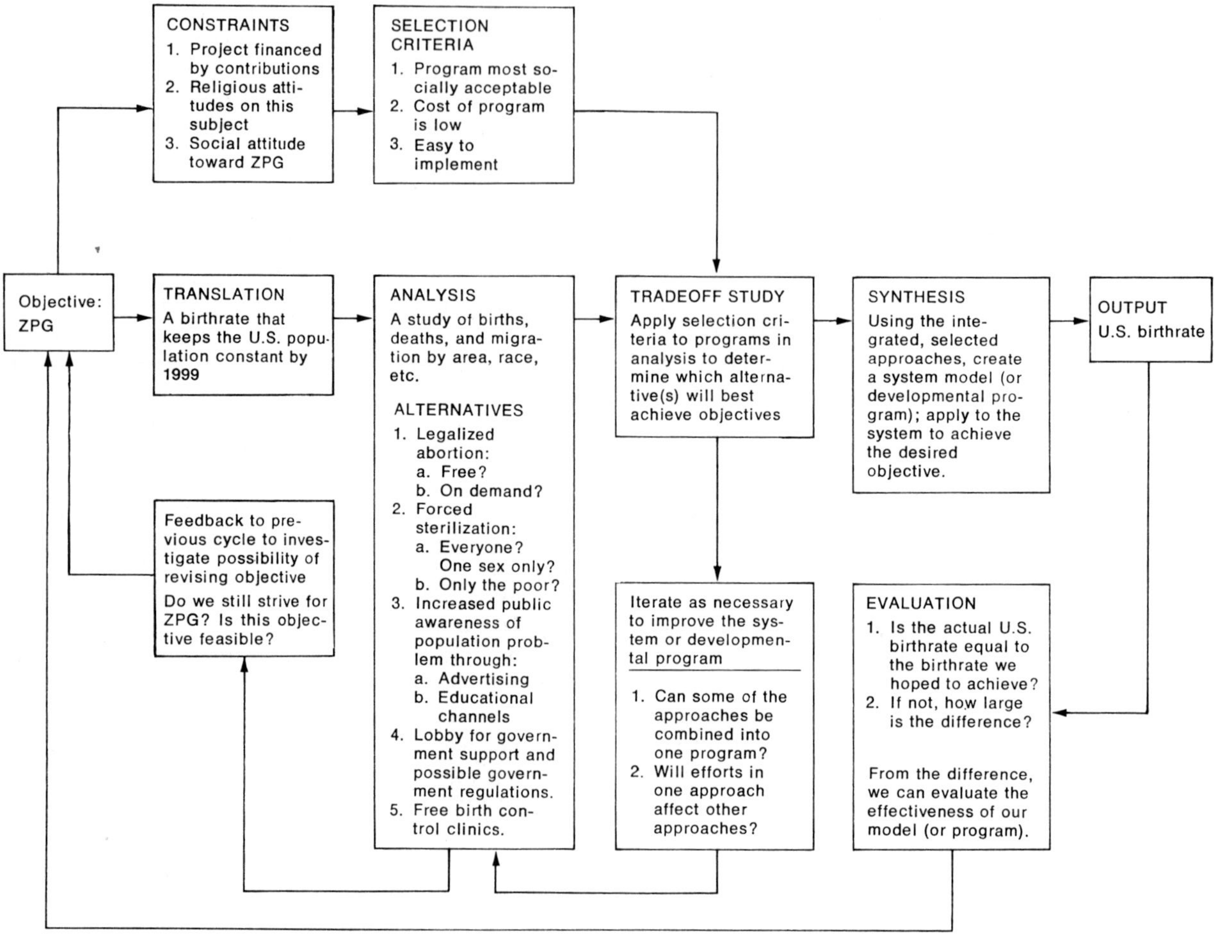

Figure 21.4 Modeling the Zero Population Growth (ZPG) Objective*

*The items used to illustrate the model are not intended to represent an exhaustive analysis.

about the initial objective, it would be possible to expand this model to create the best program to achieve the objective of zero population growth.

A systems approach demands that the analyst look at the total system. We cannot think of the population issue without considering its influence on environmental factors and sociological and technological changes. We cannot find meaningful solutions without understanding the problem as well as all factors that might affect the solution.

Defining the problem is the first order of business; determining system or organizational boundaries comes next. In national and international issues, the analyst soon becomes aware of the complexity of the problem and the degree of cross-relationships that exist. To consider an environmental issue, for example, an analyst must recognize the economic, social, and political characteristics of the issue. Even in the planning or decision-making activities of an organization, the complexity has increased. Whereas decision makers of the past were able to direct solutions primarily toward the technology objective of the organization, today's managers must include in their analyses relevant factors about the organizational climate and the pertinent environmental systems.

To illustrate, a large paper mill must consider (1) the economics of the operation so as to make a profit, (2) the factors that will maintain and develop a satisfied and productive work force, and (3) the cost of processing wastes so as not to pollute the environment. All this will add constraints and further variables to be analyzed and weighed in the decision-making process. Moreover, the decision criteria will change as social and political influences become more significant to the decision maker, perhaps eventually assuming equal weight with economic factors that were of prime importance in the past.

A decision maker must deal with those issues or problems over which he or she has an ability to influence or change. Thus, we would be well advised to consider the population problem within the context of this country, for we have more ability to influence the social, and perhaps religious, mores of this society than of another. In trying to eliminate pollution, it might be more successful to confine our efforts to a particular section or geographical area of the country or, in some instances, to a particular natural resource such as a lake, a river, a harbor, or a park that has become polluted. The problem can be identified, the boundaries are easily established, and we have the ability to influence the outcome. All of this suggests that complex problems can be solved provided that once the total has been described, the decision maker systematically analyzes a segment of that problem that lies within his or her ability to influence.

SUMMARY

The systems approach may be the most promising means of analyzing and solving many of the large-scale problems that confront our society. First of all, this approach helps identify the cross-relationships that exist among the economic, political, and social systems. Next, it provides an orderly and systematic process for proceeding from determination of the objective, mission, or issue through the analysis and to a logical and feasible solution.

A system operates on the inputs it receives. The inputs are transformed by some process that is designed to produce output that will accomplish stated objectives. Understanding how to use this framework will help the analyst relate the various subsystems to the whole. We have illustrated the orderly process for achieving the stated objective through applying the systems model to a specific issue. Although the analysis is not complete, this example does show the extent to which a study can be organized within the systems approach context.

DISCUSSION QUESTIONS

21.1. In what way will consumer demand affect the pollution problem?
21.2. Will the level of pollution increase in direct proportion to an increase in population? Why or why not?
21.3. How would a change in constraints affect potential solutions to the zero population growth issue?
21.4. List some of the consequences occurring within two years of forcibly instituted zero population growth.
21.5. Do you believe the population will stabilize eventually, even without some positive program of control? Why or why not?
21.6. Prepare your own solution to the zero growth issue. Referring to Figure 21.4, add some alternatives or change those already listed. List the constraints and selection criteria that you believe are most appropriate.

APPENDIX
The Waste Machine

Many people have suggested that the United States is a gigantic "waste machine." The average American is said to throw away about 5.5 pounds of trash a day, or approximately a ton each year. In 1974, there were 48 billion cans and 26 billion bottles discarded as trash. "Use it once and throw it away" has become a way of life since World War II.

At the turn of the century, men shaved their beards with a straight-edge razor that they sharpened themselves. Then the safety razor was developed, with individual blades that could be resharpened. Some time later, the razor blade companies started to sell cheaper blades that the customer could use for one or two shaves and would then have to discard. Today, a razor in common use has a plastic handle to which the customer attaches self-contained shaving heads. Each shaving head contains shaving edges sealed inside a cartridge. The customer is really buying a new razor each time he purchases blades.

Several years ago, one of the appliance manufacturers designed its appliances so that they could not be repaired. The toasters and irons were assembled as modular units, and when the appliance failed to function, the consumer would purchase a replacement module and snap it into place. The defective module was thrown away. Not long ago, manufacturers even promoted paper dresses and paper drapes, although neither of these items ever gained general public acceptance.

Recently, there has been some change in consumer attitudes. Recycling campaigns have been prevalent since the economy was startled by heatless homes, long waiting lines to buy gasoline, shortages of some products, and inflated prices. Although recycling successfully recovers some resources, experts suggest that this is only a stop-gap measure. They believe that the best way to save is to reuse rather than recycle. According to a study conducted by the Midwest Research Institute of Kansas City, Missouri, every time you drink a six-pack of beer and throw away the cans, you have helped to waste the energy equivalent of 1.5 pints of gasoline. They figure it this way: The manufacture, filling, distribution, and disposal of a throwaway container takes about 4,200 more Btu's (British thermal units, a standard measure of heat energy) than the equivalent process required if a container is used ten times. The energy equivalent in a gallon of gasoline is approximately 125,000 Btu's, so each wasted container represents 4.3 fluid ounces of gasoline. A simple multiplication suggests that the country wastes the energy equivalent of 2 billion gallons of gasoline each year.

1. How does reusing differ from recycling?
2. How do people's attitudes affect the "waste machine"?
3. Can economic and political leaders be influenced to conserve energy? If yes, how?
4. What responsibility should business leaders assume in the conservation of energy?
5. What are some actions that you can take to save energy and other resources?

SELECTED READINGS

DeGreene, Kenyon B. *Sociotechnical Systems*. Englewood Cliffs, N.J.: Prentice-Hall, 1973.

Ehrlich, Paul R. *The Population Bomb*. New York: Ballantine Books, 1968.

Johnson, Richard A., Newell, William T., and Vergin, Roger C. *Production and Operations Management*. Boston: Houghton Mifflin, 1974.

Posner, Michael. *Fuel Policy: A Study in Applied Economics*. New York: Macmillan, 1973.

Part Four Business and the Future

The political system in which the organization operates ultimately determines its form and potential. The growth of the American capitalistic economic system was encouraged and aided by the predictability and cooperation of our democratic political system. Today, managers face the problem of how to manage organizations effectively in a political environment in which governmental control, intervention, and regulation impose increasing constraints on managerial decisions.

One recent development is the emergence of the multinational corporation. Its influence is now significant in world economics, since it affects nearly all political systems throughout the world. Thus it is no longer possible to operate economic systems in isolation from other social systems because the world's political economy has become inextricably interwoven.

The future of management holds great promise and challenge. Management is a developing profession that needs the talents of outstanding young people. What may happen in society as a whole, and in organizations in particular, during the next decade is difficult to predict. However, if we predetermine what we wish to occur and then set forth to see that it happens, we can help shape our future.

Amtrak

Chapter 22
Capitalism and Its Future

The managers of corporations do not make decisions in a vacuum. The corporation and its managers exist in a given geographical place and work under prescribed governmental laws and regulations that have been developed over time. Thus, every business around the world must operate within the ground rules of its economic-political system. A relevant question to ask is whether the business operates under some form of **capitalism**, such as mixed-market capitalism, welfare capitalism, or national capitalism; or some variety of **socialism**, whether market socialism, state socialism, democratic socialism, or authoritarian socialism? Authoritarian socialism is often called **communism**, which also has variations—Soviet communism, Chinese communism, and Cuban communism. There are also the political systems of **fascism**—Nazi fascism, Italian fascism, fascism in Argentina under Péron, in Spain under Franco, in Portugal under Salazar, and in numerous other small African and South American countries.

The quantity of economic-political mixes in the world is extremely large. The difficulty for **multinational corporations**, particularly, is that they must adjust to each separate and different system if they are to survive. Indeed, simply to survive (let alone, prosper) under a *single* type of economic system is difficult enough. The ground rules in various countries are dramatically different, and even within a single economic-political system (like our own) they may be conflicting and contradictory. The manager must know what the ground rules, laws, and regulations are and be able to find his way through this governmental maze that stipulates how business can operate.

Corporations find that they must spend more and more time understanding and responding to the continually changing ground rules concerning environmental controls, affirmative action, and consumer regulations. This, of course, increases costs and decreases productivity. To offset this expense, managers try either to pass such costs on to the consumer or perhaps to decrease other costs of the firm. When such measures fail, the profits of the firm fall, and it becomes harder to raise the money for new plants and equipment—either from the firm's own savings or from borrowing in the capital markets.

The special interests and values of power groups in a society continue to cause new laws and regulations to be passed. This points out quite simply that while the corporation has its own ends and goals (profits being one of them), society can—and does—impose other goals and constraints on the firm. In a democratic political system, public demands are usually voiced by electing responsive politicians who, in association with representatives of other interest groups, reach bargains about what the laws and regulations will be. Business has fewer votes than other groups such as labor and has made large financial donations to political campaigns in an attempt to compensate for this disadvantage. Recent legislation has restricted this practice, however, and may have reduced business' political power in this country as a result. In a system under an authoritarian rule—such as the USSR, China, Cuba, and parts of Africa—the decisions are made by a small, inner group that controls the government. In such circumstances, business is merely a part of the government and serves the basic function of meeting the requirements of government planners and the government elite.

These two extreme systems can be represented on a continuum: At one end of the continuum, democracy enables the decentralized decision making of firms to respond to consumers and managers; at the other extreme, an authoritarian government permits a small group of planners to decide for both the consumer and the manager what should be produced and at what price. Falling between these two points are numerous combinations of centralized and decentralized decision-making organizations.

THE DEFINITION AND EVOLUTION OF CAPITALISM

Traditionally, *capitalism* has been defined as "a system in which political democracy exists, along with private property, a free market, and independent business firms seeking profits." In theory at least, this suggests a decentralized political and economic system. In contrast, communism as practiced in the USSR represents centralized political and economic power in the hands of the government elite.

Capitalism developed in full form in Great Britain early in the last century parallel with the industrial revolution. The simultaneous combination of private property, capital, entrepreneurs, and a relatively free market occurred to exploit the new industrial technology of cotton manufacturing, steam engines, railroads, and iron smelting. Although the various ingredients of capitalism had existed historically in Venice, Florence, and the Netherlands, it was in Great Britain that the coalescence of both democratic capitalism and a new technology took place to create the first capitalist industrial superpower.

It is a paradox that England, the world's greatest economic and political power in the nineteenth century, should have fallen into virtual bankruptcy by today. The reasons for this rapid demise are complex but not difficult to analyze. They include low productivity caused by obsolete plant and equipment, a divisive union and an uncooperative work force, and lackluster management by both business and government. Thus, the country that led the world in the last century trails the industrial world at this time. Does this suggest that capitalism is doomed? The United States may provide the answer to this question. The future of capitalism as a viable economic system—and perhaps even democracy as a viable system —may be tested in America during the last quarter of the century. At least many economists think so.

Features of Capitalism

There are several significant features of capitalism, but the three most important are the rights to ownership of property, a free market, and profits.

Ownership of Property

Many rights have traditionally been associated with the ownership of property. However, several of these rights have been reduced substantially over the years. Taxation, zoning regulations, eminent domain, health, fire and sanitary codes, and rent controls have all reduced the absolute power the individual has over his private property. This reduction in the individual's property rights is counterbalanced by an increase in the rights government may exercise over private property. This shift has occurred because the public has increasingly ruled that certain values involving the condition of the community, for instance, were more important than individual rights. Or, put another way, many people are saying that the choice in our society between private use and public control of property should shift toward the public interest.

Free Markets

The free market as the mechanism for economic decision making and distribution of income has been another sine qua non (essential feature) of capitalism—at least in theory. Economists and the public have found the theoretical advantages of the market so appealing that it has fostered an ideology, even a cult, of true believers. If indeed free competitive markets can exist (and whether or not they can is a real question), then decisions about what the economy can produce most efficiently and at the lowest price can be made better by the free market than by any other known alternative mechanism. The major theoretical problem is what economists call "externalities," or the effect of production (or pursuing of a firm or an individual's own interest) on innocent bystanders. For example, the smoke from a factory may blow over your yard and dirty the laundry or kill the plants or in some cases make people ill. There are ways to cope with these problems of externalities but they usually require governmental intervention.

Another question is whether or not a large company, if broken into smaller segments, could compete in the international market, and indeed, whether size (and indeed what size) might not be a requirement for an efficient scale of operation. There is a controversy among economists about whether the large firms in the U. S. should or should not be broken into smaller units. Basically, most of the products we buy today are made by firms in industries that do not operate in a real competitive market. In fact, if a free, competitive market were to be put into effect in the U. S., most major industries would have to be broken up or regulated—and the result of this action on the consumer and the country is unknown. In essence the real question is, "How competitive do we want industry to be, and which industries should be the most competitive?" Obviously, we tolerate public utilities; they manage to rank among the industries with the highest productivity levels in the country. Thus, some argue that if the concentrated industries are carefully regulated instead of broken up, it would be best for the public, since these firms would be able to produce more at lower prices, which also is the goal of a perfectly competitive economy.

Profits

The corporation's main focus on profit is yet another distinctive feature of a truly capitalist system. There is considerable evidence that profit as a goal in a public organization may either not rank as high, or at least not be as effectively pursued, as in an owner-controlled firm. However, profits tend to be the results of either efficiency or monopoly power or both. Without profits the firm faces grave difficulties in expanding or maintaining itself. Thus, Marx argued that a declining rate of profit would destroy the capitalist system. In recent years in the U.S., low profits have been held as the major reason for reduced investment in new plant and equipment and consequently, decreasing levels of productivity.

These three features of capitalism—private property, the free-market mechanism, and the profit motive—plus the human factor of management, that is, the entrepreneur (whose job it is theoretically to combine resources to produce products and, one hopes, profits), have been the components of the simple model of capitalism taught by economists for two centuries. During the past half-century, however, these basic pillars of the capitalist edifice have been remodeled, so that today the house of capitalism has a new look. To many it is hard to recognize, and to others the remodeling has only begun. Thus we should ask the question of whether we are now in the twilight of capitalism or in its full flower.

From the New Deal and World War II

Traditional American capitalism changed substantially under the influence of the Great Depression of the 1930s and the rise of the social philosophies of President Roosevelt's New Deal. Income taxes had never been a major economic factor until the late 1930s, and then with the need to pay for World War II in the 1940s, large "temporary" tax rates were passed. These "temporary" taxes became a permanent part of the tax structure. Only those businessmen who were entitled to use the tax advantages of fast depreciation or oil depletion, or who owned growth stocks and did not have to sell them and pay taxes, were able to accumulate great wealth. Otherwise, income taxes made it difficult (if not impossible) to accumulate a fortune. However, through the use of trusts, fortunes that had been made previously could still be passed on to the heirs. After World War II, the businessmen were subject to different ground rules than before and were at a comparative disadvantage in wealth accumulation. The period since World War II, though, has marked a more favorable era for the lower-income individual and the blue collar worker.

Private Property

Corporate stock represents private property. What changes have occurred during the last half-century regarding the corporation? Actually, more regulations have been passed during this period than during all the rest of the nation's history. The last fifty years represent an era in which business passed from a tradition of little governmental interference in business affairs to an accelerating trend toward more and more governmental regulation. Almost every aspect of business activity now involves government. This process has meant that the individual has much less control over his own property—whatever his business—than he had in the past. And parallel to this has been a gradual redistribution of income. (The redistribution of wealth has been less rapid).

The reasons that the public mood has dictated a more restricted role for business were undoubtedly influenced by the traumatic experience of the depression (during which twenty-five percent of the work force were without jobs and could not find them), the perils of World War II, the continuing cold war, and the general worldwide move toward socialism and communism. To attempt to deduce the reasons for the change in public opinion toward business, it might be helpful to summarize the findings of public opinion polls measuring the public's confidence in business since its high point in the 1920s. The 1930s marked a disastrous drop to an all-time low in our history. Thereafter, confidence gradually moved upward, until a new peak was reached in the 1950s. Then, business experienced another drop in public confidence in the late 1960s, a partial recovery in the early 1970s, and moved back down with the major economic recession of 1974–75. In the 1960s, the consumer, civil rights, and the environmental movement had created a crescendo of demand for more governmental regulation of business.

All of these factors have changed the nature and use of private property and the amount of free-market decision making that exists. Therefore, as the government responded to various public issues, it placed new constraints on decentralized managerial decision making as well as on the individual's rights concerning the use of private property—and eventually, of course, on the amount and rate of profit after taxes. The result is that American business and government are now partners. Indeed, one-half of all profits earned by firms go to the government, and the dividends paid out to stockholders are taxed again. Thus, the capitalist in our contemporary society has difficulty assembling capital (if he did not inherit it) and has continual difficulty maintaining a reasonable return on his capital after taxes. It is only a slight overstatement to say that the day of the private capitalist in this country has passed.

A New Kind of Capitalism

If we refer to the essential feature of capitalism we discussed earlier—the free market—we would probably conclude that capitalism may exist at present, but in a vastly changed form from that which existed before the Great Depression and the New Deal. The nature of private property, the market, profits, the entrepreneur, and public values have all changed. The corporation is no longer the dominant force in our society. The government is now first among its less-than-equal partners of capital, labor, and management.

A corporation's annual report can vividly illustrate the change in American capitalism. As recently as the late 1940s, it was not unusual to see members of the boards of directors of American corporations identify themselves by occupation in these reports as "capitalists." This title now seems obsolete. Now such directors, even if they are very large stockholders who have supplied great sums of capital to the enterprise, merely list themselves as an "investor" or often as a "director" of a firm. Thus, even the capitalists in America no longer publicly admit to this title. Obviously, if the once-proud capitalists are hesitant to admit their role, capitalism is not what it used to be. It is doubtful that children of the 1970s suggest that when they grow up, they want to be "capitalists" rather than lawyers, salesmen, doctors, teachers, or cowboys. Perhaps few ever did. If that is true, the society has prepared itself for its present situation.

Today, the professional manager is much more conspicuous than earlier in the century and the function of supplying capital is more and more carried on by savings from profits, pension trusts, and mutual funds. The vast size of corporations has made the demands for capital too great to be met by a few capitalists. As a consequence, the function of a capitalist has been progressively institutionalized. As this process has occurred, criticism of the corporation from minorities, consumers, and environmentalists, has also provoked demands that the business perform in different ways and put other goals ahead of the profit motive. Thus, management finds that it must attempt to maximize a number of conflicting goals about which it has little to say and which are imposed by various governmental agencies. The traditional assumption of capitalism—that the goal of the firm was to maximize profits, and in so doing, provide the most goods at the lowest price and thus maximize total welfare for society—has been superseded. The pressure of "externalities," monopolistic markets, and new and changed public expectations has legislated a whole new set of goals for the corporation. Today, profits are only one goal among many for the manager to pursue. This change implies the major change in society's values and attitudes, which has altered the ground rules for management.

Capitalism and Other Systems

In concluding our discussion of the definition and evolution of contemporary capitalism, it may be helpful to contrast it with some of the other basic economic systems operating today. For instance, communism in the USSR operates in a political environment in which there is only one political party, and its members are limited to a few thousand people out of a population larger than that of the United States. The real competition for power occurs within this relatively small, but elite, group. Both politically and economically, the USSR is highly centralized. The market is seldom used as an economic decision-making tool. Most economic decisions are made by planners responding to the goals of the political hierarchy and carried out by the members of the various firms. Private property exists only in the form of personal goods, occasional summerhomes for the elite, and tiny plots of land for growing vegetables and other crops for private consumption by farmers who spend most of their time working on state-owned lands. Profits are not an official goal. Rather, managers and workers are evaluated according to "production targets" of units produced, determined by the planners with the assistance of the firm. Profit as a criterion was tried recently in a modest way, but there seems to be little follow-up of this idea.

Communist China differs economically from the USSR mainly with respect to certain ideological goals: For example, everyone works part of the time with their hands, in contrast to the emphasis on efficiency and specialization in the USSR and the United States. Thus, ideology strongly influences China's economy. One might argue that the lack of effective governmental planning over resources in the United States may be due to the ideological opposition to planning held by part of our population. Many people in America regard government planning as a feature of communism that might create a more controlled and less capitalistic economy.

THE FUTURE OF CAPITALISM

Economic ideologies are often the rationalization and justification of basic values, which are seldom thoroughly examined or understood. As a result, businessmen are often forced to function in a world of unknowns because societies are almost never organized and run on the basis of rational and consistent values. In a real sense, then, managers must develop rational strategies for an irrational "Alice in Wonderland" world. Values and ideologies seem to owe as much or more to historical accident and emotion than to rational economic or political facts. It is little wonder that the logical manager may find himself exasperated by the system. Interestingly, both Soviet and American managers appear to suffer frustrations. Both systems place the manager in the untenable position of having to maximize different and opposing goals. Public goals shift rapidly. To survive in any system, a manager must be flexible and adroit—and essentially nonideological.

Growth or No-Growth Economies of the Future

In recent years, American capitalism has found itself beset by major crises that may alter its future. The traditional basis for capitalism (and indeed for all Western industrialized societies, including socialist and communist) has focused on *economic growth*. The idea of economic progress took hold in the eighteenth century with the very beginning of industrialization. This concept has been built into economies such as ours. The idea that every year one would see a real increase in gross national profit and in per-capita income—and hence, in the standard of living—has become the "accepted wisdom" in this country. The political consequences of failing to achieve this economic goal in our society have yet to be faced. Suddenly, however, the specter of energy and resource shortages, a population explosion in the Third World, and a worldwide monetary crisis and depression have served to dramatize the fact that the world has changed and that it may no longer be reasonable to pursue a policy of continuous growth. If energy and resources come to be rationed or allocated permanently, the future growth strategy of American business will be quite different from that of the past. This discussion suggests that we may have reached the end of an era of unlimited resources and affluent styles of living, which were associated with the past decade. The kinds of products people will buy will change, and the future of many firms may be short.

Alternative Futures of Capitalism

The changes that have occurred in our economic system in recent years continue to transform the way business will be operated in America. The use of wage and price controls during peacetime is unprecedented and suggests the broader role being played by government. The dramatic change in the relative cost of energy and placing the industrial Western bloc under the economic pressure of the Middle Eastern nations obviously mark a new period in Western economic history. Thus the future of capitalism is clouded. Two factors alone may suggest a change in the economic and political future of the United States (1) the very low level of productivity in the U.S. compared to that of the other major industrial nations, and (2) the fall in real per-capita income in the U.S. These two factors, if not corrected, could result in the fall of the U.S. from its position as the world's leading economic power. Further, the potential political unrest these two factors could create—when the dream of continued economic growth and affluence is shattered and people realize that drastic measures are needed to correct them—is immense. A public demand to change many of the basic ground rules in our society is likely to occur unless these two economic trends are reversed in the near future. To accomplish this may, in itself, create a major change in American capitalism. Without prescribing, however, let us look at some of the alternative business futures that could conceivably occur within the next decade.

Roll Back the Clock

While the probability is not high that our society would decide to move back to the era before the 1930s when the market and business made most of the nation's economic decisions, it is at least one possibility. Under the Nixon and Ford administrations, attempts were made to rely more on market forces and decentralized decision making. Such attempts were not notably successful and various phases of wage and price controls had to be introduced, as the economy hit new lows in a period of **stagflation**—a period with both high unemployment and high inflation. Thus, if the most avidly conservative administrations in forty years were unable to move the clock back toward less governmental regulation of business, business cannot realistically expect that it will be possible to reverse the recent tide of regulative legislation or prevent more laws from being passed in the future. Despite some firms' alleged purchase of favor from the Nixon administration to avoid antitrust actions, many antitrust suits have been filed by the Justice Department that could make some markets much more competitive. Even if some parts of our domestic market became more competitive, the international cartels in oil and other world commodities would still be less vulnerable to antitrust legislation.

More of the Same

In most forecasts, the straight-line projection is always noted as a possibility, and then is usually afforded the highest probability among alternative scenarios. This possibility is popular because few people are capable of forecasting intrusions into the systems—such as wars, depressions, inflation, interventions, or even a political coup—not to mention a general change in the public's mood or values. However, if one reviews history, these "intrusive" events happen so frequently that it would be unreasonable to forecast even a decade ahead without anticipating events of significant proportions to occur and change the vector of the present.

Let us consider for a moment what the implications for management would be if we were to move ahead with merely "more of the same." A continuing high level of regulative legislation on business will make business decisions increasingly dependent on government guidelines and approval. Also, if wage, price, rent, and energy controls are applied to the economy again, business will find a profit crisis and eventually a capital crisis developing. This forecast, therefore, suggests a highly regulated business economy. Creative management is difficult to achieve under such restriction, because management's decision-making arena is so very confined.

Nonetheless, the continuing trend in the United States over the past forty years has been for more government control of both the market and the participating industries so as to satisfy the demands of the various public interest groups. Business, its critics argue, has not acted responsibly, or has been unable to solve economic crises such as recessions and inflation. The result is that business has decided it must be increasingly active in the political scene to counter these public charges. The fact that so much regulation has been passed indicates that business as a whole has failed to understand the basic value changes and trends of our society. Further, too few businesses have understood how to get their arguments across to Congress. If "more of the same" continues, the ultimate effect will be a totally regulated economy. Such projections must indeed be disconcerting to business.

The Firm as a Government Agency

Another alternative scenario is that the economic problems of energy, unemployment, inflation, and bankruptcy will create situations in which the government may find it necessary to take over certain large firms, like Lockheed—and perhaps even whole industries, such as the railroads or the airlines. Eventually, whole segments of the economy may be run as government agencies, not unlike the socialized sectors of the economy in Great Britain. Managers in such firms become civil servants. This kind of sequence could develop from some national crisis or simply from maturing industries running down and no longer being able to provide the level of service demanded by the public or the government. For example, the U.S. Merchant Marine fleet is already so heavily subsidized that it could easily be taken over by the government.

Huge investments of capital to meet foreign competition or to provide services necessary for the economy, or environmental costs that cannot be absorbed by private concerns are other justifications for government's absorbing major firms or whole industries. These changes could be initiated by public demand for continuation of services (affecting transportation), governmental needs for military security (affecting the Merchant Marine or defense companies), or the continued decrease in public trust and confidence in certain businesses (as has affected the oil industry). Indeed, all these factors are present now; it would take only a confluence of circumstances and the "right" mix of politicans to promote them to change the system toward a form of mixed socialism. All this could be done by a reluctant Congress or president who could suggest no other alternative.

This scenario has a reasonably high level of probability of happening. The crucial question is the speed and extent to which it might occur. It becomes quite possible, therefore, for large segments of the economy to pass from private to governmental hands, with only small change in our present circumstances. One might argue that for many sectors of the economy, capitalism is not a viable economic system. This is not to say that such nationalization will increase efficiency or improve service without substantial public subsidization. In many cases, though, governmental regulations (usually at the behest of business) to protect certain industries from competition—as with the railroads and increasingly with other industries—are often destructive to private ownership and, hence, to capitalism.

The Business/Government Firm

Another alternative for the future of business and capitalism in this country is a partnership between government and private business. In the United States this has already been tried with Comsat, a communications satellite created through a partnership between American Telephone and the U.S. government. In Sweden recently, the government has bought large blocks of stock in that country's largest firms to give additional capital to the large firms, since the firms are having trouble raising capital in depressed capital markets, and to give government the opportunity to represent the interests of the public directly through its involvement with these firms. There are criticisms from both liberals and conservatives about this Swedish policy. Liberals complain that the government is wasting public monies and does not need to buy its way onto corporation boards—it can merely pass a law requiring its inclusion on the board. Conservatives criticize the idea as a first step toward socialization. In either case, the results will be interesting to follow.

The injections of large amounts of capital into new technological developments might actually require government's partnership with business. For instance, an all-out attempt to make the United States self-sufficient in energy would require massive capital investment in nuclear power plants and perhaps in solar generating plants, which would be too great and too risky for private capital markets to supply. Putting a man on the moon was a tremendous achievement and accomplished only by a partnership of private business management and governmental capital. A major technological project such as solving the energy crisis or some other unanticipated event of similar magnitude would require this kind of partnership. Also, as some of the major industries such as the automobile industry and the airlines mature and need redirection, the government might find it necessary to enter into a partnership rather than merely to subsidize or regulate. The probability is lower that this partnership method might be used in the latter case than in the former, but much could depend on the programs of a charismatic politician.

A Major Political Change?

Although the probability is slight, it is necessary to consider the potential effect on capitalism of the possibility of a military coup or the electing of a president who is a demagogue. After all, in the closing period of the Nixon administration, the secretary of defense took measures to see that no orders to the military chiefs from the White House might circumvent him, which suggests that he was worried that a coup might be considered by some. Chile had had a long history of democratic government before the Chileans elected a Marxist, Allende, whose policies pushed the country into the arms of a right wing military coup. Under both Allende and the military, traditional rights and freedoms were suspended. Thus it is possible, although we hope it is unlikely, that circumstances might occur that would alter our own political system.

Since the American "military establishment" could hardly be accused of being Marxist or even far left in ideology, one assumes that two variations might occur with a military coup. One possibility is a form of fascism in which private ownership continues but the government plans and controls the economy to meet military goals. It would be a nondemocratic government with a system somewhat like our wartime economies. Typically, under fascism the government uses industrial groups to assist in planning and coordinating the economy. Business is often quite happy at first, for monopolistic profits are guaranteed and the market is divided among firms and prices agreed on, so that risk is reduced. However, as the government bureaucrats (the military) come to realize that they hold the real power and do not have to answer to the public in elections, the private property drifts increasingly into the hands of the government and the new military elite. Thus even most businessmen, who at first might support such an antidemocratic system, eventually come to regret it, since they have no checks against the abuse of power by the government.

A second possibility is a form of socialism that has occurred in many of the military takeovers in developing and underdeveloped nations. Whether this would occur in the United States would depend on the military's need to gain support from lower-economic interest groups. Many military leaders find the planning and nationalization of certain key defense industries appealing. In underdeveloped nations, of course, there is a great need for more equitable income distribution and for many welfare measures. Most of these measures already exist in the United States. In some Latin American countries such as Brazil, the need to control inflation is a justification for a military takeover (whether or not its programs for relief are successful). Once this happens, there is a need to correct income injustices toward the aged and the poor and often the middle class, resulting in a form of wealth redistribution. Some of this is already happening in the United States through taxes and inflation. But the role of business under military socialism is precarious because of the government's proclivity to take over and nationalize the nation's basic industries. One might argue that this tendency is accentuated when those businesses have foreign ownership involvements. Obviously, neither of these results can be considered capitalism, for capitalism by definition requires a democratic political system that can diffuse power and make abuse of power answerable to the citizenry.

SUMMARY

Capitalism, as traditionally defined, means an economic system that includes private property, a free competitive market, firms seeking profits, and entrepreneurs operating in an environment with few governmental restrictions. The entrepreneurs bring together capital, labor, and land to produce the maximum number of goods at the lowest price. This political system is assumed to be democratic, with free elections, competing parties, and a free press. Democracy and capitalism have generally been considered to extend the greatest freedom—both politically and economically—and to provide the most incentive and opportunity for personal achievement.

The criticisms of capitalism are that it creates and maintains great inequalities of wealth and income and permits concentrations of power for the wealthy and for industries in which competition is less than perfect. Further, it is argued that the "externalities" of the environment are not adequately solved under capitalism.

There are several alternative futures for the corporation and capitalism. (1) Roll back the clock to the era of little governmental regulation in the affairs of the corporation. (2) More of the same, or continuing governmental regulation of business so that the scope of business decision making is increasingly reduced. (3) The firm as a governmental agency as a result of economic or political crises creating circumstances in which the government finds it necessary to take over certain large firms and even major industries. (4) The new business/government firm, as with Comsat. A major technological thrust for the economy, such as putting a man on the moon or becoming self-sufficient in energy, requires great capital and manpower necessitating business-government partnerships. (5) Finally, with a major change (such as a military coup) in the political system, democratic capitalism could vanish into a variation of authoritarian fascism or authoritarian socialism. Capitalism without democracy becomes fascism, and democracy without a competitive market and a relatively free profit-seeking business structure becomes either socialism or some new political system.

The only thing that can be certain about the future of capitalism and business in this country is that the ground rules will change—in what direction and how fast are the key questions. Therefore, the successful manager will need to anticipate what the future holds in order to succeed and remain in business. The guiding principle for the modern manager is not just to react to the problems of the past but to anticipate the problems of the future.

DISCUSSION QUESTIONS

22.1. Is capitalism "the businessman's favorite economic system?" If so, why?
22.2. Without private property, can capitalism still exist? Can private business still exist?
22.3. What role does the market play in a capitalist system?
22.4. How has our capitalist system changed since 1930? In what way does this affect business?
22.5. What are the current weaknesses of American capitalism? Can these problems be corrected? How?
22.6. Do you think that all countries should be capitalistic? Why or why not? Discuss.

APPENDIX
The Obsolete Car

Harrison put down his coffee cup and watched as Bill Evans approached from the doorway of the coffee shop. Harrison and Evans usually had coffee together in midmorning. They were both employed as accountants in different departments of the Miller Clothing Company.

"Hello Bill," Harrison said.

"Hi, Bob. You beat me to it this morning. What's with you guys up there? Nothing to do?"

"Yeah, but I wanted to walk by Johnson Motors and leave some papers," Harrison said.

"Buying a new car?" Evans asked.

"Yeah. Jane thinks the old buggy is ready to turn in. I guess I agree."

They sipped their coffee in silence for a few minutes. Finally Harrison said, "This is a funny country, Bill. That car of ours only has 40,000 miles on it. Runs like a top. And yet, here we are turning it in for a new one because it *looks* old. Hear what I said? Because it *looks old*. We do this every three or four years!"

"Yeah, we do the same thing. So does nearly everyone else I know," Bob said.

"I dunno what it is. We're just a bunch of sheep, I guess. It all seems so wasteful. Those guys in Detroit know what they're doing, all right!"

"You think they're to blame?"

"Of course! Every year they come out with a new model. A little piece of metal here, a little different curve there. My car has the old, 'square' look. The new ones have the 'contoured look', whatever that is. Basically, the two cars are the same. They just look different."

"The ads don't help, either," Bob said.

"You're absolutely right!" Harrison said. "This is supposed to be a free country. Our whole system is built on the idea of choice. Prices are supposed to keep things in balance. But Detroit leads us around as if we had rings in our noses! It's a deliberate, planned obsolescence. Wasteful as hell! Same thing happens with nearly everything you buy. A friend of mine who sells beer told me once that all our applicances—refrigerators, dryers, TV sets, and things like that—have a built-in life span. So many years, and Pow! It's ready for the junk heap."

"Well, I dunno," said Bob. "Our refrigerator is ten years old and still going strong."

"Look at what happens in Europe," Harrison said, ignoring his friend. "Car manufacturers over there stick to the same model year after year. My boy has a 1962 VW and you can't tell it from one made in 1975. Same way with all their cars. And they make them smaller and cheaper to run, too. Why can't we do it like that in this country?"

1. Do you agree with Harrison's assessment of the automobile economy? Is it fair? How about his comment on major appliances? What are the implications for management?
2. Should this country shift over to the European example of fewer model changes and economy? Why or why not?
3. In a country where democratic capitalism is the dominant economic system, how could such changes away from planned obsolescence be brought about? What part could or should business management play in such a change?

SELECTED READINGS

Bjork, Gordon C. *Private Enterprise and Public Interest: The Development of American Capitalism.* Englewood Cliffs, N.J.: Prentice-Hall, 1969.

Edwards, Richard C., Reich, Michael, and Weisskopf, Thomas E., eds. *The Capitalist System.* Englewood Cliffs, N.J.: Prentice-Hall, 1972.

Friedman, Milton. *Capitalism and Freedom.* Chicago: University of Chicago Press, 1962.

Grossman, Gregory. *Economic Systems*, 2nd ed. Englewood Cliffs, N.J.: Printice-Hall, 1974.

Holm, S.N. *Economic System.* New York: Holt Rinehart and Winston, 1955.

Monsen, R. Joseph, Jr., *Modern American Capitalism. Ideologies and Issues.* Boston: Houghton Mifflin, 1963.

Silk, Leonard. *Capitalism the Moving Target.* New York: Praeger Publishers, 1974.

Chapter 23
The Multinational Corporation

The Roman Catholic church and the House of Rothschild have both maintained large interests in foreign countries for centuries. Britain's East India Company, however, may be closer to the prototype of the modern multinational corporation. The East India Company ruled one-fifth of the world's population for nearly two-and-a-half centuries in its day, exceeding in economic power General Motors—currently the largest multinational corporation, with annual sales of nearly $30 billion. During the 1930s, United Fruit Company (now United Brands) had become so large and powerful, with its holdings of four million acres of land stretching from Cuba to Ecuador, that it could—and did—topple governments in Latin America. Not many years ago the country of Liberia in Africa was known as the "Firestone Republic."

Although the multinational corporation is not a new idea, only lately has it become so widespread and potent that it has alarmed many major world powers such as France. The vast majority of the five hundred largest U.S. corporations have extensive overseas investments and representation. These corporations now play a significant role in foreign trade and in the movement of foreign capital: One need only think of the role the large oil companies have played in the last few years in upsetting economic and political systems around the world. Some recent currency devaluations (even of the supposedly mighty U.S. dollar) have been precipitated by decisions made by multinational financial managers.

FROM NATIONAL TO MULTINATIONAL

National corporations become multinational because such expansion provides growth of sales and profits and, at least in the beginning, an outlet for investment. A **Marxist view** would argue that the multinational corporation is merely the next phase of capitalism turning into economic imperialism. Marxists have traditionally asserted that capitalism would eventually run out of profitable channels for investment in developed countries; therefore, such organizations would be forced to invest their profits abroad to keep their capital busy earning the highest possible rate of return.

The non-Marxist view contends, however, that investment opportunities have not run out, or even run down, in the United States. Rather, investment opportunities often pay more abroad because there is more risk—and the greater the risk, the higher the rate of return. This line of argument suggests that as firms reach the point where they cannot expand their sales easily at home—antitrust laws being one major limitation—an active and aggressive management will naturally seek out places where they can sell more of their product.

Initially foreign ventures may be treated as mere sideshows. This attitude finds its financial explanation in the gambling theory of foreign earnings, in which the firm regards its foreign earnings much as a man does his winnings at a race track. There is a strong tendency to let the money ride; that is, to reinvest it in the firm that generated the earnings without thinking that there may be even more profitable prospects elsewhere. At some point, however, the home office recognizes that the foreign sideshow constitutes a fair percentage of the total company's business, at which time top management moves to set up more central controls of its international subsidiaries. It moves from joint ventures (which are often difficult to control) to wholly owned operations. The corporation may begin saying somewhat self-consciously, "We are an international firm." Progressively, however, this self-consciousness disappears and the firm finds it is involved in the multinational game as an operational strategy.

Severe organizational problems may develop during this transition, not unlike those the company faced during its national growth. First, as a company expands its national market, its all-purpose management (generally quite small) may be replaced by a more complex organization broken into functional areas, such as finance, marketing, and production. This organization works until the firm has to diversify further. Then with a number of product lines to serve, great strains develop. Each product line has its own manufacturing and marketing problems. The corporationwide functional managers cannot solve all the problems. Eventually a new organizational structure may take shape around divisions that are profit centers built on specific products. A central office and staff are created to serve all the product lines.

Risk and Rate of Return

As mentioned, the high rate of return may often be illusory; that is, although the rate of return may appear much higher in a foreign country, in practice it may actually result in a lower percentage of return. For instance, political risks for the security of business are often much greater abroad than in the United States, and the danger of either expropriation or war loss is often far from remote. The Singer Company's experience in operating around the world gives an excellent example. In the sixty years up to 1965, Singer had gross earnings from its foreign subsidiaries of about $670 million. In the same period, however, total assets with a low book value of $145 million (spread over nearly forty countries) were either confiscated, nationalized, or destroyed because of wars. To compensate for these risks, a rate of return (on the surface) that is expected in a foreign country must be equal to the rate of return in the United States *plus* an amount equal to insurance against the risk of war and confiscation.

Inflation presents an additional risk. The loss of investment and earnings due to currency inflation represents a significant cost in doing business abroad. Most companies attempt to use local borrowing to hedge against such losses. Still, it is necessary to consider the possibility of inflation and analyze the cost and availability of overseas local bank financing. Further, the costs of managing an international subsidiary are usually greater than in a comparable domestic business. Such additional costs relate to transportation, salary allowances for nonnationals, and communication. These extra costs vary greatly from one country to another.

All of these factors must be considered in determining a realistic expected rate of return in a foreign country. Yet, this does not even figure the costs or problems in taking profits out of a country and the political problems at home and abroad that may confront the firm. Nonetheless, most major U.S. firms today have become multinational and think of themselves as corporate "citizens of the world."

STRATEGIES

The advantages of using systems analysis before deciding to enter international operations may seem logical; however many firms have become involved in these operations almost by accident. The decision may be based on nothing more than, "Our competitors are going overseas and we are afraid that they will get a head start in a potentially rich market or will acquire a cheap source of supply for possible reimport into the U.S., thus threatening our domestic market position," or, "Antitrust limits our growth at home, so to continue to grow and expand we have no choice but to go abroad." Sometimes, the strategy for going abroad is to link a competitive advantage—which might be in technology, reputation, cheap capital, brand name, or skilled management—with the advantage peculiar to another region, for example, cheap labor. Often, the competitive advantage is seen as the superiority of American professional training of management. Thus, just as countries export products, so can they export managerial skills. The export of American managerial skill is particularly evident in cases where the American manager has a competitive edge due to long experience in organizing the sale and servicing of high-technology products.

Thus, a complete analysis of a company's assets, including its management, may suggest that a vastly higher rate of return may be possible overseas. The systems approach to management suggests, then, that the entire globe is the potential market, and that the rational multinational firm will constantly shift its resources between countries, as political constraints allow, to maximize its rate of return.

Information Flows

Before a firm can develop a detailed strategy, however, it must have resolved the information flow problem, which is also a measurement problem. For instance, what is meant by the concept of "return on investment"? Is it local currency return or dollar return? If it implies dollar return and the local currency is devaluating rapidly, what exchange rate should be used, and when? At the end of the month? The end of a quarter? The end of the year? Even if these questions can be answered, there remains the question of what management means by "investment." Is it original dollar investment, the current dollar value of investment, or the remittance value of the investment? Such considerations, along with the problem of constantly shifting exchange rates for both the dollar and foreign currencies, suggest that a firm may make money selling its products abroad only to lose it all through devaluation or inflation. To avoid this possibility of loss requires the service of highly skilled financial managers who constantly review financial and political activities throughout the world and understand the value of the multinational firm's liquid assets in all currencies—which is no small task.

There is often another problem concerning the accuracy of economic data in small, underdeveloped nations. In other words, when a decision is to be made on whether or not to invest in such a country, it may be difficult to get valid economic data about what is really going on in that country. As a consequence, the multinational corporation may soon find that it needs its own international economic department and even its own "state" department. Multinational firms come to resemble sovereign nations as they deal on a global basis with countries on both an economic and a political level. This is a new experience for most managers trained in the context of a single domestic firm.

Management

Although the period of time required for personnel from one country to function effectively in another seems to be diminishing due to increasingly similar managerial practices around the world, the need for frequent contact with local governmental officials and for keeping constantly abreast of local news developments make it essential for multinational management to speak the local language. Furthermore, strong nationalistic pressures in many countries are making it more difficult to rotate management routinely into middle- and upper-managerial posts. For political reasons, the management of one nationality may become unacceptable to the government of another country—which, of course, reduces the flexibility and certain economies of management that the multinational corporation has enjoyed. Such factors are making it all the more important that international firms plan their manpower needs on a long-term basis, with the nationalistic pressures of foreign countries in mind.

For years, Singer has followed the practice of relying on local managers around the world. They are given broad autonomy—unless local operations go sour. Then the headquarters-based vice-presidents take matters into their own hands. For example, in 1967, Singer's International Consumer Products Group responsible for all consumer goods outside North America and Europe, had overextended credit to customers in South America, Africa, and Asia. Delinquent accounts had become perilously high. Singer's president replaced a top layer of management, installed nine executives from New York in key spots on each continent, and personally ran the group for nearly two years before gradually releasing authority to a group vice-president.

Marketing

Of all the managerial functions, marketing is difficult or impossible to centralize at the headquaters of a multinational firm. This difficulty is due to variations in social patterns and the subtlety of local conditions, which have the most pronounced effect on basic business strategies and tactics. Thus, the responsibility for marketing planning must be assigned as close to the local level as possible, and it must be administered by those executives most familiar with local tastes. Information about local values and tastes may be difficult to acquire in many foreign countries. As a result, managers must often make decisions based on less information than the parent firm may desire, which is all the more reason that such decisions are carried out by personnel most conversant with local consumer tastes.

Because of the need to stay in close touch with local demand, few companies that have had prior foreign experience centralize the marketing function. Even the most product-oriented firms usually administer marketing on a national or regional basis, with only a small amount of intercompany coordination of local advertising and promotion. The result is that in the marketing area, there is generally little value in integrating into a single, uniform system corporate long-term plans for international marketing.

Product Planning

Cultural differences among nations have implied additional constraints on centralized product planning. Some authorities use as a rule of thumb: The closer a product is to the consumer and the less technically complicated it is, the less useful it is to determine detailed product specifications outside the specific market in which it is to be sold. For example, it may be possible to plan a line of laboratory instruments at the central corporate headquarters for worldwide distribution, but it is less feasible to forecast the demand for electric stoves in various world markets. Most extreme would be a centralized staff's attempt to plan what styles of home furniture should be designed and marketed in certain overseas regions. Local laws, trade restrictions, and pressures from local groups have resulted in a maze of regulations that make it necessary to develop the product as closely as possible to the actual purchaser. Indeed, the U.S. Department of Commerce lists dozens of nontariff barriers—including special packaging, labelling, and marketing restrictions. It is easy to see that product specifications must be geared closely to the actual market, and that even when there are products that have worldwide demand and where taste factors are not significant, the problem of meeting local specifications—even current safety regulations—reguires some individual tailoring of each product for its specific market. Needless to say, this greatly complicates the whole production cycle.

PROBLEMS FACING MULTINATIONAL CORPORATIONS

The most typical problems facing the managers of multinational firms are overcentralization, money management, local politics, and local reactions to foreign business.

Overcentralization

The complexities of international activities suggest that corporate planners for multinational firms, unlike their counterparts in domestic firms, find themselves restricted in the amount of detailed programming they can undertake effectively. This restriction leads to smaller and more compact planning staffs at the international headquarters of multinational firms. Consequently, a majority of the planning takes place in the overseas branch offices.

Just as American business has tended to swing from overcentralization to overdecentralization, so the international firms tend to swing back and forth between these two poles in attempting to reach the correct functional balance for their activities. Such experimentation is to be expected and even encouraged to avoid organizational stagnation and to cope with continually changing circumstances.

There is a general tendency to centralize whenever possible. For example, when the head office looks at the company's worldwide operations and identifies few differences between one country and another. To illustrate, for years the Singer Company had no separate organization for Canada but merely considered it an extension of its American activities. In time, their Canadian profits seriously declined. When they created a separate Canadian division, profits began to rise almost immediately. Similarly, a mistake made by many firms is to lump Argentina and Bolivia together because they have the same cultural heritage and to assume, therefore, that market requirements are the same for both countries. However, the political environments and levels of sophistication between the two nations are quite different and require different marketing and supply strategies. Occasionally, marketing strategies for different countries may be combined successfully. For instance, Canada and Australia, although in different hemispheres, are much more similar both politically and culturally than either Japan and Korea or Mexico and British Honduras.

Despite the need for decentralization, centralization is still the rule rather than the exception. The forces causing centralization are related to the problems of overlapping markets, diversified supply, and production. The tendency toward centralized control of product and market planning is not evident among the heavy equipment manufacturers, however. For instance, International Harvester has a joint venture located in Turkey, with engines produced in Germany, the chassis made in the United States, and a local Turkish source to supply other components. This firm implemented a degree of centralization only for coordination. Concerning the design, product development, purchasing, and financing, the heavy equipment manufacturers tend to be highly centralized. Yet from an operations standpoint, the overseas branches are run with a high degree of decentralized authority.

Centralizing tends to increase as trade and tariff barriers decline; decentralization increases as the reverse occurs. Massey-Ferguson is an example of a multinational firm that has carried the interchange of components and products all the way to its home and foreign consumer markets. It uses a French-made transmission, a British-made engine, a Mexican-made axle, and American-made sheet metal parts—all of which combine to produce in Detroit a tractor sold in Canada. Chrysler, Ford, Sperry Rand, and IBM are also contemplating increasing the degree of "internalization" or "commonization."

Choosing the Best Alternative

As a firm weighs the decision to integrate its production internationally, one of two opposing models of international operation must be chosen: (1) to design or adapt products to each specific marketplace with decentralization of engineering and product planning, or (2) to opt for substantial quality and cost control by standardizing products and processes.

Which alternative is chosen obviously depends to a large extent on the product. Automobiles must be adapted to local driving conditions and pollution standards, and motors must be made so that they will run on local power supplies. For many products, such concerns are less important. In a relatively underdeveloped country, if the product is the first in a particular market, and particularly if it involves high technology, the local industry will sometimes accept it and adjust it to the standards it sets. An example of this adaption has occurred in the television industry, in which a product developed for sale in the United States became the standard for Latin America.

Money Management

Whereas centralized production planning can be debated in multinational firms, coordination of finance at the headquarter level is generally accepted as essential. United Shoe Machinery (USM) is an example of a firm that operated its subsidiaries in virtual isolation ten years ago. Now, its European headquarters at Lausanne, Switzerland, handles the financial flows and reserves of all of its subsidiaries to maximize the return on its liquid resources. The following case illustrates the firm's financial strategies.

One of USM's Danish subsidiaries had excess cash on hand, which it lent to another Danish subsidiary that was receiving goods from the Swedish subsidiary. By having the Danish unit prepay its account with the Swedish subsidiary, it was possible to use the money to finance the movement of Swedish goods into the Finnish subsidiary. What did this accomplish? Simply, that if Finland had been required to pay for the goods, it would have had to borrow at a rate of nearly fifteen percent, the going Finnish interest rate. Further, if the Swedish subsidiary had had to finance the sale, it would have had to borrow at about a nine-percent interest rate—but the cash in Denmark was bringing in only five to six percent. Thus, a considerable savings was made in interest. Moreover, since the Danish currency was then in a weak position in relation to the Swedish currency, by speeding up payments to Sweden it was possible to hedge against a devaluation of the Danish krone. Hence, good money management accomplished two goals.

USM, like many multinational firms, starts a foreign subsidiary with a limited amount of equity and a generous supply of debt. This is possible because foreign lenders often provide far more money as a percentage of equity than American lenders. For many firms, this added leverage is a particularly attractive feature of international business because it increases the effective rate of return on their foreign investment. Leverage, however, works two ways. In subsidiaries that are thinly capitalized, troubles may occur. This happened with USM's fledgling Portuguese operation, requiring headquarters to ask the British and German subsidiaries to ship components and finished goods for inventory to the Portuguese subsidiary on open account, taking no payment for over a year. Such procedures become much more difficult when various countries impose mandatory controls on capital movements. The continually changing governmental regulations, not to mention interest rates and currency positions, make international money management one of the most sophisticated, but potentially profitable, operations of multinational firms.

When the United Kingdom's pound sterling was devalued in 1967, instead of being hurt, USM actually came out ahead. It had not only made use of the forward exchange market (where you can buy and sell currency at some time in the future), to hedge against devaluation, but it had also encouraged subsidiaries in countries with solid currencies to defer payments for goods sent them by the British subsidiary for as long as six or seven months. Thus, when USM did pay its subsidiaries, they could buy the pound sterling cheaper and save a considerable amount. The problem with this method is that it could trigger a devaluation of a country's currency if many money managers were to adopt the same strategy. This could cause major political problems, both for the government and for the multinational firms. Massey-Ferguson, also, was able to gain from devaluation of the pound because of its worldwide production flexibility. Since many of its subsidiaries produce a large line of interchangeable parts, it could switch subsidiary purchasing immediately from France to the United Kingdom, thus taking advantage of the fall in British export prices caused by the devaluation of the pound. Exchange-rate planning is important to multinational firms since the ability to predict fluctuations in the exchange rate can cause them to make or lose tremendous sums of money.

Local Politics

Closely associated with devaluation or revaluation are all the political problems of local countries and their currencies. Although corporate planners may devote a great deal of time to quantifying the effects of economic changes in the U.S. business environment, domestic business has traditionally taken little account of political shifts in this nation. The American government has been highly stable, and major political-economic changes have not occurred often or unexpectedly. In many areas of the world, however, this is certainly not the case; politics have much more effect on business and are apt to change more rapidly and perhaps more violently. Thus, corporate executives used to dealing with domestic plans that have been condensed to one or two pages of statistical data often cannot see the need to take the time to plow through the voluminous reports that may be required to understand the effects of overseas political trends, or to bother spending the time with key managers to review local politics and its significance for the subsidiary. Yet in many cases, the profitability of overseas operations and crucial decisions pertaining to overseas investments rest on the accuracy of assumptions about local political trends. As more and more firms have faced problems of nationalization, this fact may come to be recognized.

Reactions of Foreign Governments

There appears to be considerable ambivalence among host countries to the investment overtures of multinational corporate groups. On the one hand, they recognize the ability of multinational companies to provide vital capital and the technology for development and also superior marketing knowledge and market access. On the other hand, there are many potential problems associated with the operating plans of the multinational firm. For example, the possibility exists that the international firm will not bring new capital and technology into the country but instead will simply muster much of the scarce domestic resources, capital, and labor, which only serves to create competition and deprive the national firms the use of such resources. The recent history of fund raising in Europe by U.S. corporations seems to suggest that this situation occurs in many instances.

Furthermore, host countries realize that a giant firm that can provide export markets for their products can also cut off such markets and eliminate the jobs that depend on these exports. If Korea becomes a cheaper producer than Italy for a given product, the multinational firm can decide to switch its operations out of Italy and into Korea. The fear also exists that the international firm that has the ability to raise foreign capital for a host government's economy may decide to drain away the capital for use elsewhere. This is the typical accusation of "decapitalization," which is a common point of contention between foreign investors and Latin American countries.

The host government might also suspect that the multinational firm may be in a position to avoid the jurisdictional reach of the sovereign country. In the field of taxation, problems of avoidance are often expressed by host countries. Such business practices as transfer pricing, royalty payments, and headquarters cost allocations among members of the multinational corporate groups are often sore points. Similarly, the subsidiary's more subtle ability to borrow from its parent company or other branches may put it beyond the power of the host nation's monetary regulations. Thus, its critics argue that the multinational firm has many unfair advantages in business over its local counterparts.

But the fears mount to near hysteria if the suspicion exists that the multinational firm is acting in partnership with some foreign sovereign. In the underdeveloped nations, for instance, there is a constant fear that the subsidiaries of United States firms, even though they are nominal nationals in the host country, could always call in the marines to support a point, or more realistically, may be able to influence U.S. foreign aid and thereby dictate the outcome of a potential dispute between the country and the corporation. Canada and some European host countries are sometimes concerned that the subsidiaries of American firms are disciples of misguided U.S. policies with regard to East-West trade and prescriptive policies regarding military technology. It is all too easy to document many recent situations of foreign suspicion toward American business.

The responses of multinational corporations to these various fears and manifestations of hostility are often positive. In many cases, they may make an effort to be as unobtrusive as possible, particularly by becoming partners with a national firm in the host country. Joint ventures are also used to quell fears. In fact, in some countries such as Japan, it has become very difficult for foreign firms to do anything other than to assume a minority interest in a joint venture with a local Japanese firm. The Japanese government is often accused of being "a corporate state" where business and the government work hand-in-glove to exploit foreign firms and investors.

One can also recall the hostility that General de Gaulle offered foreign, and particularly American, firms investing in France. Yet he reluctantly accepted the offer of General Electric to absorb Machines Bull, the French electronics concern, when no other option seemed feasible. In like manner, Canada was finally persuaded to permit American automobile manufacturers to consolidate their position in the Canadian market.

Such cooperative actions, however, do not end the suspicion and hostility of nationalist politicians who are encouraged by the rise of nationalism in many countries around the world, and who know that the large, U.S. based multinational firm is often fair game for **nationalization**, or at least greater control. Thus, the economic logic that is stimulating the growth of multinational corporations has countervailing political pressures that make the future a complex period for multinational managers.

MULTINATIONALISM IN MODERN PERSPECTIVE

In the early stages of its expansion, a business firm tends to export its products abroad. Similarly, the company is likely to import needed materials to manufacture its products. We have pointed out the increasing tendency, even necessity, on the part of firms to go beyond the export-import relations and become involved in international business. The problems are no longer confined to how to conduct foreign trade but have expanded to include problems of foreign investment climates, including economic, cultural, social, and political dimensions.

At the present time, United States foreign direct investments, mainly represented through investments in branches and subsidiaries abroad, amount to around $90 billion. The combined foreign direct investments of all other foreign nations, including Japan and Canada, amount to some $70 billion. It is clear that the United States has had a major influence on the international business scène because of its multinational, even global, enterprises, the largest of which include such companies as General Motors, Exxon, Ford Motor, General Electric, IBM, and Mobil Oil. The largest multinational firms located abroad include Royal-Dutch/Shell (Britain–Netherlands), Unilever (Britain–Netherlands), British Petroleum (Britain), Philips (Netherlands), and Volkswagen (Germany).

The same reasons that originally attracted United States firms to invest their money in foreign countries are now encouraging many foreign companies to review their investment policies regarding United States markets. The devaluation of the dollar, the relatively low price-earnings ratios for companies quoted on the New York Stock Exchange, the rising costs of labor and transportation for goods manufactured abroad, and the fear of protective tariffs and quotas have led to direct foreign investments in the United States. Thus, Sony decided to begin manufacturing television sets in California; Kikkoman Soy Sauce manufacturer located a new plant in Wisconsin; Michelin Tire decided to produce radial steel tires in Carolina; and the German chemical and pharmaceutical concern, Bayer, acquired Cutter Laboratories. These and other foreign companies representing leading firms in their countries and, for that matter, in the world, are investing in the United States at the rate of $1 billion a year. Totally, their foreign direct investments in the United States amount to some $15 billion.

Simultaneously, the multinational firms have come under attack from their own legislators, governments, and interest groups. In contrast to past patterns, the U.S. Congress appears to be oriented decidedly against multinational firms, in large measure due to the pressure exerted by labor groups. Labor contends that the multinational firms are treated too generously by governmental legislation. As a result, they are encouraged to invest abroad, with the effect that employment opportunities for American workers are lost to foreign labor. In turn, the manufactured goods are likely to be imported at inexpensive prices, leading to further deterioration in levels of employment in certain industries.

Congressional legislation has been introduced which would curtail the right of United States companies to defer taxes on earnings until they repatriate these earnings to the United States, to claim deductions for foreign income taxes paid against United States income taxes, to lump together foreign earnings from all nations, thus writing off losses in one country against those in another—all apparently aiming at discouraging multinational firms in their foreign direct investment activities.

Multinational firms contend that employment in the United States has continued to rise, and that they have contributed to a net inflow from earnings on previous investments, thus aiding the United States balance of payments. Furthermore, they assert that they have generated more in new exports than in new imports. Of course, the problem is that the new jobs they may have generated are not necessarily in the same place as were the lost ones. Unions in such industries as steel, rubber, textiles, and shoes have been particularly hostile against multinationals' expansions abroad.

SUMMARY

Although multinational corporations have existed for centuries, in recent years the power of conglomerates has become so widespread and potent that they play a significant role in trade and the movement of capital throughout the world. National corporations move to multinational activities in order to increase sales and profits, and, at first, to provide an outlet for investment. Because of the greater risk, investment opportunities abroad often have a greater return. When firms begin foreign ventures, they may consider them a sideshow. Later, however, as the share of total business increases, the home office makes certain organizational changes—setting up more central controls of international subsidiaries and moving from joint ventures to wholly-owned operations.

It is important to consider all factors in determining the rate of return in a foreign country. Even though the rate may appear high, the risk of political change, expropriation, and war loss is often present. Inflation is another problem, and the costs of managing an international subsidiary are usually higher than in a domestic firm. Managements using a systems approach in deciding on overseas activity will carefully consider their strategy—including competitive advantages, which may include the greater experience of management. The rational multinational firm that plans carefully may shift resources between countries to maximize the return rate.

The development of strategy requires consideration of several areas. Accurate information must be obtained about the country and terms carefully defined. Managers imported from the U.S. must understand the local culture and language. In some cases, local managers may be used, with supervision from the central office where necessary. Marketing is one area that must be handled as close to the local level as possible because of variations in social patterns, and should be administered by executives familiar with local tastes. Most companies do not centralize marketing operations. In product planning, the general rule is, the closer a product is to the consumer and the simpler it is, the more important it is to determine the specifications of the product in the specific market where it is to be sold.

The major problems facing the managers of multinational firms are overcentralization, money management, local politics, and local reactions to foreign business. Most international firms tend to swing back and forth between the two poles of centralization and decentralization in attempting to reach the correct functional balance. There is a tendency to centralize wherever possible, despite the need for decentralization, and errors can be made because of this. Centralization is often related to the problems of overlapping markets, diversified supply, and production. A decline in trade and tariff barriers will tend to increase centralization. In deciding how to organize its production internationally, the firm may choose one of two opposing models of operation: (1) to decentralize engineering and product planning and adapt products to each specific market place, and (2) to seek quality and cost control by standardizing products and processes. This decision often depends on the product involved.

In money management, coordination of finances at the international level is generally considered essential. Financial strategies can maximize the return on liquid resources by switching funds among countries. Many multinational firms start foreign subsidiaries with limited equity and a generous debt because of the varying practices of foreign lenders. However, problems may occur in subsidiaries that are thinly capitalized. International money management requires knowledge of government regulations, interest rates and currency positions—all of which constantly change—but it can be a highly sophisticated and profitable operation in a multinational firm.

The creative challenge for multinational firms in the future may lie in negotiating contracts and licensing arrangements that gradually transfer ownership of physical facilities to natives of the host country, or that create joint ventures, already stipulated by law in a number of countries. The growth of multinational firms has been one of the major business events of the past quarter century. Today, although still expanding rapidly, they face a future requiring that political awareness become a crucial factor in their decision making if they are to avoid outright nationalization or punitive regulations.

DISCUSSION QUESTIONS

23.1. What are some different situations that tend to get a company into international business?

23.2. Why would a company expect to earn a higher return on its investment abroad?

23.3. What would qualify a company to refer to itself as a "multinational" firm?

23.4. In organizational relationships, what consequences follow from centralization or decentralization as the company's prevailing managerial philosophy for foreign operations?
23.5. In what ways would a multinational company appear to be independent of the rules of its host country?
23.6. What are some difficulties multinational firms face in operating in a foreign culture?
23.7. Why do multinational firms meet with resistance in expanding their worldwide operations from their home country? From the actual or potential host country?

APPENDIX The Swedish Multinationals in Trouble

Sweden has a number of multinational firms, such as Volvo (automobiles), SKF (ball bearings), Sandvik (steel products), L.M. Ericsson (telephones), ASEA (electronics), and Electrolux (vacuum cleaners). The Swedish Social Democratic government has proposed a new law that would require these multinational firms to submit any plans for expansion abroad to the government for approval. This would enable the government to decide whether or not the planned expansion would affect employment in Sweden.

The investments of foreign firms in Sweden must also be approved by the government. Thus, when Saint-Gobain, a multinational French firm, wanted to expand its glass-making activities to Sweden, the approval was contingent on the plant's being established in the small city of Emmaboda and on taking over the town's glassworks, which happened to be the town's largest employer.

The trade unions in Sweden indicate that employment has increased seventy percent in the plants of Swedish multinationals outside of Sweden, while it has increased only five percent in plants in Sweden. The Swedish firms disagree with this conclusion, however. A spokesman for L. M. Ericsson indicates that his firm has built eight plants abroad in eight different countries during the last ten years, but during this time Ericsson's exports from Sweden have increased from $20 million to $100 million, thus creating more employment in Sweden.

The Swedish government contends that the multinationals are currently enjoying the kind of freedom that most industrial firms experienced during the beginning of industrialism. To regulate the multinationals and still preserve the independence of small, industrially advanced nations, some experts believe that it may be necessary to develop international regulations that would become an organ of the United Nations.

1. What factors appear to have led to new constraints on multinationals in Sweden? How do these factors apply to multinationals in the United States?
2. If you were a trade union spokesman, how would you defend clamping down on investments abroad by the multinationals?
3. If you were a representative of the management of a multinational firm, how would you defend your freedom to invest abroad?

4. What factors might lead to multinational firms' having to become responsible to international cooperation through an international organization such as the United Nations?

SELECTED READINGS

Fayerweather, John. *International Business Management: A Conceptual Framework*. New York: McGraw-Hill, 1969.

Kolde, Endel. *International Business Enterprise,* 2nd ed. Englewood Cliffs, N.J.: Prentice-Hall, 1973.

Salera, Virgil. *Multinational Business*. Boston: Houghton Mifflin, 1969.

Stopford, John M., and Wells, Louis T., Jr. *Managing the Multidimensional Enterprise*. New York: Basic Books, 1972.

Chapter 24 Managers of the Future

In much of the Western industrialized world managers are men of influence in their communities. They have control over resources and decide on their allocation within the characteristic constraints of that society—government regulations, market forces, legislative framework, and values and attitudes. Managers therefore provide models for society's members to imitate and follow. They set standards of behavior. Their values and attitudes are important for the future development not only of the organizations they head but also of the society and economy in which they live.

Managers of the future will have the responsibility to make long-range plans for investing billions of dollars in a constantly changing environment. In addition, multinational corporations with home bases in the United States will influence most other parts of the world, while marshalling resources and transforming and distributing output to customers. And for better or worse, the political and business systems of the United States have become inextricably intertwined, each influencing and sharing the fate of the other. Despite the broad horizons of the modern corporation, the most critical function to be performed in the organization is still the proper management of people.

MANAGING PEOPLE

Since the industrial revolution, the employee has been viewed as a means of achieving the objectives and goals of the organization. During the Middle Ages, man had identity through membership in a community, whether a feudal manor, a guild, his family, or the church. The communion of the church cemented the relationships of the community. Later the Reformation, and particularly seventeenth-century Calvinistic thought, replaced the values of the community with the values of rationality, which became the hallmark of the industrial revolution and the Puritan work ethic. Now in the last half of the twentieth century, there appears to be emerging a new desire for a sense of community. But this time, a sense of community is not only sought within the context of the work environment, but also outside the organization. A counterculture has appeared in society that seeks to return to a simpler mode of living, closer to nature, with renewed emphasis on the importance of personal relationships and the support and security of a community for its members.

From a foundation of high productivity and the resultant high standard of living and education, the present generation of employees has developed new expectations of the work environment. In the Western industrialized world, these expectations are interpreted as an increased involvement and increased participation in establishing and carrying out job assignments. During the 1950s and 1960s, recognition of these expectations was demonstrated through feeble attempts at job enlargement, at times resulting in more work of the same kind without altering or responding to the underlying issue. Today we see developing progressively stronger emphasis on job enrichment, which by design strives to fulfill human values and needs. Meaningful participation and involvement in the work experience for both managers and workers may, in turn, contribute to the organization's objectives and goals. The modern organization is pivoting from a punishment- and incentive-oriented approach to motivating workers toward positive rewards and a recognition that the quality of work can directly affect the quality of life.

What, specifically, can organizations do to give more attention to employees' goals and objectives and thus to support the individual in his search for a satisfying life?

1. Employees can be considered as "job customers" just as organizations consider the public as "product customers."
2. With the changing values in society, organizations will be called upon to recruit and hire persons potentially trainable and qualifiable instead of selecting only those already trained and qualified. This emphasizes "including in" rather than "screening out" the potential employee. Anyone who presents himself for an available job opening may have to be accepted for probationary employment until proved untrainable.
3. Selection will become a two-way street. Why should it remain a one-way street in which the employer alone decides whether an applicant is suitable for job openings? Why not let the applicant view the various jobs which are open, even try out several to determine which is most suitable for him? If the applicant tries out the job in advance and then takes it, the chances of the reality matching his expectations are much improved.
4. Organizations need to ascertain the desires and goals of employees and maintain up-to-date inventories of this information.
5. Use of flexible hours and schedules will increase. Employees who can work fewer than eight hours a day because of home commitments will be accommodated. Some firms already provide "split schedules" where two workers agree to keep one job filled and arrange between them which one of them will work at any particular time or day.
6. Perhaps the concept of life-time employment will gain greater

acceptance as one point of stability in an individual's life. Long a norm in Japan and frequently in Germany, a core work force at least is assured of permanent jobs. If a recession occurs, the total payroll is reduced as payments are cut back. Surplus employees go into training status to acquire new skills and new knowledge, giving them a versatility of later value to the organization.

7. Employees in the future must have alternative choices related to their lives and careers. For some there may exist a need to intersperse other activities with work, such as service work, or community work.
8. The creative, innovative talents of management must be directed toward making jobs more meaningful for those employees who are satisfied with nothing less than that. One way is to redesign jobs to allow and encourage greater participation by employees in decision making. Instead of having responsibility only for the doing of the job, employees will have responsibility for the planning and inspection or control aspects as well.
9. Employees in the future will demand more control over their own fortunes and destiny. Jerry Wurf, President of the American Federation of State, County, and Municipal Employees has stated: "White collar workers will not turn down higher salaries, but this is not their main reason for unionizing. The main reason is so that they can be a part of the process that determines their destiny." This statement is equally relevant to other groups in our society, e.g., women, blacks, students, and others.

    Most union agreements today call for grievance procedures culminating in arbitration. This is another step in the "humanizing process" since arbitrators tend to be especially sensitive to actions of management which are arbitrary, capricious, or discriminatory.
10. Organizations will increasingly work toward humanizing their operations. Some recent illustrations point to the effects of the difference between a traditional mechanistic approach and a humanizing approach in treating employees.
11. Organizations will reduce, or be forced to reduce, time spent on the job for routine jobs which cannot be redesigned. Perhaps not every job can be made satisfying, interesting, and meaningful. We face some situations where there are more employees seeking meaningful work than there are meaningful jobs. In these cases the impact of work on the individual and his life must be minimized.[1]

The contingency approach to management has served as a reminder that there is no single recipe to apply to management and

[1]Borje O. Saxberg and Robert A. Sutermeister, "Today's Imperative: Humanizing the Organization," *Personnel Administrators* (January–February 1974), pp. 57–58.

organization. The choice between a mechanistic or organic approach or a combination of these depends on the situation—the internal and external conditions governing the organization. But developments in a society with changing priorities and changing personal values relating the manager to those managed are tending to unite the two in a common view that work need not be a drudgery, unpleasant, and onerous. If the company, which until recently has believed that efficient production must rely on the tedium of assembly-line work, discovers that there are no workers to hire at any price to do such work as presently designed, then the company must make changes to enrich the job. Although this approach may seem to increase costs and bring a high price for consumers, it may turn out to be more productive for the organization if the workers accept more responsibility. The potential savings gained from eliminating a separate inspection unit, from improved attendance, less turnover, and improved employee attitudes may more than offset any additional costs incurred by changing the nature of work assignments.

We must introduce an element of caution into this analysis. To this point, all of the efforts at job enrichment have been achieved within the context of experimentally operated units and in the setting of a large corporate organization. These experiments have often reported great successes, but a closer examination reveals that the experiment involved a limited and somewhat sheltered operation within the organization, while the larger part of the firm continued to operate in a traditional manner.

A small enterprise is in a better position to implement many elements of job enrichment because of the advantage of face-to-face contact between managers and employees and the possibility for greater communication among all members of the firm. The variability of human nature is such that the larger the organization, the more difficult it becomes to find a management team and employees who have a common set of values. The future may well require a serious reevaluation of the organization's objectives and goals as they relate to a desirable balance between ownership profits and human happiness. So far, no job enrichment plan appears to have accepted the possibility that management will settle for lower production in return for creating a better work life. If such demand gains support, we can expect further efforts in job enrichment design for both managers and workers.

Not all workers desire the responsibility of participatory management. Some people prefer dependent relationships; others strive for interdependence in working together with their peers, both on and off the job. Still others who possess strong motivation to achieve or a need for power seek independence and possess the physical stamina, have the intelligence, and are willing to accept the responsibility and loneliness to become the decision makers. The challenge for managers of the future is to maintain a contingent, flexible frame of mind, making possible an organizational environment in which those who work may find a measure of fulfillment and thus be self-motivated to contribute their best efforts toward the organization's missions. Indeed, the very survival of organizations of the future will depend on the intelligence and sensitivity of leadership to remain alert to the human, social, economic, and political requirements of the internal and external environments. Today, this involves not only the technical and motivational problems of the firm but also includes consideration of the domestic and international global environments. The impact and importance of global events must be foreseen by management because every organization will be directly or indirectly affected by them. The jet age has reduced the time-lag impact of change in another part of the world and each operating system must be flexible enough to adjust to changes in the worldwide environment if it is to survive.

MANAGING CHANGE

One of the most significant tests of managers of the future will concern their ability to manage the effects of change on the human and social aspects of organization. Writing in 1945, Elton Mayo expressed the opinion that World War II would never have happened if society's social skills had kept pace with its technical skills.[2] Although the gap between social science and physical science has been narrowed somewhat since then, the fact remains that a substantial lag still exists. One measure of this situation is that in the United States today, only one of every thirty dollars spent on research is directed toward social research. The rest is spent on the physical sciences.[3] This bias has helped create a superb technological machine from which we have gained great benefits; but, as we have pointed out elsewhere in this book, it has left us behind in the discovery and application of information designed to cope with the human and social problems of our society.

In looking toward the future, Toffler describes in vivid terms what he calls the "economics of impermanence."[4] This concept is directly related to our swiftly advancing technology. In such an economy, the costs of manufacturing an item are lower than the costs of repairing it, because in manufacturing an item we have efficient machine production whereas in repairing it, we are obliged to use costly hand work. Rapid technological development makes frequent model changes mandatory. We therefore build for the short term rather than the long term. As change reaches more and more people, uncertainty about our future needs increases. The net result is that we have become a "throwaway culture," in which the only certainty is the inexorable, accelerating advance of change.

[2]Elton Mayo, *The Social Problems of an Industrialized Civilization* (Boston: Division of Research, Graduate School of Business Administration, Harvard University, 1945), p. 33.

[3]Paul Hersey and Kenneth H. Blanchard, *Management of Organizational Behavior*, 2nd. ed. (Englewood Cliffs, N.J.: Prentice-Hall, 1972), p. 3.

[4]Alvin Toffler, *Future Shock* (New York: Bantam Books, 1970), pp. 57–58.

Change imposes severe strains on our ability to manage the technical aspects of industry; however, this is of relatively minor importance compared to the effect of change on the human organization and the human being. To cope with an economy of impermanence, the organization must become more responsive, more creative, and more flexible. There is already evidence of a trend toward less rigid organizational structures designed to complement the demands of a dynamic environment. This is particularly true in the aerospace and electronic industries and in large government projects, where change has long been a fact of life. As the change phenomenon begins to erode the bureaucratic structures of these industries, other industries will follow suit.

Organizations based on the "temporary-group" concept place new strains on the human personality. People need increased capacity and encouragement to adjust and adapt to transitory relationships. This method runs counter to a basic human need for certainty and stability and creates new stresses all along the organizational structure. In a bureaucracy, the members enjoy some sense of permanence and constancy in their associations with others. Work rules, procedures, and policies largely define established ways of doing things and of relating to others. In organizations in which the task force or the project team is the dominant form of working groups, group members must learn to work with new people, new supervisors, and new relationships revolving around the task rather than the job or the position they occupy. They must learn to adapt quickly and effectively or performance will suffer and work will slow down.

People do not adjust easily to temporary groups. Individuals are used to finding satisfaction of important needs as members of more-or-less permanent groups such as the family or the department, within which they may enjoy recognition, status, and affection as persons who have earned their way. But it takes time for groups to establish themselves and to achieve the degree of cohesiveness and interdependence needed to promote mutual satisfaction. Thus, managers involved with temporary groups must learn ways of speeding up group processes and accelerating group development, if the organization is to cope with the human needs of its members.

In organizations based on horizontal relationships rather than the traditional vertical ones, people must learn to communicate effectively with their peers, without the structural and traditional supports of status and hierarchy to pull them through the trouble spots. Horizontal organizations require a form of give-and-take that differs sharply from the more traditional superior-subordinate relationships of a bureaucracy. People must listen more effectively since collaboration in peer groups is not based on a communion of "explainers." Instead, it requires exchange in both an informational and a psychological sense. Effective listening, alas, is not easy. It demands empathy, respect for the ideas of others, and an absence of threats. If managers are to promote these conditions, they must set an example. Knowles and Saxberg point out that in many present-day organizations, traditional patterns of competitiveness and impersonality have created what is described as the **derogatory syndrome**, or the tendency to disdain and disparage ideas and sentiments that differ from one's own.[5] The manager of the future must understand the nature and effects of this tendency to disparage and be able to combat it, while at the same time promoting a sound motivational climate for effective interpersonal communication at all levels.

Authority in the fast-moving organization of the future must be located where the knowledge is, if adaptability and flexibility are to be maintained. It will no longer be possible to wait on the protocol of an "executive decision" or a rubber-stamp approval by the board of directors on matters that must be decided now. This redistribution of formal authority will be accompanied by alterations in the formal communication structure, removing some of its time-consuming, formalistic character while ensuring that necessary information is preserved at the appropriate places of control. The traditional authority structure and formal lines of communication will see a shift in the essential character of management, from decision making and directing to coordinating and controlling. Control, however, will be motivational rather than regulatory or restraining.

Conflict in the bureaucracy is easily resolved by "someone in authority" or by written rules. In the transient organization of the future, this method of settling disputes will not be good enough. As we have pointed out earlier in this book, conflict can be healthy as well as destructive. Settling conflict by edict merely files the conflict away for future reference, or worse, causes objectors to resort to underground methods of resistance that obstruct and delay. For conflict to be constructive, it must be faced squarely and resolved. The effective manager must have special skills to make disagreements a positive experience. (Some methods for developing such skills were discussed in Chapter 11.)

Defensiveness in organizations, which tends to be widespread in bureaucratic structures that foster a win-lose climate, is a serious barrier to the flexible and adaptive behavior requisite for the organization of the future. Defensiveness always reflects a return to the past, to earlier, more comforting ways of coping with experience, a clinging to rigidities of mind and personality that we learned at an earlier time of life. Defensiveness can be reduced when its causes and its manifestations are understood and taken into account in the actions of managers toward their subordinates and their co-workers. A defensive climate in a problem-solving group leads inevitably to slowing down progress, to hang-ups in information exchange, a reduction in participation and communication, and a stunting of creativity.

[5]Henry P. Knowles and Borje O. Saxberg, *Personality and Leadership Behavior* (Reading, Mass.: Addison-Wesley, 1971), p. 145.

We have described some of the stresses and strains that affect people in organizations undergoing rapid change. We believe that the age of bureaucracy and the "organization man" have come face to face with the future, and that managers should be exploring alternatives. As the future quickly becomes the present, individuals at all organizational levels will need to be more flexible, less rigid; more creative, less complacent; more adaptable, less conforming; more open, less defensive; more the leader, less the follower.

Beyond managing change is actually effecting change. Although specialists are available for this purpose, managers cannot avoid the role of change agent. Technical and economic changes in organizational life depend on the possession of technical skills and/or material resources. To effect social and psychological changes, however, one must draw heavily on information from the behavioral sciences and intervention skills. This all seems to point to the fact that the manager of the future must be better trained in the behavioral sciences than he is today. A corollary is the need to encourage and provide support for social research proportionate to that provided for technical and physical research.

PREDICTING CHANGE

In 1972, a White House conference was held on "The Industrial World Ahead." That meeting of 1,500 top business, labor, and academic leaders was organized to determine the economic troubles that were likely to emerge between 1972 and 1990.[6] This group of "experts" was confident that the period under study would forecast years of spreading abundance. For example, they concluded that: (1) more than one-half of all families would be earning the equivalent of $15,000 in 1972 dollars by 1990; (2) unemployment would be about four percent; (3) more women would be working; (4) a smaller portion of income would be spent for food and other necessities; and (5) more young people would be attending college. We should note, however, that this group forecast continuing problems of unemployment, poverty, inflation, crime, and environmental pollution, although confident that a healthy economy could direct sufficient resources toward research and development to solve these problems.

Prediction of Output

The group predicted an output of $2.4 trillion by 1990, using 1971 dollars as a base. This is more than double the 1971 output of goods and services. The conference speculated that a greater portion of the output would be used to provide additional services for people and require less for the manufacture of goods. The growth predicted for the United States is illustrated in Figure 24.1, which shows the dramatic increase forecast for total spending, with the largest increase predicted for governmental spending.

[6]"The America of 1990," *U.S. News & World Report* (February 14, 1972), pp. 42–43.

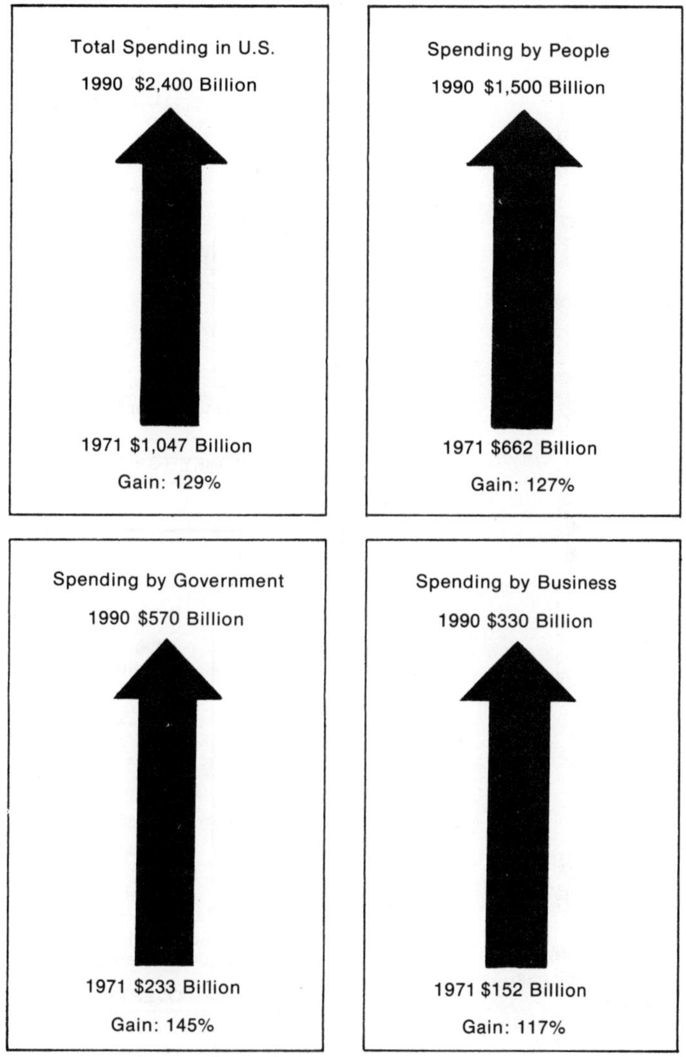

Figure 24.1 Growth Ahead for the U.S.—As Top Economists See It

Note: 1990 estimates are based on 1971 prices to eliminate effects of inflation.
Source: "The America of 1990," (February 14, 1972), p. 42. Reprinted from *U.S. News & World Report.* Copyright 1972, U.S. News & World Report, Inc.

Benefits for People

The conference participants pointed out that children from the Baby Boom of the 1940s and 1950s would be entering the labor market, which would increase the work force about thirty percent, to approximately 110 million by 1990. This, they suggested, would cause the problem of unemployment to continue and to remain a problem for years to come.

Figure 24.2 shows that the average person in 1990 will have more money to spend, almost twice as much as in 1971. The number of families earning more than $15,000 a year would more than double during this period. Although there will be fewer

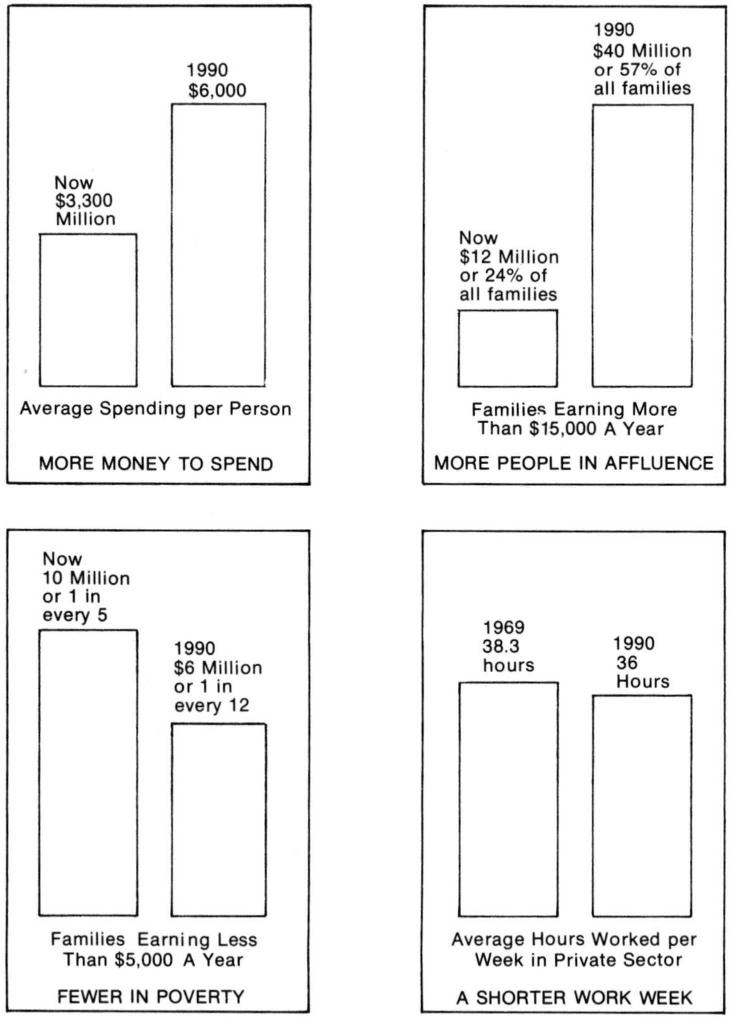

Figure 24.2 Better Times Coming for Americans—Some Measures

Note: 1990 dollar figures based on 1971 prices to eliminate effects of inflation.
Source: "The America of 1990," (February 14, 1972), p. 43. Reprinted from *U.S. News & World Report.* Copyright 1972, U.S. News & World Report, Inc.

families earning less than $5,000 a year by 1990, one in every twelve families, or approximately six million families would continue to earn below this amount. The work week would decline, but perhaps not as rapidly as some people expect.

Other predictions included a reduction in the percentage of income that would have to be spent for food—a decline from eighteen to fourteen percent; and that the percentage of eighteen- to twenty-four-year-olds enrolled in colleges or universities would increase from thirty percent in 1970 to about forty percent by 1990.

The conference also suggested that the United States will probably become more dependent on imports, although there will be a continual investment in new plants, equipment, and construction by business in this country. Electricity was expected to provide one-third of the nation's energy by 1990, compared with about one-fourth in 1972. Land will become increasingly scarce as the population increases.

Evaluating the Forecast

Someone has said that it is relatively safe to make long-range forecasts or predictions because by the time the forecast date arrives, people will either have forgotten about the forecast or will no longer be available to make an evaluation. From a present perspective, changes have occurred so rapidly that some of the predictions made in 1971 should already be modified. It was interesting that the 1972 forecast made no mention of an energy shortage and predicted an increase in the number of students attending colleges or universities. These predictions were upset by activities in the Middle East oil-producing countries, which proved to have a greater impact on our economy than expected. In addition, the changing attitudes of young Americans seem to have reduced the growth rate of college and university enrollment, at least temporarily.

Such unforeseen changes suggest that our country is very vulnerable to worldwide economic and political actions. In addition, people's decision criteria, their priorities, may cause significant alterations in our way of life. Assessing attitudes forms an appropriate basis for forecasting future change. As an illustration, let us look at the attitudes of college students on some private and public issues.

Attitudes of College Freshmen

Table 24.1 is a survey of the attitudes and goals held by entering college freshmen in 1973–74. These students believe that the United States isn't controlling pollution or protecting the consumer. They also believe that the wealthy should pay a larger share of taxes, and that women should have job equality. Unfortunately, about 41.1 percent believe that an individual can do little to change our society.

A little over 50 percent of the students regard themselves as middle-of-the-road, with a liberal group (32.6 percent) being in the next largest category. Sixty-nine percent of the students believe that developing a philosophy of life is their prime objective, with helping others who are in difficulty (64.3 percent), and becoming an authority in one's field (62.5 percent) following in that order.

Although there are differences in answers among men and women and among students from different kinds of schools, it is interesting that these differences are not often statistically significant. Such attitudes as illustrated here, assuming they represent those of a major interest group, can influence the direction our society will take in the future.

Table 24.1 Their Attitudes and Goals*

	All Institutions			2-Year Colleges		4-Year Colleges				Universities		Predominantly Black Colleges	
	Men	Women	Total	Pub.	Priv.	Pub.	Priv. Non-sect.	Prot.	Cath.	Pub.	Priv.	Pub.	Priv.
Student agrees strongly or somewhat that:													
U.S. isn't controlling pollution	86.7	89.7	88.1	87.3	86.7	88.3	89.5	87.9	88.2	88.9	89.4	87.7	87.6
U.S. isn't protecting consumer	75.6	80.8	78.0	79.0	76.5	77.5	79.4	75.7	79.9	76.8	78.8	81.8	82.7
U.S. isn't desegregating quickly	46.3	51.2	48.6	50.6	47.8	45.0	53.6	48.7	50.6	45.6	51.9	67.7	71.6
Courts are too concerned for rights of criminals	55.7	43.7	50.1	53.3	53.2	49.8	44.3	48.7	48.2	48.5	42.7	39.6	36.4
People should be paid equally	27.7	25.8	26.8	31.8	29.9	25.1	23.5	25.2	25.8	21.9	21.3	47.5	44.1
Women's activities best confined to home	40.9	18.8	30.4	35.5	38.6	28.4	25.0	33.8	29.2	25.0	22.0	40.7	36.7
Wealthy should pay larger share of taxes	75.8	69.3	72.7	73.0	68.1	74.6	71.6	71.3	73.7	72.5	70.2	74.5	73.6
Marijuana should be legalized	50.9	45.2	48.2	48.6	41.1	46.1	54.7	38.5	45.7	50.0	55.7	42.9	45.8
Large families should be discouraged	67.1	63.5	65.4	61.0	64.4	66.3	70.0	68.0	48.8	72.3	68.7	45.2	38.9
Women should have job equality	88.0	96.2	91.9	89.6	90.2	93.0	94.1	91.8	92.5	93.7	94.6	88.7	90.6
Individual can do little to change society	44.6	37.3	41.1	43.8	40.4	39.9	41.1	36.7	37.5	39.9	38.8	44.5	40.7
College has right to regulate off-campus behavior	12.5	10.1	11.4	12.7	16.5	10.7	9.5	16.7	10.2	8.6	9.1	25.6	24.6
Chief benefit of college is monetary	61.6	49.4	55.8	67.0	61.1	54.5	44.0	46.8	45.2	49.0	37.0	75.7	70.4
Students should help evaluate faculty	74.4	75.3	74.8	71.9	71.6	76.0	77.6	73.8	74.6	78.4	76.7	69.2	70.2
College grades should be abolished	35.1	34.6	34.8	36.5	35.2	33.6	36.4	34.2	36.2	33.3	30.6	35.6	35.5
Organized college sports should be deemphasized	26.1	23.4	24.8	25.4	23.1	21.3	28.0	27.0	24.9	24.5	29.4	29.6	27.0
College should regulate student publications	31.2	30.4	30.8	36.6	44.0	29.9	23.0	37.6	28.2	23.0	19.5	52.8	46.3
College has right to ban speakers	25.7	20.3	23.1	26.0	30.3	23.6	18.3	27.0	21.5	18.0	18.7	27.9	24.1
Disadvantaged should have preferential treatment	39.5	38.1	38.8	42.4	49.5	37.5	36.1	43.2	34.5	34.6	29.6	64.3	64.9
Colleges have been too lax on student protests	40.3	32.9	36.8	40.6	44.8	37.2	31.2	40.5	34.7	31.3	28.1	41.9	38.3
Public colleges should have open admissions	36.7	33.9	35.3	50.4	37.1	27.4	28.8	30.4	34.1	22.5	22.8	52.8	46.3
Standards for degrees should be same for all	79.0	76.9	78.0	79.2	74.6	77.8	77.2	76.6	76.4	77.4	79.4	74.2	73.4

	All Institutions			2-Year Colleges		4-Year Colleges				Universities		Predominantly Black Colleges	
	Men	Women	Total	Pub.	Priv.	Pub.	Priv. Non-sect.	Prot.	Cath.	Pub.	Priv.	Pub.	Priv.
Student's political views:													
Far left	2.7	1.6	2.2	2.5	1.8	1.6	3.3	2.2	1.6	1.8	2.8	5.3	4.5
Liberal	33.2	32.0	32.6	30.2	27.4	31.5	40.2	31.2	33.8	34.1	42.3	36.5	44.9
Middle-of-the road	48.2	53.5	50.7	53.7	54.1	51.3	43.4	47.4	51.3	50.0	40.9	40.3	37.6
Conservative	15.0	12.6	13.9	12.9	16.1	15.0	12.5	18.5	12.9	13.7	13.4	14.7	11.2
Far right	0.3	0.4	0.6	0.8	0.7	0.6	0.5	0.7	0.3	0.4	0.6	3.2	1.8
Objectives student considers essential or very important:													
Achieving in a performing art	16.7	23.9	20.1	17.8	20.5	20.0	25.1	22.8	20.4	20.4	25.4	24.4	23.2
Becoming an authority in his field	67.0	57.6	62.5	60.4	61.9	64.4	64.6	61.9	61.2	63.4	66.1	75.6	77.5
Influencing the political structure	17.9	11.1	14.6	12.8	13.0	14.6	17.3	15.6	17.5	15.0	21.0	27.1	30.0
Influencing social values	29.2	33.1	31.1	28.9	35.4	32.0	32.0	37.1	36.9	29.5	34.2	46.0	46.6
Raising a family	54.8	58.2	56.4	58.7	61.4	56.9	54.4	56.3	60.3	51.7	53.9	55.2	58.9
Having administrative responsibility	32.0	21.9	27.2	30.4	28.5	26.8	25.3	23.3	26.2	24.1	24.1	43.8	40.2
Being very well-off financially	63.8	45.5	55.1	60.7	51.8	53.8	52.9	43.6	49.1	52.5	51.9	71.7	73.0
Helping others who are in difficulty	56.0	73.4	64.3	61.5	69.3	66.5	64.9	71.4	73.7	61.5	66.7	74.7	78.5
Being successful in his own business	51.6	31.3	41.9	45.7	45.7	37.4	42.1	38.1	39.5	40.7	40.9	59.0	60.8
Helping to clean up environment	35.8	32.6	34.3	35.4	35.8	33.1	36.4	32.9	33.2	33.6	31.5	42.8	41.6
Developing a philosophy of life	64.7	73.7	69.0	63.1	68.8	70.2	74.7	76.1	75.7	71.3	76.3	70.1	75.9
Becoming a community leader	27.8	34.0	30.7	28.2	32.7	32.2	32.7	34.4	36.4	30.4	32.3	45.8	49.8
Keeping up with political affairs	43.7	40.8	42.4	36.4	36.5	44.1	47.7	42.8	48.9	46.1	55.1	48.1	54.7

*All figures represented as percent of those responding.

Source: *The Chronicle of Higher Education* (February 11, 1974), p. 8 (as adapted from the American Council on Education figures).

SUMMARY

Managing people rather than things will be the most important challenge managers face as we move toward the twenty-first century. Not only will human intelligence and skill have to be allocated wisely and precisely in relation to the technical demands of industry, but people must also be trained and motivated to deal effectively with each other and with changing organizational requirements if the fullest measure of the human potential is to be realized.

During this century, man has emerged from a mechanistic organizational society in which he was the servant of bureaucracy into a transitional phase based on organic, adaptive ideas of organization and work. It is now possible to conceive of the organization as an instrument of man, rather than the other way around. As this trend continues, managers will need the knowledge and skill to understand the goals and desires of employees, giving them more meaningful jobs, a wider choice of career patterns, and more freedom to control their own activities and futures.

As the future unravels, we are sure to see an acceleration of the effects of change on the human organization. New information technology, the advances of knowledge that affects man's work, the rapidity with which new machines and methods displace the old, the greater mobility of the work force, and the swiftly changing values of society will have significant and lasting effects on social customs and relationships in organizational life. Since human beings must learn to accommodate to change rather than resist it, future managers must prepare themselves not only to handle technological change but also to cope with organizational problems relating to social change and its effects. Thus, managers of the future will more and more assume the role of change agents and will need to be aware of developments in the behavioral sciences that will help them understand and deal with the human needs of their employees.

Other aspects of change are projected economic growth, the expansion of spending throughout the economy, the spread of affluence into new sectors of the population, and the prospect of more leisure time during the next twenty years. These changes will affect the nature of markets, products, and social customs and will therefore be primary elements in managerial decisions, both now and in the future.

At no time in our history has the future provided a greater challenge for managers. As we move through the next few crucial decades, their actions will affect the eventual fate of industry, of new generations, and perhaps of the human race.

DISCUSSION QUESTIONS

24.1. What is the difference between "job enlargement" and "job enrichment"?

24.2. Why do you think this country has spent more money on technological research than on research in the social sciences?

24.3. Explain the reasons for Toffler's concept of the "economics of impermanence."

24.4. In what ways does the project, or systems, type of management affect individuals?
24.5. Will authority and communications change under systems management? If so, how?
24.6. Why is the future so difficult to predict?
24.7. What changes do you foresee for this country by 1999?

APPENDIX The Cowdominium Concept

We're living in an age of miracles. In the past decade, more of the things "they" said couldn't be done have been done than most of us would ever have believed possible. And miracles keep right on happening. Only this one—the Cowdominium Concept and all of its revolutionary ramifications—is just beginning to happen. Just what is the Cowdominium Concept? Well, it's a large, far-reaching, and highly beneficial concept that has been proposed to revolutionize the dairy and beef cattle industry. That's an industry that hasn't had anything of a revolutionary nature happen to it for a long time—actually, not since the days of Louis Pasteur. It's an industry that sorely needs some revolutionizing.

The dairy Cowdominium (see Figure 24.3) is a 120-acre, wheel-shaped structure with about twenty "spokes" radiating from a central processing plant. This processing plant converts animal waste and other cellulose materials (sawdust bedding or straw) from the 6,000 resident cows into a crude high-protein animal foodstuff. Converting waste materials not only solves pollution problems but also creates protein which is so badly needed worldwide.

Each of the spokes of the dairy Cowdominium is a 5-acre individual operation that includes an open area, a feed storage barn, covered corrals, and feed stalls for 300 cows, a wash barn, a milking parlor, and a milk house. Both the wet and the dry wastes are channeled to the processing plants through the installation's slotted floors. The animal and vegetable waste solids are then subjected to heat, pressure, and chemicals. The solids eventually emerge as a fine-textured, sterile substance having a molasseslike aroma. This substance is called "Cowmel" (Cowmel Process: patent and trademark pending).

Individual dairymen who lease the land on which their individual, uniformly constructed barns are built may market their milk however they choose. They have access, however, to a variety of very specialized services. One of these services is a computerized veterinary center that has complete control of the environmental complex.

The Cowdominium Concept is a breakthrough in caring for cattle in a controlled environment, in addition to giving the individual dairyman the ability to compete on an equal footing with large, efficiently operated farms while still being able to maintain his independence. He can be his own man and still manage to run a

Registered trademark, U.S. Patent Office, patents pending. This case is published by permission of FarmEcology Corporation, Enumclaw, Washington.

Figure 24.3 Dairy Cowdominium

better dairy farm and make a better living than he ever could do on his own.

Each Cowdominium has computerized control in all phases of the operation. The control concentrates on clinical sanitation, care and feeding of the animals, climatic and environmental control, and the complete control of bacteria, parasites, insects, and so forth. In fact, the plant has computerized operations throughout, resulting in the complete elimination of one of the dairyman's greatest problems—namely, the disposition of animal waste.

Despite the fact that a cow has multiple stomachs, its actual food-processing efficiency is only 40 percent. The remaining 60 percent of the cow's intake becomes waste, which ultimately forms a large pile in the farmyard. Composting is only one of the means of disposing of the waste. Spraying it over the acreage is another, but when the same section of land is sprayed continually, tiny fibers in the animal waste form an impenetrable mat that seals off the soil. This kills the vegetation and prevents new grass from growing. It also allows the waste to pollute rivers and streams. The diagram below illustrates an integrative systems approach to the problem of recycling waste.

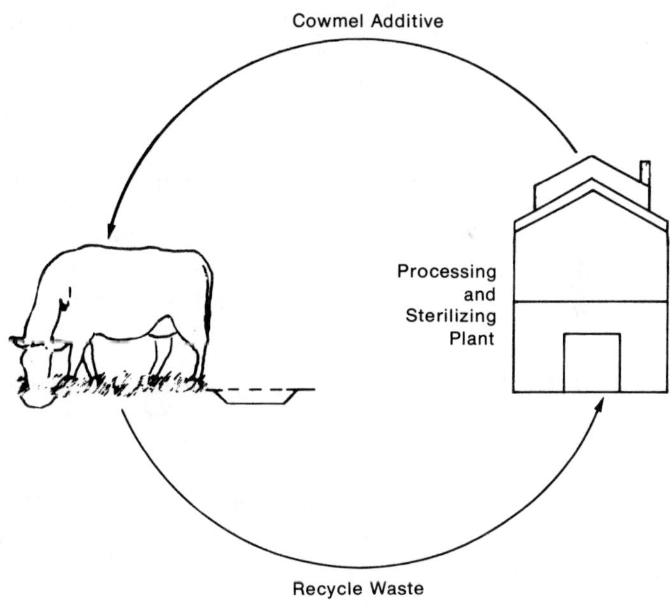

Cowmel, a highly nutritious crude-protein feed, is equivalent in food value to alfalfa, soybeans, and cottonseed meal. Cowmel is a product of the completely reprocessed waste solids, resulting from the separation of the waste solids and liquids; the waste liquids are purified by the total process. The result of the processing is material that has crude-protein value in the controlled range of 20 to 60 percent, and total digestible nutrients in the controlled range of 35 to 80 percent.

Economic Considerations

The total capital cost of a 6,000-head dairy Cowdominium including land, plant, and all improvements is approximately $5 million. It has been estimated that a dairyman would save $42 per head per year by using the recycled feed material. Thus a 250-head milking operation could save $10,500 per year on feed alone. A dairyman using the Cowdominium Concept operation does not need to own animal-waste handling equipment, which also represents a significant saving.

FarmEcology Corporation owns the land and installs off-site improvements such as the water lines, waste conversion lines, power lines, streets, fences, veterinary clinic and laboratory, administration building, and security building. Each of the twenty Cowdominiums are to be built on five unimproved acres and then leased to a dairyman for ninety-nine-years. This lease is subordinated to a mortgage to facilitate financing the building on each of the five-acre plots. The value of the unimproved five acres of land is approximately $20,000. The diary building is constructed according to FarmEcology Corporation's specifications, with a certain amount of latitude available to the dairymen for interiors and design, including the milking-parlor arrangement. The estimated cost of such a building is approximately $180,000, or a total value of $200,000 for land and buildings per each five acres. At eighty-five-percent financing of such evaluations, a dairyman can own his own building on this leased land with approximately $10,000 down, and the balance on a term mortgage.

1. If you were an independent dairyman, what additional information would you need to determine your interest in leasing acreage in a Cowdominium?
2. Do you believe the government should require all dairymen to join a project of this type?
3. Assuming this concept is successful, will it have any effect on the value of farm land? The small dairyman? The consumer?

SELECTED READINGS

Bell, Daniel, and Kristol, Irving, eds. *Capitalism Today*. New York: New American Library, Mentor Books, 1971.

Bennis, Warren. *Beyond Bureaucracy*. New York: McGraw-Hill, 1973.

Jenkins, David. *Job Power*. New York: Doubleday, 1973.

Kahn, Herman, and Bruce-Briggs, B. *Things to Come*. New York: Macmillan, 1972.

Likert, Rensis. *The Human Organization*. New York: McGraw-Hill, 1967.

Meadows, Donella, Meadows, Dennis, Randers, Jorgen, and Behrens, William W., III. *The Limits to Growth*. New York: New American Library, 1972.

Schon, Donald A. *Technology and Change*. New York: Dell, 1967.

Toffler, Alvin. *Future Shock*. New York: Bantam Books, 1971.

Glossary

Action research. When the investigator observes or changes or intervenes in an ongoing organizational or group situation.
Activator. The element in a control system that provides or releases new inputs into the system as required to return the system to expected output.
Affirmative action legislation. Legislation requiring that business and other institutions take positive steps to remedy minority imbalances in their employment.
Alternatives. Different courses of action that may be taken. Each possibility should achieve the objectives as outlined.
Analog computer. A machine that creates a physical analogy, or model, by way of electrical voltages, as a way to analyze and solve the problem being studied.
Authority. The right to command, decide, and act by virtue of the position a person holds in an organization.
Autocratic. One-man rule with minimal participation by others in the organization. This is the concept of a king whose decisions are above question.
Behavioral scientists. Investigators and researchers drawing on, and contributing to findings in psychology, social psychology, sociology, cultural anthropology, political science, and psychiatry.
Behavioral subsystems. Systems that emphasize the activities of humans in organizations, for example, their output, how they are motivated, their group interactions.
Brainstorming. The process of generating innovative ideas through group dynamics; of building on each others' ideas to solve problems.
Bureaucracy. A formally planned organizational structure whose rules and procedures are in writing. There is little, if any, modification of these plans by operating managers.
Capitalism. Assumes democracy in addition to an economy with private property, free markets, and a profit motive by business. Capitalism implies free entry and independent action.
Centralization. A managerial philosophy denoting a minimum delegation of authority and responsibility. The subunits are managed by established and often detailed guidelines.
Charisma. Meaning, "by grace of God." Personal magnetism in leadership. The ability to command the respect and cooperation of the people.
Checklist. A method of getting idea-clues by proceeding through a prepared list of ideas and relating each to the specific problem at hand.
Choice. The selection from among two or more alternatives. The decision among two or more possible courses of action.
Closed-loop control. When changes in input are exercised

automatically as the system recognizes a change in operation that would cause the system to vary from a predetermined plan.

Committee organization. An arrangement in which decisions are made by participating groups of people.

Communication. The sending and receiving of thoughts, letters, directives, data, and so forth, by any means. The transfer of meanings between persons and groups.

Communism. A planned economy under centralized authority and having no privately owned means of production. The political system is controlled by a small fraction of the population.

Comparator. That element of a control system that performs the function of comparing the system's performance with the established standard.

Component. The smallest unit or segment of a system that performs a part of the transformation or processing function. Determining what is the smallest unit may be an arbitrary decision.

Compromise. Obtaining agreement through mutual concessions. Each party modifies his or her point of view to achieve a workable solution.

Concentration. The ability to direct complete attention to a particular mental task or physical activity and thereby improve the chance for objective achievement.

Conflict. A feeling or expression of hostility between persons or groups. This may vary from "bad feelings" to open hostility.

Consensus. A group problem-solving process that results in solutions to problems with which everyone in the group can at least partially agree.

Constraints. Those factors that limit the number of feasible or workable alternatives. The reasons why certain alternatives could not be applied.

Consumerism. A sociopolitical movement demanding greater protection of consumers in the quality and safety of products purchased. A grouping of power by those who buy.

Contingency. A condition or uncertainty in a situation that creates a potential for a changing response.

Contingency approach. The theory that the best way to lead, organize, etc., is not fixed but depends on the situation.

Control. The function of providing the direction of a system in conformance to the operating plan and the stated objectives; also the ability to make adjustments when necessary.

Coordination. Assuring through designated positions and subdivisions a planned contribution by individuals in the organization to its objectives. Getting individuals or groups to work toward the same objective.

Core values. The individual internalizes attitudes and beliefs in early childhood which become highly important for the subsequent development of personality, and change only with difficulty; similarly an organization or a society may possess certain attitudes and beliefs as a reflection of members' shared values which have a long tradition behind them.

Creativity. The act of producing new ideas or new knowledge. Creativity includes the process of testing combinations of information or knowledge to achieve unique results.

Culture. The way of life of a society, group, or organization based on shared values. It includes history, social mores, and so on.

Decentralization. In current usage, a managerial philosophy implying maximum delegation of authority and responsibility. Each subunit is given great freedom to act independently.

Decision making. The choice from among alternative courses of action. The process of deciding what to do.

Degradable pollutants. Environmental impurities that tend to disintegrate by natural processes over time. Those impurities that recycle themselves in the environment.

Democracy. A government freely elected by the majority, with each person having an equal vote. This should result in equal representation for all.

Derogatory syndrome. A societal and organizational emphasis on destructive competitiveness and impersonality.

Design. The bringing together of materials, energy, and information to create an organized system to accomplish a particular objective.

Differentiation. An organizational approach implying that departments and their professional staff employees are governed by unique objectives that differ from those of other departments.

Digital computer. A tool for computation based on the manipulation of symbols and counting, using the binary number system, which is based on only two digits—0 and 1.

Divisionalization. A decision by a company to establish autonomous organizational units with responsibility for performance resting with a divisional manager left to run the division as if it were a separate organization.

Domination. Obtaining results through direct command. Using power or influence to impose one's will on others.

Drive theory. A psychological principle that attributes behavior to primary (biological) and secondary (learned) motives.

Ecology. The relationship between living organisms and their environment. It encompasses the lives of people, animals, and plants in the world.

Effectiveness. The ability to accomplish the objectives of the system. A direct relationship exists between output and objectives.

Efficiency. The measure of output relative to resource input. The degree of efficiency indicates how well we have managed our resources.

Elite. The "choice" or most carefully selected part of a group or society. Usually a group having special privileges.

Entrepreneur. A person who organizes and manages a business and assumes a risk for the possibility of making a profit.

Equality. Having the same rights, privileges, and rank. No discrimination because of race, religion, national origin, and so on.

Esprit de corps. Positive spirit of community. The feeling of unity and working together toward a common objective.

Ethnocentrism. A belief in the inherent superiority of one's own group.

Euthrophication. The mineral enrichment of water caused by phosphates. This, in turn, accelerates certain plant growth.

Evaluation. A study of a system relative to its stated objectives. An analysis to see whether the system is properly designed.

Fascism. A centralized authoritarian political government with a private property economy and in which business is usually controlled by the government. Frequently implies a military regime.

Feasibility. That which is possible to accomplish within given resources and constraints. It may be one of several alternatives.

Feedback loops. Self-correcting mechanisms that control systems by furnishing information about the effects of system activity.

Flexible system. A system whose structure (design) adjusts to changing environmental requirements. Sometimes called an "open" system.

Flexibility. The ability to adjust to meet changing conditions. This is particularly relevant to the design of a system.

Force-field concept. The theory that behavior or level of activity is the result of multiple forces in the relevant energy field of the environment.

Formal communication. A communication that follows organizationally prescribed lines of authority.

Formal organization. The planned organizational structure as a means for achieving organizational objectives and goals.

Free-form organization. No recognized formal organizational structure. An organization that modifies itself as conditions change.

Frustration. Not being able to overcome some obstacle in trying to reach a goal. Also the mental state induced by such blockages.

Functionalism. Pertains to the ability to do a job; in organizational terms functionalism refers to division of work and thus organizational structuring based on functions such as production, purchasing, sales, and others.

Functional organization. Based on specialized activity. Differentiation or distinction among the duties in an organization.

Fusion. Joining together. The combination of two or more materials, individuals, or groups to form a new identity.

Goal. An object one strives to achieve. Often considered to be somewhat more specific than objective.

Goal oriented. Actions aimed directly at reaching a specific end. The goal-oriented system is designed to achieve specific goals.

Grapevine. Underground, informal channels of communication. These are channels not designed by the organization but that occur out of the interactions of the people in the organization.

Group dynamics. Knowledge about the nature of group behavior; also, the interdependent forces and processes in groups that influence behavior.

Group problem solving. A group process aimed at clearing up a problem or making a choice. There may not be an identified leader.

Hardware. The equipment (such as a computer) that is used in an information system. It includes all of the computer's mechanical and electronic devices.

Hedonism. Also known as **psychological hedonism**. A theory of motivation that explains human behavior as a desire to seek pleasure and avoid pain—to do what pleases one.

Hierarchy. Positions ranked in importance within an organization by degree of authority. This forms the structure of the organization.

Holistic thinking. The whole is more than the sum of its parts and therefore needs separate attention.

Horizontal strain. Competition among the various functional areas of an organization. An effort to optimize subunits rather than the total system.

Human relations movement. An emphasis on individuals and groups in the organization and their objectives and goals.

Individualism. Leading one's life in one's own way. In economics, the doctrine of laissez-faire.

Informal organization. The formation of groups or cliques superimposed on the formal organization based on authority relationships. The informal organization exists for its members and their protection and survival.

Information. Data that have been processed or analyzed for a specific purpose.

Input. That which is introduced into the system in the form of information, energy, and materials. This represents the resources to be used.

Integrated data processing (IDP). This approach implies data and information in a continuous flow from input to use without human intervention in between.

Integration. Providing a unity of parts. Obtaining results through agreement on the common interest and the factors that govern this interest or mission.

Intelligence. The ability to identify the important information, to assess, and to understand. Common sense is also associated with intelligence.

Interest groups. Those groups organized around common interests. Such groups are usually aimed at affecting political action or gaining economic sanctions.

Interface. Relationships between organizations or systems suggesting cooperation or collaboration but no loss of separate identity, objectives, or goals.

Interpersonal communications. Includes face-to-face, verbal and nonverbal sending and receiving of messages with emphasis on the transfer of meanings.

Interventions. In this sense, any activity designed to help an organization change some aspect of its attitudes or behavior, usually with reference to the activities of a change agent in an organizational development context.

Iteration. The process of repeating a logical sequence of steps in order to improve or refine.

Johari window. A simple model of human communication. It shows the communications relationships that exist under various circumstances.

Laissez-faire style of leadership. "Let alone" leadership style involves abdication of leadership and letting the members of an organization get on with assignments as they see fit without providing direction or guidance.

Leadership. Gaining commitment from followers by their acceptance of an individual and his or her definition of the objectives and goals to be achieved.

Liberalism. Today, refers to a position of reform or tolerance.

Line organization. The formal organization's relationships between superior-subordinate. This pertains to individuals in the operating units of the organization.

Lobbying. Actions by individuals representing interest groups or organizations to influence government agencies or legislation.

Maintenance function. A leadership action that has to do with supporting and sustaining effective human relationships in the group.

Management. The art and science of making appropriate use of resources to accomplish the objectives of the organization.

Matrix organization. Members in the organization have dual allegiance—to a project and to a department.

Marxist view. View held by Karl Marx, father of Communism, or his followers. This includes an economic and a political philosophy.

Mechanistic view. Based on the concept of a machine, to think toward optimizing mechanical devices with little regard for people.

Mission. The task to be accomplished, or the objective to be achieved. This implies a very specific objective.

Model. A representation of the real world in some other form that will make it easier to analyze the situation.

Model building. The representation of reality in a form that can be understood and analyzed. The development of analogous forms.

Motivation. Energy generation from inner tension that incites or prompts a person to behave in a way that will reduce tension. Motives are internal, as distinguished from external forces inciting a person to act.

Multinational corporation. A corporation that engages in worldwide business activities—producing, marketing, and financing its operations in many different political units throughout the world.

Nationalism. Putting an individual nation's interests above the interests of other nations. Devotion to national interests, unity, and independence.

Nondegradable pollutants. Impurities that will not disintegrate by natural processes. Examples would be certain detergents, discarded glass bottles, some industrial wastes, and so on.

Objective. The purpose or mission that is to be accomplished. The target toward which action is directed.

Object program. The instructions that the computer follows. These instructions are developed by systems analysts and are then programmed into a machine language.

Open-loop control. When the control function is exercised by an individual(s) outside the operating system. For example, when you continually adjust the setting on the house thermostat you are exercising open-loop control over what is otherwise a closed system.

Operational control. The control of inputs and activities in the day-to-day operation of an organization. This is one of the functions of a manager.

Optimize. To maximize the utility of a particular function. This is usually accomplished as a tradeoff or underutilization of another function.

Organic organization. An organization characterized by flexibility and wide distribution of power and authority, in contrast to the rigid structure and hierarchical control of mechanistic organizations.

Organization. The relationships and/or structure that provide a means for coordinated effort. The arrangement of the interdependent functions required to perform a complex task into a unified system of interrelated activities.

Organization chart. A graphic symbolization of the formal information and decision-making network of the organization.

Organizational climate. The trend of fundamental concepts and attitudes toward people that characterizes an organization.

Organizational control. Procedures, rules, organizational structures that incorporate all of the plans on how the organization will operate and also includes a follow-up to see whether or not the system has been a success.

Organizational development. The application of behavioral science to learning, changing, and the growth processes in organizations. A professional change agent is used in this process.

Oscillation. The fluctuation of a system's output beyond the limits of allowable variation; abnormal swings above and below designed output levels.

Output. That which is produced or transformed by the system. Automobiles are part of the output of General Motors, refrigerators of General Electric, etc.

Particulates. In air pollution, refers to solid matter or particles in the air. Auto exhaust smoke, for example, consists in part of carbons and other solid matter.

Performance. Achievement of planned goals and objectives, frequently measured through an emphasis on efficiency, profitability, or return on investment.

Planning. Deciding in advance what is to be done and how it should be done. Planning may result in a formalized program, schedule, or scheme of action or it may simply be an informal mental process of looking ahead and deciding what is to be done.

Pollution. To make unclean, impure, or contaminated. For example, a river may reach an unacceptable level of pollution through the discharge of human waste into its waters.

Power. The ability to control others through personal ability, through the use of rewards and punishments, through position, or through the use of superior knowledge.

Pressure groups. Economic and political interests that organize to influence government. The "labor lobby," the "milk lobby," and consumer organizations are examples of pressure groups.

Productivity. The relation of the output to the input of resources (efficiency). Productivity may also be measured in terms of an absolute or relative standard, for example, a monthly quota, or one person's production compared to another person's.

Profit. Difference between gross revenue (sales) and costs measured over a given period of time. It is also expressed as the ratio of this difference to capital invested. For example, "I made a 10 percent profit on the stock I purchased."

Progress. Idea of improvement or betterment. Became a popular idea and goal of society in the eighteenth century with the introduction of the industrial revolution.

Psychoanalysis. The theory of personality and psychotherapy advanced by Sigmund Freud and his followers. It is based on the idea that mental illness originates in unconscious motives related to the conscious repression of desires.

Psychological systems. Such systems refer to various operations within the individual's mental activities contributing to the individual's ability to function normally in human society—processes of thoughts, feelings of hostility or security, fear of unknowns, perceptions, and features of personality, for example.

Psychosocial systems. Organizational and group relationships involving individuals and their personal idiosyncracies, emotions, and behavior patterns.

Racism. Evaluating individuals on the basis of race. The assumption of superiority of one race over another leads to racial domination and discrimination in organizations affecting such things as hiring practices, promotions, etc.

Ranking. Rating from high to low. May also be used to describe a person or thing that is at the top of a hierarchy, such as a social order or a number of alternatives. For example, to rank steel above aluminum for certain purposes.

Rapidity of response. The speed with which variations of a system's output from operating plan are corrected.

Real-time data processing. The updating of the master file of information immediately after each transaction occurs.

Recycling. Literally means to begin a production process again. For example, wood is initially made into paper; then paper is made into pulp from which it can be made paper again.

Reliability. The mathematical probability of system survival or nonfailure for stated environmental conditions and time periods. The consistency with which operations are maintained. This may vary from a predictable output to a complete stoppage.

Response. The nature of an individual's reaction to a stimulus. For example, we pull our hand away in pain from touching a hot stove, or we react negatively to an overbearing leader.

Responsibility. The requirement or duty to carry out an assigned task or function. This is often but not always accompanied by a feeling of obligation to perform an assigned duty or task.

Rigid system. A system in which the basic design does not adjust to the changing needs of the environment in which the system exists. Opposite of a flexible or open system.

Role analysis. Clarification of the expectations and requirements on a person's behavior in a particular organizational position and in a particular situation.

Satisficing. Refers to the business pactice of striving for a "satisfactory" profit rather than a "maximum" profit or achieving roughly adequate, rather than wholly satisfying, behavior.

Scalar principle. Ranking of duties according to the degree of authority and responsibility. This establishes an organizational hierarchy.

Scientific management. The use of science in management. As a proper noun, a term associated with the work of Frederick W. Taylor.

Selection criteria. The list of priorities that are used in determining the best alternative. In buying a car, for example, price, make, and style may be important in that order, indicating that price will be given the greatest weight.

Sensitivity. Capacity to respond to stimulation. Also, the degree of such responsiveness. For example, "He had great sensitivity to others' feelings."

Sensor. A means to identify and/or measure a characteristic that simulates the desired output of a system.

Simulation. A series of test activities to determine how a system behaves under different conditions of variables and/or values. Usually characteristics not identical to those of the actual system, but analagous to it, are used.

Smog. Air pollution manifest in a visible form. We usually associate a noticeable mixture of air and smoke hanging over a city or region with smog.

Socialism. A democratically elected government with a centrally planned and controlled economy. It can exist with varying degrees of government control over business.

Sociocultural subsystems. Such systems refer to various operations within the setting of groups, organizations, or even nations, which have the effect of creating conformity or compliance of behavior with shared values or norms—education and training, recruitment and socialization, institutionalization of values through school, church, and religion; also groups, institutions, organizations, and others consisting of members who share in a common set of values or attitudes and beliefs.

Software. Includes all of the programming or instructions used in the application of the computer in information systems. The program tells the computer what steps must be followed in reaching the solution.

Soldiering. Withholding effort in the work environment, as used originally by F. W. Taylor. It refers to the pegging of production output both in manufacturing and office situations whereby employees retain a measure of control over their own work effort.

Solid waste. Garbage.

Special interest group. A group representing the particular economic and political views of a subgroup of the total population.

Stability. The tendency to operate without great fluctuation; to maintain a relatively constant pattern of output or behavior. Stability may also refer to a routine situation that is characterized by little change.

Staff authority. Specialists available to advise managers in the line organization. Frequently specialists may wield authority derived from the status and position of the line manager to whom they report.

Stagflation. An economic condition in which both high unemployment and inflation exist.

Standard of living. A society and its members will exist at a certain level of material affluence frequently measured through per capita income which can be used for comparing the standard of living in various regions of a nation, or between countries. This measure is also intended to reflect nonmaterial aspects of living but these are measurable only with difficulty.

Standards. An established level or norm of behavior or output to which current performance can be compared. A standard may reflect the outcome of decisions on operating policies and procedures.

Stimulus. Something in the environment that incites one to activity. Stimulus can be equated with an influence that brings about certain behavior or results.

Strategy. Refers to long-range planning. It is the general philosophy and plans made by an organization to accomplish its mission. A strategy refers to plans for the future; strategy enables short-run decisions to be made in relation to a defined organization image.

Structure. The arrangement of the various functions or activities performed in an organization. In the systems context, it is the arrangement of the subsystems and components.

Subsystem. A subset of the defined whole which also can be defined by its inputs and outputs. The subsystem can be analyzed as a system in its own right; however, each is a subsystem in a larger system.

Synthesis. A combination of alternatives into a course of action that accomplishes the objective according to the stated assumptions. It refers to a prior investigation of data and facts, an analysis of their composition and relationships. A synthesis represents a complete statement of all the relationships viewed as a whole.

Systems. An organized or complex whole: an assemblage or combination of things or parts forming a complex or unitary whole. In general, the whole—or the system—is more than the sum of its parts.

Systems analysis. The application of modeling and problem solving in a logical or systematic way. This may rely heavily on some type of feedback or two-way communication model.
Systems approach. The total science of systems where the assumption is made that all phenomena can be discussed at a level of abstraction which can use a system approach. The systems terminology is the same across the various sciences.
Systems management. The design and operation of an organization as a system. It reflects the need to consider interrelationships involving both definition of task and human behavior.
Systems philosophy. To view situations, problems, or organizations as systems in order to identify and understand the interrelationships among the parts.
Systems theory. The development of a systematic, theoretical framework for describing general relationships of the empirical world. It contains a common conceptual scheme and terminology.
Task function. A leadership action that supports and encourages effective work activity in the group. Task performance is an organizational expectation and requirement.
Technical system. Systems that emphasize machines, nonhuman components, or information development. The technical system tends to refer to the task of the organization. It is generally judged on efficiency or costs.
Technique. A methodology for solving a problem or performing a task. Technique is a prescription for producing results; it may be incorporated into the design of a machine.
Thematic Apperception Test (TAT). A psychological test that requires the subject to make up stories based on series of pictures. The subject "projects" his characteristic ideas, attitudes, hopes, fears, etc., in his stories and thus reveals them to others.
Thermal pollution. River, lake, or sea water heated to the point where it tends to destroy marine life. It is particularly relevant today with construction of atomic power plants using water as a coolant.
Time sharing. The utilization of a computer system by two or more users with each user having current access to the computer.
Tokenism. In this sense, a seemingly genuine but actually false gesture of friendship, inclusion, cooperation, etc., by a majority group toward a minority group or a member of such a group.
Tool. An instrument to aid in the accomplishment of a task.
Tradeoff. A comparison between the advantages and disadvantages of alternatives relative to established priorities or selection criteria.
Translator program. Also called a compiler. The instructions prepared to convert the general instructions of the programmer into the detailed machine language instructions for the computer.
Trickle-down process. Refers to a method of housing distribution whereby each group in income scale follows the group above it in moving into the available housing. Thus, new housing for the wealthiest results in better housing for all.
Value analysis. A technique used by purchasing departments to measure the functional usefulness of items purchased and to get the greatest value for money spent.
Value approach. The introduction of external ethics and values in the decision-making process. Values provide a frame of refer-

ence in which to judge and to evaluate human progress in a larger sense; but also in relation to decisions and results in the immediate sense.

Values. Ideas (or ideals) that provide means for making choices and ordering preferences. Values tend to reflect family and educational background. They also tend to be shared in an organizational or group setting or for a society or culture as a whole.

Vector. Something possessing both magnitude and direction, i.e., an arrow in flight.

Vertical strain. Competition between the different hierarchical levels in an organization.

Vitalist theory. The importance of thinking in terms of the whole and in integrating the total system relative to the parts.

Welfare capitalism. Responsibility for an individual citizen's welfare is superimposed upon the traditional capitalist system.

Welfare state. The state or government assumes responsibility for the citizens' total welfare.

Zero population growth (ZPG). A balance in population between increases and decreases. Births and immigrations would equal deaths and emigrations.

Index

PHOTO CREDITS

Front cover: Gordon Menzie, Everett Peck; Half title: Gordon Menzie; Title: Everett Peck; Part One: Library of Congress, Photophile; Chapter 1: Everett Peck; Chapter 2: Everett Peck; Chapter 3: Everett Peck; Chapter 4: Everett Peck; Part Two: Southern Pacific Railroad; Chapter 5: Debra McQuiston; Chapter 6: Everett Peck; Chapter 7: Everett Peck; Chapter 8: Debra McQuiston; Chapter 9: Everett Peck; Chapter 10: Gordon Menzie; Chapter 11: Everett Peck; Chapter 12: Everett Peck; Chapter 13: NASA, Boeing Aircraft; Chapter 14: Boeing Aircraft; Part Three: UPI, Photophile; Chapter 15: Photophile; Chapter 16: UPI; Chapter 17: San Diego Gas & Electric Company; Chapter 18: Lorenzo Gunn; Chapter 19: UPI; Chapter 20: UPI; Chapter 21: Reserve Synthetic Fuels, Inc.; Part Four: Black Star; Chapter 22: Everett Peck, Debra McQuiston; Chapter 23: Debra McQuiston; Chapter 24: Photophile